DESIGN & POLITICS

The Public Promotion of Industrial Design in Postwar Belgium (1950–1986)

Design & Politics

The Public Promotion of Industrial Design in Postwar Belgium (1950–1986)

Katarina Serulus

Leuven University Press

Published with the support of
Universitaire Stichting van België

ISBN 978 94 6270 135 9
D/2018/1869/33
NUR: 656

Layout: DOGMA
Cover design: Griet Van Haute
Cover illustration: View of the *Industrial Design show* at the sixth Liège International Fair in 1954. Fonds Hannoset © AML (Archives et Musée de la Littérature)

GPRC
Guaranteed
Peer Reviewed
Content
www.gprc.be

Contents

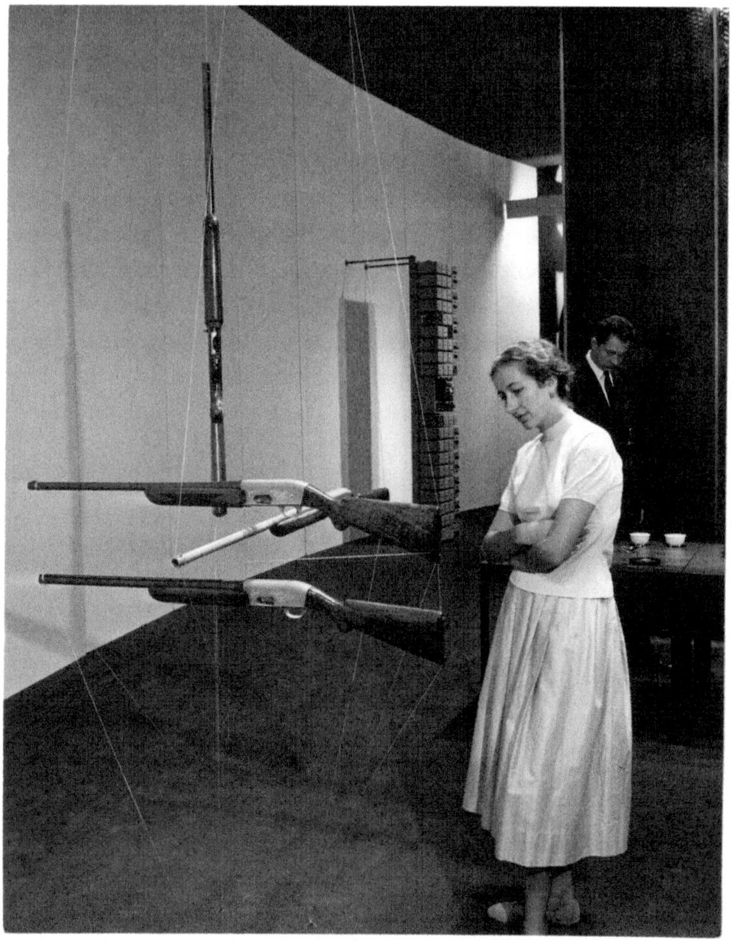

Fig. 0.1. Double-barreled FN Browning rifles on show in the Belgian section of the 1957 Milan Triennial. Source: Archivio Fotografico © Triennale di Milano.

Introduction

1. GOLDEN GUNS

In October 2011 Colonel Gaddafi was captured and killed by rebels in Sirte, Libya. His gold-plated gun was held up to the camera as a symbol of his fall. It was a customized, luxurious High Power MKIII pistol with elegant curling engravings in gold and a carved wooden stock. The engravings on the golden barrel frame an open space where the words "MADE IN BELGIUM" are neatly inscribed in a sans serif typeface. This Belgian product, which became one of the many symbols of the Libyan conflict, was manufactured by the Fabrique Nationale in Herstal (FN, founded in 1889 and now part of the Herstal Group). It is a notorious design by the American firearms designer John Browning (1855–1926) who became famous for his role in the development of semiautomatic and automatic weapons for civilian and military use.

Although the involvement of FN firearms in controversial armed conflicts is now considered a blot on Belgium's escutcheon, half a century ago the Belgian government consciously held these weapons up to the camera. A picture of the 1957 Milan Triennial shows four double-barreled FN Browning rifles floating in the air, mounted with invisible thread in a three-dimensional composition, attracting the attention of a female visitor (fig. 0.1). The Belgian entries of the X (1954) and XI (1957) Milan Triennial let the international design-minded public become acquainted with fine Belgian industrial design: technical equipment such as microscopes, lenses, surgical instruments and the famous FN Browning rifles, but also furniture and domestic appliances. The highly lauded Belgian attendances were the first initiatives supported by the Belgian government to promote industrial design. It was the Ministry of Economic Affairs, headed by Jean Rey, that organized the Belgian participation and from 1954 incorporated industrial design in its policies. The aim was to stimulate the dynamics of the national industry and to strengthen the image of Belgian products on foreign markets. Thus, in this case, the FN Browning rifles did not illustrate bloody conflicts, but demonstrated Belgian excellence in the production of highly technical equipment.

Advertised as reliable and sound, Belgian guns were successful export products in the postwar era.[1] In the 1980s, the British secret agent James Bond also used one.[2] The FN weapons even entered the museum in 1984 as one of the thirty objects that played an important role in the development of Belgian and international design. Indeed, that year, the Belgian Royal Museum of Fine Arts opened a new permanent section: "Design in Belgium".[3] The aim was to outline chronologically the history of Belgian design. The museum displayed Browning rifles next to Willy van der Meeren's furniture, Val-Saint Lambert's glass, Nova's electrical household appliances, Michel Olyff's graphic design, and the Brussels subway train of Phillipe Neerman. The rifles on show in the museum were the latest design of the American John Browning and were considered one of the highlights of "Belgian" design. The museum catalog explained that the engravings on the steel and walnut rifles illustrated the famous craftsmanship of the FN workers.

This two-faced image of Belgian weapons that developed over time — recently regarded as an offense to human rights, before as a highlight of Belgian design — brings us to the observation that as time goes on objects acquire different narratives and public discourses. These are constructed by a complex network of actors such as, in this case, the press, international human rights organizations, writers, the entertainment industry, cultural institutions, industry, companies and governmental bodies. This study will especially focus on the latter: How does design become entangled with state politics and ideologies? How does the state participate in shaping a certain design culture? How do some objects become the exponent of national idiosyncrasies? What role does the state play in the discursive embedding of design? But also, can the study of these design objects clarify, enlarge or alter our understanding of postwar Belgian politics?

2. THE PUBLIC PROMOTION OF INDUSTRIAL DESIGN IN POSTWAR BELGIUM

The above FN weapon was only one of many products promoted by the Belgian government at international and national events to enhance its economic and cultural image. The array of promoted industrial products ranged from an industrial milking machine and children's toys to a tennis racket and interior furnishing. To explore the modality of the relation between design and politics — here understood as state-related activities — this study examines the three most important governmental and quasi-governmental bodies for industrial design promotion of that period in Belgium: the Signe d'Or (1956–1966), the Institute for Industrial Design (1956–1971), and the Design Centre (1962–1986). These institutions are considered in this book as intermediary platforms where the shape and form of design in Belgium was negotiated by different stakeholders such as the state, industry, educational organizations, businessmen, objects, consumer groups, financial institutions, design professionals and social and cultural organizations. Through a close historical investigation of these institutions and their activities, this book aims to provide new insights into the wide range of fantasies, agendas, intentions, narratives, manifestations, wishes and ideas of industrial design in postwar Belgium and its official promotion.

The institutionalization of industrial design in Western Europe dates from the decades after the Second World War and was largely directed by government-backed events.[4] The interest of official circles in engaging with design was mainly guided by a political and economic rationale. In 1957, "the Six" — i.e. Belgium, France, Germany, Luxembourg, the Netherlands and Italy — signed the Treaty of Rome which stipulated, in a three-phase plan to last for 12 years, the establishment of a free trade area between the members, the creation of a European customs union and eventually a political union.[5] A couple of years later, in 1960, the European Free Trade Association (EFTA) was established by the "Outer Seven" — Austria, Denmark, Norway, Portugal, Sweden, Switzerland and the United Kingdom — as an alternative free-trade bloc. In spite of its small size, Belgium played an important role in international politics as one of the driving forces of the European Economic Community (EEC). Brussels took on the role of Europe's unofficial capital, crystallizing in the 1958 Brussels World's Fair, and became the diplomatic and symbolic heart of Europe, positioned between the Cold War protagonists: the United States and the Soviet Union.[6] Given this context, it is then hardly surprising that the early promotion of Belgian design was characterized by transnational and European initiatives.

The same postwar period is marked by the Cold War between the two superpowers emerging from the ruins of World War II: the United States and the Soviet Union. The power struggle between the two rivals dominated the geopolitical stage and divided the world into two ideological spheres of influence: the capitalist West and the communist East. The role that design played was not insignificant: in fact it was an important ally in the game of superiority for both blocs.[7] Indeed, the Cold War conflict itself was very much "designed." The chasm between East and West was made visual and material on various occasions in the international arena through design objects. The most famous example cited by many historical accounts on the subject, is known as the "kitchen debate" between Richard M. Nixon and Nikita S. Khrushchev at the 1959 American National Exhibit in Moscow. A lemon-yellow American model kitchen at the American National Exhibit in Moscow in 1959 formed the backdrop to a discussion between the two leaders about the differences in living standards and consumer regimes in the two countries. Khrushchev assured Nixon that the Soviet Union would soon overtake American consumer culture, with the famous words: "When we catch you up, in passing you by, we will wave to you."[8] Design was thus a major weapon in the political jousting of postwar international politics.

Another high spot in ascertaining the political meaning of material culture in the Cold War was the first postwar world fair, the legendary Brussels Expo 58 under the rubric: "balance of the world for a more humane world." At this international high-profile fair, the pavilions of the United States and the Soviet Union were juxtaposed, illustrating almost literally a miniature version of the Cold War being waged.[9] The monumental Soviet pavilion with replicas of Sputniks and a huge sculpture of Lenin represented its technological capability, while simultaneously exhibiting leisure and home culture. The American pavilion focused instead on displaying the American way of life, showing modern domestic objects. In this sphere of European unification and Cold War tensions, high-profile international fairs became a stage on which different nations and regions promoted their production, but which also served as a cultural and ideological battleground. Even for a small country like Belgium these international appearances were carefully choreographed without underestimating their diplomatic power. For example, the Belgian section at the X Milan Triennial in 1954 was internationally praised and rewarded for its high quality. It was one of the first promotional exhibitions on Belgian "industrial design" and showed, among others, microscopes, objectives, cameras, surgical instruments and guns.[10]

Against the backdrop of the European liberalization and Cold War tensions, the Belgian government started to show interest in such exhibitions as a tool to stimulate the dynamics of the national economy and as an instrument of modernization. The first to formulate a firm design policy was Minister of Economy Jean Rey (1902–1983) in 1954. His primary concern was to turn Belgium into a competitive force on the European common market as a manufacturer of consumption goods. In 1954 he organized the first Belgian manifestations of industrial design in Liège and at the X Triennale of Milan and in 1956 funded two new design bodies: the Institute for Industrial Design for Belgium and the Grand Duchy of Luxembourg and the Signe d'Or prize.

The alliance between politics and design received new impetus in 1962 when a Design Centre was established in Brussels with the aid of the Belgian Office for Foreign Trade. The new center was the brainchild of two design organizations already in existence. The former chairman of the institute Pol Provost (1907–1990) and the secretary of the Signe d'Or Josine des Cressonnières (1926–1985) joined forces and acquired the roles of, respectively, chairman and director of the brand new initiative. The purpose of the Design Centre was to add prestige to Belgian products in foreign markets after the successful examples of Italian, Scandinavian and German design.

The decision to promote Belgian design by means of a design center was nothing new: it was a tried and tested formula all over the world, and the majority of European countries already had one.[11] The Belgian version was modeled on the well-known British institution that was established in London in 1956: The Design Centre. The formula consisted of a permanent exhibition of jury-selected products that would change every three months, supplemented by temporary themed exhibitions and a library. After two years of preparations, the Brussels Design Centre officially opened its doors in 1964 in the galerie Ravenstein, which was strategically located in the heart of Brussels between the commercial and the business district. This was to highlight Belgian design for the next twenty-two years, establishing one of the longest track records among such centers. Over two hundred exhibitions were held on diverse themes, from workers' attire and souvenirs to children's toys and designs for the state. Parallels can be drawn between the design reforms of the nineteenth century when the promotion of "good" taste was similarly employed by the Belgian state to assert its position in the arena of trade and international politics.[12]

The already existing Institute and the new Design Centre were supposed to function in "tandem."[13] The Institute was to promote industrial design in the Belgian industry, and the Design Centre to make sure it was fruitful through the export of Belgian design. Yet, the foundation of the Design Centre in fact eclipsed the Institute that eventually discontinued its activities in 1971.[14] The Design Centre became the protagonist, organizing a permanent exhibition that showcased a constantly changing selection of "good" Belgian products as well as temporary thematic exhibitions, award competitions and conferences. To communicate with a large audience, the institutions made use of press coverage and radio and television broadcasting, and published their own design magazines.[15] The activities of these two institutes were not purely restricted to promotion but included advice, training and education. These bodies not only served the interests of the state, but functioned also as representatives and defenders of the modernist discipline of industrial design, keeping close contact with the international design scene.

Since the position of these institutes was at the intersection between public and private interests, economics and culture, national and international design networks, materiality and political discourse, these are interesting subjects to help us discern which meanings, storylines and political actuality were engrafted on industrial design in postwar Belgium. In that way, this study contributes to the correct understanding of postwar Belgian design and stimulates critical awareness of the political undercurrents of these a priori innocent, trivial and frivolous objects.

3. HISTORIOGRAPHY

Although "Belgian design" is now a hot topic and design research in Belgium has gathered more and more steam in the last decades, this episode in the Belgian history of design is still surrounded by obscurity. Little has been written on the actual relationship between the Belgian State and industrial design in the postwar area.[16] Most studies on postwar design in Belgium focus on designers,[17] companies,[18] domestic furniture or design for the home.[19] In addition, furniture design is largely featured in these accounts, while industrial design, such as investment goods, consumer goods, vehicle design or design for the public interest, is mostly left undiscussed. There are some exceptions. Architectural and design historians Mil De Kooning, Iwan Strauven and Fredie Floré were the first to give a considerable amount of attention to the Institute for Design and the Signe d'Or in their overview of 1950s furniture in Belgium that was published in 2000.[20] Furthermore, there is the work of Fredie Floré on the government-backed Belgian participation in the Milan Triennials and the design historical accounts of Javier Gimeno-Martinez of industrial design initiatives in the 1950s.[21] Along with these introductory accounts, this study owes much to Gimeno-Martinez's work on the role of creative industries in constructing regional identities in Flanders and Valencia after 1968.[22] Gimeno's research emphasized the role of mediating actors and the importance of governmental plans, exhibitions, publications and awards in the construction of design identities. Similarly, this study considers the Belgian design bodies to be mediating agents that forged links between design and Belgium, but focusing on the four decades after the war, when design promotion in Belgium was still a national concern and was not yet a regional competence.

The foreign counterparts of these Belgian institutions have, on the contrary, received considerable attention in recent academic literature. [23] In the international discipline of Design History, many accounts deal with how postwar nation-states endorsed modern design for various reasons: the creation of a national design identity, Cold War propaganda, the improvement of the national economy and its position in global markets.[24] The importance of design institutions is generally accepted in these studies as they prove to be interesting study objects.[25] This intense scholarly attention to design institutions fits the recent "mediation" turn in the field.[26] As design historian Kjetil Fallan noted in 2014:

> Recently, design history has toned down the conventional focus on persons (designers), objects (artwork), styles, movements, periods, etc., and is instead becoming increasingly concerned with

other aspects of and actors in design culture. There is a growing interest in the roles of mediators, critics, curators, educators, consumers and users, as well as arenas like journalism, exhibitions and education.[27]

Already in the late 1980s, prominent design historians emphasized the importance of the institutional dimension of design. Australian design historian Tony Fry states that "design has no identity and is given coherence through institutional orchestration."[28] Design historian John Walker devotes a little heading in his 1989 book *Design History and the History of Design* to "Design institutions as an object of study."[29] For him, the study of such institutes is quite relevant for design history because it "shifts the emphasis away from products and individuals, though this is not to deny that certain forceful individuals have a strong impact upon the direction of policies of institutions." Moreover, he thinks that "the overriding interest of design institutions is their role as ideological battlegrounds, as the sites where different ideas about design are made explicit and struggled over. Institutional studies are to be welcomed because they add another layer of complexity to the object of study [...]." Design historian Anna Calvera in the mid-2000s also underlined the importance of this topic by indicating that the design practise and design culture was spread in a range of countries through the founding of design promotional institutions such as design centers and professional organizations.[30] According to Guy Julier, the pattern of institutionalization can be found in all capitalist or proto-capitalist countries and is a productive starting point for design historians to question why design arrived in a given nation.[31]

The political dimension of design is equally addressed from the 1980s in design historical studies. As historian Patrick Maguire phrased it: "Politics and design might appear to make uneasy bedfellows but, like sport and politics, they usually make inescapable ones. However much contemporaries, or subsequent chroniclers, might have wished otherwise, the two were inextricably linked."[32] Design historian John Heskett devotes an entire chapter in his book *Industrial Design* (1980) to "Design and Politics", admitting the importance of this relation between design and politics: "The subordination of design to political ends makes the need to consider the context in which designs are produced and used nowhere more relevant than in the realm of government intervention."[33] The link between politics and design is elaborately described with regard to Cold War design when, as mentioned above, design was infused with political aspirations.[34] According to design historian Alison Clarke, "design has been systematically under-represented in its role in 'making' the world, due to its assumed depoliticized status."[35] From a similar observation, many recent events give evidence of demystifying "design" from its dominant progressive modernist fashion, i.e. the well-designed and commercially successful design that improves one's life. Instead, the multiple personalities of "design" are emphasized, including the "evil" and "political" ones. The *Museum of Cruel Designs* curated by Gavin Grindon at Banksy's Dismaland showed, for example, design objects that are used to hurt, exploit and exclude people.[36] It can be read as a critique of the neoliberal systems of control and obedience. Analogue visions have been the subject of several other exhibitions. The 2014 V&A show *Disobedient Objects* in MoMA's curatorial experiment *Design & Violence* both address similar political dimensions of design.[37] This tendency is also noticeable in the academic field of design history. The 2014 Design History Conference was devoted to "Design for War and Peace" where, for example, Kjetil Fallan discussed the Swedish production of military weapons that contrasts highly with the usual suspects associated with the popular brand of "Scandinavian" design.[38] At a recent workshop in Oslo on "Design History in Practice" Carlos Bartolo raised questions on how design historians should approach the dark political periods of design history, such as the "fascist" work of Portuguese designers that actually laid the foundations of the current Portuguese design discipline.[39]

The Belgian case is not discussed in most of these historical accounts of international design. Even in large overviews or "global" histories, it is hardly mentioned. One can argue that this blind spot is a result of the lack of secondary literature and the poor state of the Belgian design archives.[40] However, one can also argue that Belgium is a peripheral or marginal country in design matters. Design historian Anna Calvera defines these peripheral countries as economically and culturally dependent on a center: "There is a delay, sometimes, subtle, sometimes quite large, which is culturally speaking the main attribute of the peripheral character."[41] According to Calvera, national or local design narratives of peripheral countries have a similar structure. The starting point is that the nation is only known thanks to a key event or personality. In the case of Belgium, this could be Henry van de Velde — although many doubt that he can be considered Belgian since he spent most of his career in foreign countries — and the 1958 World's Fair in Brussels. However, as Calvera concludes, hardly anyone knows anything about the historical background that explains these highlights. This then serves as the main argument for carrying out such research.

The danger, however, is to create an attractive "Belgian" design history with famous designers and amazing stylish objects, but that would be plain mythmaking or — as Tony Fry put it — just fiction inspired by history.[42] This research aims thus not to list a range of Belgian design highlights or invent such glorious national design moments. This book would then be just one of the many promotional actions to put "Belgian" design on a pedestal and sell it as something outstanding to the general public, industry and researchers. Nor does it aim to measure the Belgian design history against the large narrative of modernist design to evaluate the degree of modernization. Belgian design is, except for a few examples, not included in international design canons. Should one conclude then that Belgium is just not relevant in the realm of design? Indeed, maybe Belgium was not of much significance for this linear modernistic narrative of innovation. Yet Belgium may be an interesting case because it does not quite fit into the straitjacket of the dominant narratives but rather challenges it and opens up new perspectives. So, can Belgium's rather weak national design identity be a contrast for the well-known Italian or Scandinavian cases? This research will thus try to pursue Anna Calvera's claim: "on the part of peripheral or national histories, the job to face is firstly, how to get a place, or the right place, in the global history of Design and, secondly, how peripheral research can win credibility and be attractive enough to locate its outcomes in Design History."[43]

4. DESIGN: TACKLING THE POLYSEMIC

The simple word "design" throws up a host of associations. But what practices and media are actually subsumed under the term "design" when talking about design history? The popular use of design today is vast and overwhelming, especially when looking at advertisements, the lifestyle press and marketing strategies.[44] Design seems to emphasize a certain degree of style, quality or luxury. Although the popular discourse on design and the designer culture is a fascinating object of study in itself, this understanding of design is, however, quite the opposite of how most design historians currently try to approach the subject. Kjetil Fallan, for example, describes design history as a field "where the mundane, the cheap, the amateurish, the flawed and the garish elements of material culture feature alongside the usual suspects of good design."[45]

Recent developments in design history proclaim a wide understanding of design. There is the tendency to be geographically and chronologically inclusive rather than to operate the narrow understanding of design as related to the industrial revolution of the eighteenth century.[46] The underlying aim is to transcend the modernist narratives of innovation and mass production and include marginal histories. Taking pre-industrial objects into account also avoids a solely Eurocentric discipline of design history and fits it into contemporary developments in other disciplines. In a special issue of the *Journal of Design History* entitled "The Global Future of Design History" in 2005, Victor Margolin, Anna Calvera and Johnatan Woodham elaborate on this subject.[47] Margolin argues that design historians have to supersede the limited definition of design that confined its study to industrialized regions of the world. In his vision the history of design includes all geographical areas and time periods, and thus design should rather be studied as "how different cultures have provided for their respective material needs, as well as how larger forces of politics and economics have shaped conditions for material production in those cultures."[48] Victor Margolin effectively achieved this ambition in this three-volume *World History of Design* whose first two parts appeared in 2015.[49] Edited by Pat Kirkham and Susan Weber, the large 2013 volume *History of Design: Decorative Arts and Material Culture, 1400-2000* seems to join this trend.[50] This overview of 600 years of design reflects a broad understanding of design and covers East Asia, the Americas, India, the Islamic World, Africa and Europe. Other publications followed a similar trend of inclusive design histories, albeit without offering an all-encompassing global design narrative. For example, *Global Design History* is a collection of short essays edited by Glenn Adamson, Giorgio Riello and Sarah Teasley, where design is understood without any geographical or chronological boundaries.[51] This inclusive approach must be framed in the ambitions of the discipline of design history to extend its territory, renew its aims and escape its marginal status as formulated in international conferences on Design History and Design Studies from 1999.

The past decade, influenced by a range of turns such as the cultural, the material and the visual, has seen design studies enriched with new approaches. New tendencies focusing on the everyday and the material not only widened further the understanding of design, but also secured its relevance in the academic field. Design Culture is one more recent approach at the interface of design history, cultural studies and social sciences. A leading publication in this area is Guy Julier's *The Culture of Design*, in which "design artifacts and design practice are seen as imbedded in a design culture as a whole, a range of relations to formal and informal institutions, to markets and media, to designers, to users and consumers are viewed upon as conditions for interpretation of form and meaning."[52] Although the subject of the institutionalization of design in Belgium was linked to modernistic design thinking, it is preferable to include it in these recent developments and to operate on the basis of a wide and inclusive understanding of design as a cultural phenomenon. By approaching it in this way, it is possible to include and apply theoretical insights from other disciplines and to avoid traditional pitfalls when discussing national design cultures, but rather to explore new entries to design studies beyond the dominant narratives.

5. DESIGN AND POLITICS: CONSTANT BUT UNEASY BEDFELLOWS

There is no such thing as a fixed definition of politics or the political.[53] Throughout history, many views on politics have been theorized, from Plato to Hannah Arendt. In a narrow and more popular sense, politics is limited solely to the realm of the state or government. However, many studies and theories do not consider politics as related to governments, but relate it to a much larger realm that includes families, universities and the personal. Since no consensus exists on the boundaries of the political, it makes sense that the relation between design and politics can be explored in diverse ways. Roughly, three levels of entanglement are discerned between design and politics.

The first employs a broad understanding of the political in a range of settings. The thing itself, as part of the material fabric of everyday life, is politically charged — whether linked to the state or not.[54] It is also at this level that we can situate the utopian modernistic design thinking.[55] Depending on the designer and theorist, the underlying principle of modernistic design was often infused with political aspirations. There was this belief that everyday spaces and objects affected their users in such a way that design practise could be a kind of political and social reform, or even a moral education. Greg Castillo dubbed this belief into the power of objects to influence human behavior in postwar design circles "environmental determinism."[56]A different line of reasoning, although also concerned with the political aspect of material fabric, is offered by design theorist and philosopher Tony Fry.[57] For him, design in general is profoundly political, and he argues that "design gives material form and directionality to the ideological embodiment of a particular politics." He goes even further: "design expresses power materially and in ways that shape how people interact and ontologically prefigure their material culture and economy."[58]

The other side of the spectrum is design charged politically through the explicit link with the government such as flags, guns, logos, state buildings, public infrastructure and public housing.[59] The notable reference here is French historian and philosopher Michel Foucault.[60] According to the latter, it was at the end of the eighteenth century that "one sees the development of reflection upon architecture as a function of the aims and techniques of the government of societies."[61] It was not the case that before the eighteenth century design was apolitical, but that from then on a very broad and general reflection took place on architectural questions concerning the city, collective infrastructure and housing. As Foucault describes in an interview conducted in 1982, "it was not necessarily a change in the minds of architects, or in their techniques — but in the minds of political men in the choice and the form of attention that they bring to bear upon the objects that are of concern to them."[62] This kind of relation between state politics and material culture is still a productive focus in a range of disciplines. In Science and Technology Studies, Langdon Winner argued in *The Wale and the Reactor* (1986) that technical decisions are political decisions.[63] He illustrated the political impact of material objects by his famous example of the low-hanging overpasses on Long Island designed by architect Robert Moses. These bridges were designed so that only cars, owned by the whites of the upper and middle class, and not buses, used by racial minorities and low-income groups, could easily pass and have access to areas of recreation. James Scott's *Seeing like the State* (1998) then critiques the system of "high modernism" and the belief that the state can handle its subjects and their environment through social engineering programs.[64] A similar focus can be found in materialism studies. The more recent book *Material Powers* (2010), edited by Tony Bennett and Patrick Joyce, examines, for example, how power is made up and exercised by studying the role of material infrastructures in the organization of state power.[65]

The present research can be located somewhere between these two levels. It is interested in the political aspect of design, yet does not focus on design's ability to guide behavior or on explicit design for the state or the use of design by state officials. Rather, the study focuses on how the promotion of the then fashionable international discipline of "industrial design" fits into a political strategy of the Belgian State aimed at a range of effects: to stimulate the economic dynamics and to construct a modern image out of cultural and economic concerns. Thus, studying the introduction of the modernist concept of design as a political strategy of the state does not neglect the wide understanding of the political character of the material, but it is just the analytical focus of this study. It is especially in the postwar period that design enjoyed such a political success: state-related bodies for design promotion were set up around the world. And it is this symbiotic entanglement of design and politics that will be studied here and that is often referred to as "soft power."[66] It was political scientist Joseph Nye that first coined this term in 1990. In contrast to "hard power", "soft power" has the ability to persuade people through intangibles such as culture and belief systems. "Hard power", on the contrary, is direct force exerted through plain displays of military, economic or political means. In design history, especially through the contributions of Greg Castillo and Gay McDonald, the term has now become commonplace to describe the role of modern design during the Cold War. However, it was then rather used to indicate the inscription of design in a political strategy and its diplomatic potential.

6. DESIGN AND THE NATION-STATE: AVOIDING THE PITFALLS[67]

As design historian Paul Betts explains, "the tradition of linking design with the state went back at least as far as the London Crystal Palace Exposition of 1851."[68] At the beginning of the nineteenth century, the national industrial production became the subject of large and diverse exhibitions. In the postwar era too, design historian Javier Gimeno-Martínez notes, "the nation-state is the dominant category in design promotional activities. Fairs, exhibitions, and exchanges are mostly set up according to the logics of nationalism."[69] As such, when one studies design in these periods the nation-state pops up as a self-evident category of analysis. However, empirical research in the case of Belgium shows that the official design bodies were not only linked to officials, semi-governmental actors and the idea of the nation, but had in fact many offshoots beyond the borders of the nation-state. For example, they testify to strong personal transnational networks, to close ties with the private sector operating in liberalized markets and to an intense dialog with the global developments in the discipline of design. How then should we approach the study of design promotion in Belgium if we do not want to leave out these crucial points? Many current debates in design history deal with this challenge of combining national narratives with local, transnational, supranational and global narratives. One such framework is methodological transnationalism.

Abandoning the nation-state as a key frame in historical studies was already initiated by the work of earlier world historians, such as William McNeill, Fernand Braudel and Immanuel Wallerstein, some dating back to before World War II.[70] The approach of methodological transnationalism is largely indebted to these pioneers, and was further developed in the 1970s in many different academic disciplines including cultural studies, anthropology, cross-border studies, social sciences and history.[71] The transnational approach was especially mobilized to transcend the Weberian "state-territory-people" triad known as methodological nationalism.[72] Most cases of methodological nationalism assume that one given territory or geographical space (a nation) corresponds to one specific social space (a community). Transnationalism, however, posits that a one-to-one relation between a territory and a social space does not exist, but that "the boundaries of societies, cultures or civilizations are permeable and mutually constitutive."[73] As such, it was highly successful in exploring topics such as global cities, transnational migration, cosmopolitanism, international relations, transnational corporations and global institutional restructuring.

The main advantage of this approach — as it is conceptualized in the cross-border studies — is that "[...] it provides alternative definitions of units of analysis general enough to think of social entities as territorialized and deterritorialized as well as nationalized and cosmopolitan at the same time."[74] One can continue studying "national" practises while including transnational processes. Historian Daniel Rogers puts it differently, explaining that the boundaries of the nation-state can become an "analytical cage" when they obscure "the junctures where the nation-state's permeability might be brought into view, where the transnational forces do their most important work."[75] It is these junctures that are made visible through the transnational perspective and which are also the subject of this study, namely to unravel the national and transnational imbroglios of national design cultures. One other valuable aspect is that transnationalism is concerned with movements, flows and circulations as an analytical set of methods.[76] And these movements, flows and circulations relate not only to social practices or symbols, but also to artifacts.

Transnationalism and the study concerning the material environment became good bedfellows in later years. However, many studies — old and recent — already included these transnational processes when studying material culture without labeling them as transnational histories.[77] The trend of using the notion of transnationalism with regard to design has gained considerable popularity only since the mid-2000s. Design historian Tim Putnam, for example, argued for "studies recognizing both distinctive 'national routes' and *transnational* institutions in the development of design culture" in his capacity as a guest editor for the 2007 special issue of the *Journal of Design History* devoted to "Design and Polity Under and After the Ottoman Empire".[78] The 2009 special issue of *Design Issues* on "Design in a Global Context," co-edited by Karen Fiss and Hazel Clark, similarly touches upon the possibilities of transnationalism with regard to design history.[79] Subsequently, design historian Guy Julier elaborated more explicitly on what transnationalism could mean for design history.[80] For him, transnationalism in the field of design involves "[...] the movement of ideas, styles, processes, technologies and objects between nations."[81] He subsequently described the new task of the design historian: "to go beyond borders while also paying attention to the local conditions of design cultures, itself being both ideologically constructed and embedded in the material practices of everyday life."[82] In a similar fashion, design historians Kjetil Fallan and Grace Lees-Maffei pointed out the sound value of the genre of national design histories, but nevertheless warned scholars of the possible danger of an essentialist attitude to national identity.[83] Accordingly, they "propose national studies that are attentive to cultural exchange and international trade and influence."[84]After this first acquaintance, the transnational tendency seems to be to flood the current field of design history.[85] For example, design scholars Yuko Kikuchi and Yunah Lee edited a special issue of the *Journal of Design History* completely devoted to "Transnational Modern Design Histories in East Asia."[86] They observe that design history has yet to develop a true debate on transnationalism, but account for its promising results regarding the study of modern design in East Asia. [87]

Neighboring disciplines are also responding to this tendency. In the field of history more and more attention is given to transnational processes in the making of national material cultures. For example, the research platform National Movements & Intermediary Structures in Europe (NISE) for comparative and transnational studies on national movements chose, as the inaugural volume of its monograph series, the study by historian Anne-Marie Thiesse on the creation of national arts and crafts in nineteenth-century Europe.[88] The book opens with the case of the matryoshka nesting doll that is considered a traditional Russian object, but is in fact an imagined tradition that dates from the late nineteenth century. Her claim is that objects such as the matryoshka are the result of the intensification of transnational contacts on artistic theories and practises in the light of modernity. To describe this exchange she uses the term "transnational cultural transfers." A transnational perspective, however, does not overlook the importance of nationalism when studying design promotion.

7. DESIGN PROMOTION AND NATIONAL IDENTITY[89]

When exploring a certain design history in a specific country, the focus is often on designers, artistic or creative movements, consumer groups or a connoisseurship-like attitude towards objects. The crucial role of promotional design bodies funded by industry or government is often overlooked. Yet it is these bodies that cluster and transmit storylines, which often later on become adopted as clichés, popular stories or even history. Moreover, these intermediary bodies often acted as a sieve, selecting objects or designers that could flow right on into history books. By spotlighting some products and persons while neglecting others, they more or less reconfigured the "anteroom" of design historiography. Therefore, such institutions should not be considered as a plain sample sheet of design from a certain national design culture. Instead, agendas, intentions and the criteria according to which these institutions promote "design" should be carefully dissected and contextualized.

The three Belgian design institutions that are central in this study — the Institute for Industrial Design, the Signe d'Or and the Design Centre — were charged with the promotion of "Belgian" design, or at least with improving "Belgian prestige." The most visible part consisted of the design exhibitions that showed a careful selection of objects answering to criteria informed by international debates on the discipline of industrial design. Over a period spanning four decades, these exhibitions (altogether more than 200!) differed in scale and content. Some were dry representations of Belgian production of that moment, others were exhibitions with a central theme that linked the various objects while yet others showed the *crème de la crème* of Belgian production marked by the Signe d'Or award. Different interior architects framed these selections with their designs: from sober and functional to animated and instructive. The exhibitions were shown to a national and international public at a range of events. However, the key question is how is the creation of a national identity established through these kinds of practises?

There is a tendency to assume that a national identity is inherent in objects or that somehow a national soul or characteristics are imbedded in the material appearance of designed objects. The object is then considered as a pure expression of the national spirit or a mere reflection of a culture that produced and used it. Advancing from the somewhat confusing premise that there is no such thing as a quintessential Belgian identity, studies on Belgian culture tend to describe the lack of national pride, the merging of German and Latin cultures, the individualist and surrealist character, and the preference of moderation and modesty as main characteristics.[90] This reading of objects is known as an essentialist or primordialist approach, which is quite common in popular literature but has since exploded in academic circles.[91] To bypass this pitfall, this study will apply the constructivist approach to disentangle the construction of the Belgian design image.

8. STRUCTURE

This book examines the different modalities of the relation between design and politics in postwar Belgium in six chapters, each one having its own focus ranging from design pioneers, failed attempts, national styles, Cold War exchanges to Belgian myths. The key narrative thread that knits these chapters together is the three design bodies: the Institute for Industrial Design (1956–1971), the Signe d'Or (1956–1966) award, and the Design Centre (1962–1986). The six chapters are bundled into three thematic parts that follow each other in a roughly chronological order.

The first part looks at how and why modern design was endorsed by the Belgian government in the 1950s. Instead of a top-down movement, the beginnings of the official institutionalization of industrial design can be characterized as a mash-up of divers intentions, agendas and backgrounds. By examining the different actors and discourses at play, this part maps the competing visions on industrial design in the heyday of Belgian design institutionalization. The first chapter explores the continuation and mobilization of the interwar traditions in the early industrial design events. It also chronicles the very first dubious government-backed attempts in Liège and Brussels to advance the industrial design discipline in Belgium before 1956. It was especially the prospects of the European Economic Community (EEC) and the liberalization of international trade that stimulated the Ministry of Economic Affairs, headed by Minister Jean Rey, to include "industrial design" as one of the spearheads of its policy. The second chapter examines how the instalment of "industrial design" — which was considered after all as a foreign phenomenon — was legitimized through the mobilization of Henry van de Velde's legacy. The elderly Van de Velde, who was then residing as a voluntary exile in the Swiss mountains, provided a timely and needed service to the young Belgian design institutes in the late 1950s to appropriate the new discipline. His international title as design pioneer and his claimed and ascribed spiritual fatherhood of the Bauhaus principles turned this previously politically controversial figure into a welcome but versatile national symbol of the new industrial design movement.

The second part deals with the promotion of industrial design in the light of Cold War politics. This part is especially focused on the intense personal and professional networks and practises that reached far beyond the borders of the Belgian nation and is accordingly characterized by a transnational gaze. The third chapter then explores the Belgian visions, ambitions and fantasies of promoting "European" design in the 1960s. Although the institutionalization of industrial design in Belgium was largely prompted by national interests, it was characterized, curiously enough, by a supranational and international scope. In 1961, the Institute of Industrial Design would even take the initiative in setting up a true supranational body for design promotion with the aim of negotiating the shape and form of the future "European" design. Although this "European" design was never realized, this case gives new insights into the political and economic hopes grafted onto industrial design in postwar Europe that transcended solely national scope. The fourth chapter then takes a look at the design exchanges between Brussels and Moscow in the 1970s within the framework of Cold War détente diplomacy. Key in this chapter is the leading role played by the director of the Brussels Design Centre, Josine des Cressonnières, in the transnational design community. Her important position as secretary general of the International Council of Societies of Industrial Design (ICSID) form 1961 until 1977 facilitated many exchanges with counterparts around the world. Her good contacts with the Russian designer Yuri Soloviev were, for example, one of the reasons why Brussels got the Western world premiere of its Soviet counterpart's design show in 1973.

The final part analyzes the promotion of "Belgian" design by the Brussels Design Centre through a constructivist lens. The fifth chapter focuses on how the Design Centre crafted a distinct and exclusive modernist "Made in Belgium" style in the 1960s that had to present Belgium abroad as a fully-fledged modern competitive nation on the global markets. The chapter takes a closer look at the different mechanisms behind the installation of this "national style" that was internationalist in orientation. The final chapter goes on to examine how the Design Centre continued to forge a link between the Belgian territory and design in the 1970s while the varnish of the Belgian unitary state started cracking and federalization was on its way. The growing regional tensions at the end of the 1960s between the French-speaking and Dutch-speaking communities in Belgium hindered the regional dissemination of this "national" design doctrine. Around the same time, the "good" design legacy was questioned by the introduction of new activist and environmental design paradigms. The Design Centre subtly changed its strategies. Along with its regular promotion of "good" "Belgian" products, it held temporary exhibitions focusing on the design "environment" of the elderly and children, on public urban spaces, ecological issues and public transport.

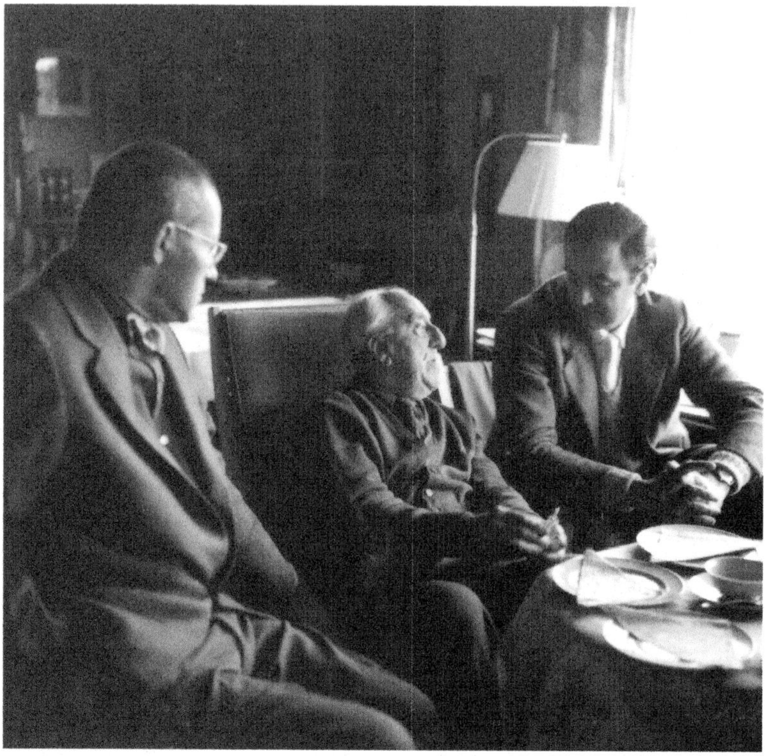

Fig. I.1. Max Bill, Henry van de Velde and Tomas Maldonado in Oberägeri (Switzerland), 1954. Source: Ulm, HfG-Archiv.

PART I
Competing Visions of Design

The phenomenon of industrial design prevailed in Belgium in the aura of postwar optimism and modernity. It was considered fresh and new, imported after foreign models. While it is certain that foreign countries such as America, Great Britain and Italy were very influential in the rise of industrial design in Belgium, and that the new postwar reality was key to its institutionalization, the new discipline did not, however, just appear out of the blue. On the one hand, it linked up through the agency of some pivotal members and exhibitions with the interwar legacy of the avant-garde movement 7 Arts, but also with educational reforms discussed at La Cambre and the Ministry of Public Education attempting to reunite the arts and industry in the atmosphere of the 1937 international Paris exhibition. On the other hand, the prewar traditions were consciously rehabilitated to give the Belgian offshoot some legitimation and body. The elderly architect and designer Henry van de Velde proved to be a very timely and appropriate national design mascot that could fit various competing economic, industrial, humanist and artistic agendas. Thus, instead of a smooth arrival, the 1950s were industrial design's heady years of doubt, experimentation and uncertainty.

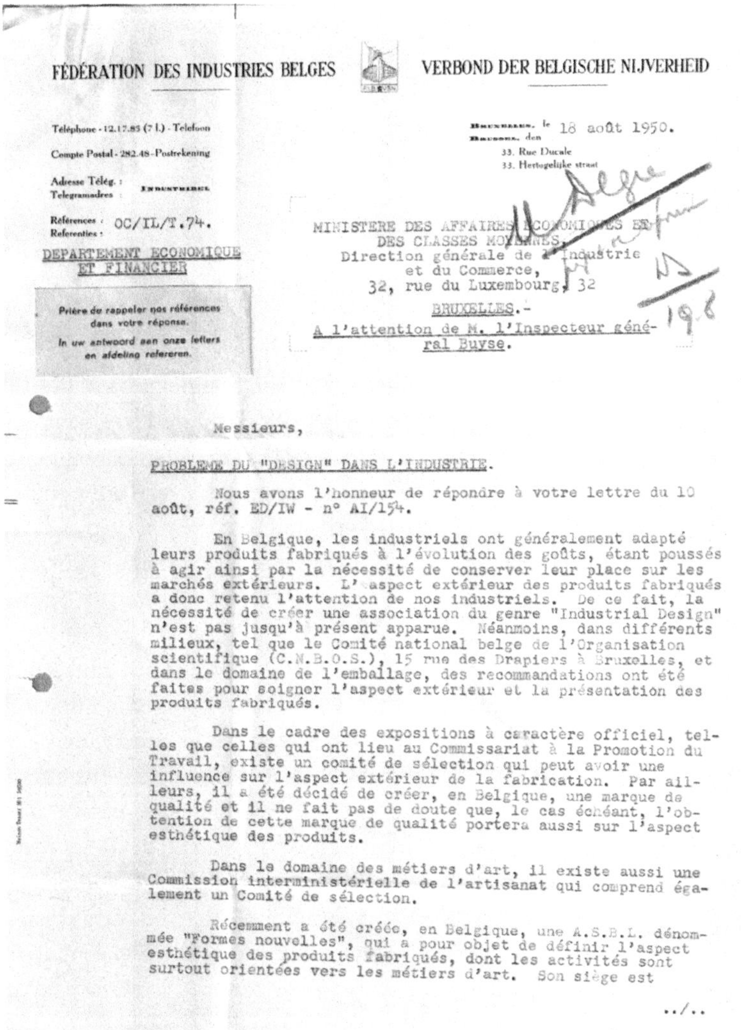

FÉDÉRATION DES INDUSTRIES BELGES — VERBOND DER BELGISCHE NIJVERHEID

Téléphone - 12.17.85 (7 l.) - Telefoon

Compte Postal - 282.48 - Postrekening

Adresse Télég. : / Telegramadres :

Références : / Referenties : OC/IL/T.74.

DEPARTEMENT ECONOMIQUE ET FINANCIER

Prière de rappeler nos références dans votre réponse.

In uw antwoord aan onze letters en afdeling refereren.

le / den 18 août 1950.

33, Rue Ducale

33, Hertogelijke straat

MINISTERE DES AFFAIRES ECONOMIQUES ET DES CLASSES MOYENNES,
Direction générale de l'Industrie et du Commerce,
32, rue du Luxembourg, 32
BRUXELLES.-
A l'attention de M. l'Inspecteur général Buyse.

Messieurs,

PROBLEME DU "DESIGN" DANS L'INDUSTRIE.

Nous avons l'honneur de répondre à votre lettre du 10 août, réf. ED/IW - n° AI/154.

En Belgique, les industriels ont généralement adapté leurs produits fabriqués à l'évolution des goûts, étant poussés à agir ainsi par la nécessité de conserver leur place sur les marchés extérieurs. L'aspect extérieur des produits fabriqués a donc retenu l'attention de nos industriels. De ce fait, la nécessité de créer une association du genre "Industrial Design" n'est pas jusqu'à présent apparue. Néanmoins, dans différents milieux, tel que le Comité national belge de l'Organisation scientifique (C.N.B.O.S.), 15 rue des Drapiers à Bruxelles, et dans le domaine de l'emballage, des recommandations ont été faites pour soigner l'aspect extérieur et la présentation des produits fabriqués.

Dans le cadre des expositions à caractère officiel, telles que celles qui ont lieu au Commissariat à la Promotion du Travail, existe un comité de sélection qui peut avoir une influence sur l'aspect extérieur de la fabrication. Par ailleurs, il a été décidé de créer, en Belgique, une marque de qualité et il ne fait pas de doute que, le cas échéant, l'obtention de cette marque de qualité portera aussi sur l'aspect esthétique des produits.

Dans le domaine des métiers d'art, il existe aussi une Commission interministérielle de l'artisanat qui comprend également un Comité de sélection.

Récemment a été créée, en Belgique, une A.S.B.L. dénommée "Formes nouvelles", qui a pour objet de définir l'aspect esthétique des produits fabriqués, dont les activités sont surtout orientées vers les métiers d'art. Son siège est

../..

Fig. 1.1. Letter from the Federation of Belgian Industries on the "problem of design", August 18, 1950. Source: Brussels, State Archives, BE AGR/ARA Prijzendienst (510-2331), 609: P. Van Lint [Secretary General of the Federation of Belgian Industries], Letter to Buyse, Inspecteur Général de direction general de l'Industrie et de Commerce (18/08/1950).

CHAPTER I

Prefigurations (1950–1956)

1. INTRODUCTION: THE PROBLEM OF "DESIGN"

Concerns about the promotion of "industrial design" as a particular discipline and method were expressed for the first time in the summer of 1950. At the request of the British Council of Industrial Design, inquiries were made at the address of the Belgian government on this issue.[1] The precise question was whether organizations existed that were devoted to the promotion of design. The issue, labeled in the correspondence as the "problem of design in the Belgian industry," eventually ended up, through the mediation of the Ministry of Economic Affairs and Middle Class and of the Ministry of Foreign Affairs and Foreign Trade, at the address of the Federation of Belgian Industries.[2] The federation replied to the Belgian officials that no such organization existed or was necessary, since the Belgian industrialists spontaneously adapted their products to the evolution of consumer taste prompted by competition on foreign markets (fig. 1.1).[3]

Six years later, however, two brand-new institutes were established to promote the new phenomenon of industrial design: the Institute for Industrial Design for Belgium and the Grand Duchy of Luxembourg,[4] and the Signe d'Or.[5] Both were granted financial aid and protection by the Belgian government. These were the first "official" industrial design bodies in postwar Belgium in that they were acknowledged and funded by the government. However, what happened between 1950 and 1956 that prompted the creation of two institutions of design? Where did this need come from and how was "industrial design" introduced? What visions existed in relation to industrial design? Who were the key figures and what were their motives? This chapter tries to answer these questions by looking at events, publications and networks of the period that preceded the creation of the first official design institutions.

These pioneering years in the early 1950s were one big jumble of ephemeral initiatives instigated by an amalgamation of artists, craftsmen, architects, cultural policymakers, economists, officials, politicians and industrialists. Each had their own motive. For many artistic minds, industrial design was a way to connect with interwar concerns on social art and foreign movements. Belgian industrialists, on the other hand, were concerned with increasing their sales and productivity to cope efficiently with the strong competition on internal and foreign markets. Officials, in turn, faced a reducing Belgian economy as a result of the outdated industry and the onset of the liberalization of international markets. The official instalment of industrial design in Belgium thus was not a top-down process, but instead emerged from various interests ranging from the arts, industry and government.

The study of this period is significant for several other reasons. First, it explores the continuation of the interwar artistic legacy in the first promotional events of industrial design. Second, it describes the initiation of a new "commercial" track of postwar design promotion. Next to the "artistic" tradition that originated in the interwar avant-garde, new developments were marked by a concern about the commercial and economic impact of industrial design. Finally, it sheds light on the first inclination of official circles towards the promotion of industrial design, and especially its endorsement by the Ministry of Economic Affairs.

2. INDUSTRIAL DESIGN AS THE TAIL END OF INTERWAR LEGACIES

While Belgian industry was, in 1950, still unsure about the notion of industrial design — as the above-mentioned letter illustrates — the artistic and cultural circles had by then already endorsed this new phenomenon. The first to pick up the early postwar signals of the international "industrial design" phenomenon were personalities previously involved in the interwar artistic circles. The avant-garde magazine *7 Arts* was one of the hotbeds for later design advocates as it preached the machine as a source of beauty and as a premise for social art. Many of the first adherents to industrial design were equally involved in prewar discussions on the reconciliation between the arts and industry that had arisen in the 1930s in the circles of the Ministry of Public Education, the La Cambre design school and the short-lived Art and Industry organization. Many of these latter discussions converged around the Belgian participation in the 1937 world exhibition in Paris. After the war, the same concerns were reiterated by Minister of Public Education Pierre Harmel in the 1952 exhibition *L'Art Monumental*, organized at the Center for Fine Arts in Brussels. While the exhibition mainly displayed interwar artworks, the show kindled a new enthusiasm that resulted in the first events of industrial design in Brussels and Liège. Thus, the rise of modern industrial design in Belgium ensued not so much from a radical break with the past, but from a historical reach backward that reiterated the prewar concerns about art and industry.

2.1 Social Preoccupations

The Formes Nouvelles (New Forms) organization was established in 1950 at the initiative of the young architect Axel Lemesre and his decorator brother Eric Lemesre in Brussels.[6] The organization gathered together urbanists, decorators, architects, economists and art critics. The association's first articles defined its aim broadly: to generate and promote new forms in the realm of art, techniques and literature.[7] However, the focus soon shifted to furniture, interior design and housing problems, emphasizing the importance of the intellectual and aesthetic aspects as well as the technical one.[8] Although Formes Nouvelles was quite active and associated with the realm of crafts, it showed interest in the new discipline of industrial design, albeit always connected to domestic culture.[9] In 1951, it attended the first International Design Congress organized by the British Council of Industrial Design (CoID) in London (fig. 1.2).[10] After this international gathering, Formes Nouvelles would profile itself as the Belgian equivalent to the famous British Council.

Formes Nouvelles's Vice-President Marcel Baugniet (1896–1995) probably introduced this industrial design focus. He was a painter, architect, designer and critic and started his career after the First World War in 1918 with abstract and cubist art works. He soon moved towards a constructive art vision.[11] Significant to his intellectual evolution was the avant-garde magazine *7 Arts* (1922–1928). The magazine was set up by a group of artists, architects and literary men, among whom were the brothers Bourgeois — Victor (1897–1962) and Pierre (1898–1976) — Pierre-Louis Flouquet (1900–1967), Karel Maes (1900–1974) and Georges Monnier (1892–1974). The publication had an international audience and was inspired by constructivist views such as Le Corbusier's magazine *L'Esprit Nouveau* (1920–1925). It eulogized the machine as a symbol of the new civilization.[12] Marcel Baugniet soon became part of its entourage. The magazine's original objects were formulated very clearly in its first issue in 1922: "Notre object est vaste. 7 Arts: tous les arts. Et par eux, la pénétration en tout la vie individuelle ou collective."[13] The critic and man of letters Maurice Casteels (1890–1962) was in charge of covering the "industrial arts" in the first years of the magazine. In his introductory account on the industrial arts, he puts it simply: "The machine is the most efficient means to achieve social art."[14] And social art was for him equal to modern art. From 1924 onwards, Marcel Baugniet would advocate, in the same vein as Maurice Casteels, the beauty of the machine in the pages of *7 Arts*.[15]

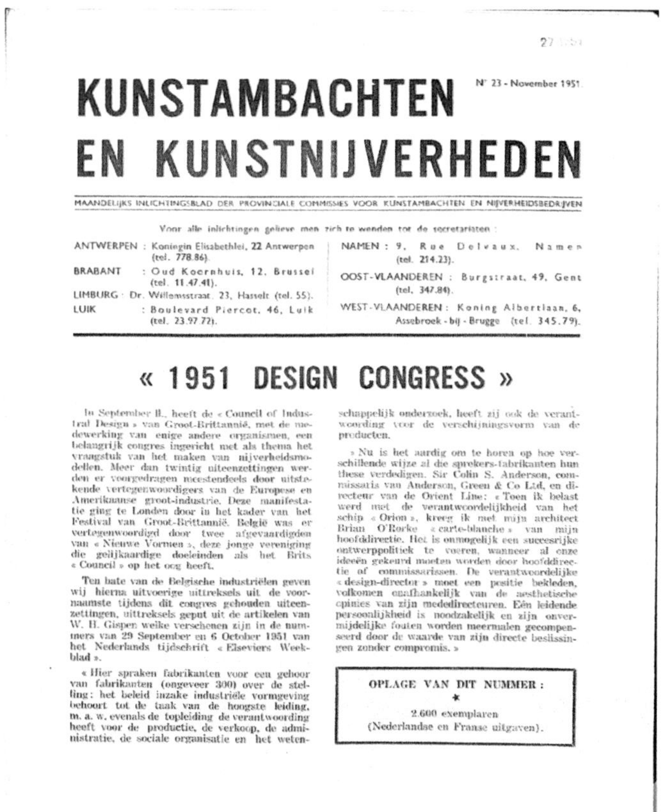

N° 23 - November 1951

KUNSTAMBACHTEN EN KUNSTNIJVERHEDEN

MAANDELIJKS INLICHTINGSBLAD DER PROVINCIALE COMMISSIES VOOR KUNSTAMBACHTEN EN NIJVERHEIDSBEDRIJVEN

Voor alle inlichtingen gelieve men zich te wenden tot de secretariaten :

ANTWERPEN : Koningin Elisabethlei, 22 Antwerpen (tel. 778.86).
BRABANT : Oud Koornhuis, 12, Brussel (tel. 11.47.41).
LIMBURG : Dr. Willemsstraat, 23, Hasselt (tel. 55).
LUIK : Boulevard Piercot, 46, Luik (tel. 23.97.72).
NAMEN : 9, Rue Delvaux, Namen (tel. 214.23).
OOST-VLAANDEREN : Burgstraat, 49, Gent (tel. 347.84).
WEST-VLAANDEREN : Koning Albertlaan, 6, Assebroek-bij-Brugge (tel. 345.79).

« 1951 DESIGN CONGRESS »

In September ll., heeft de « Council of Industral Design » van Groot-Brittannië, met de medewerking van enige andere organismen, een belangrijk congres ingericht met als thema het vraagstuk van het maken van nijverheidsmodellen. Meer dan twintig uiteenzettingen werden er voorgedragen meestendeels door uitstekende vertegenwoordigers van de Europese en Amerikaanse groot-industrie. Deze manifestatie ging te Londen door in het kader van het Festival van Groot-Brittannië. België was er vertegenwoordigd door twee afgevaardigden van « Nieuwe Vormen », deze jonge vereniging die gelijkaardige doeleinden als het Brits « Council » op het oog heeft.

Ten bate van de Belgische industriëlen geven wij hierna uitvoerige uittreksels uit de voornaamste tijdens dit congres gehouden uiteenzettingen, uittreksels geput uit de artikelen van W. H. Gispen welke verschenen zijn in de nummers van 29 September en 6 October 1951 van het Nederlands tijdschrift « Elseviers Weekblad ».

« Hier spraken fabrikanten voor een gehoor van fabrikanten (ongeveer 300) over de stelling: het beleid inzake industriële vormgeving behoort tot de taak van de hoogste leiding, m. a. w. evenals de topleiding de verantwoording heeft voor de productie, de verkoop, de administratie, de sociale organisatie en het wetenschappelijk onderzoek, heeft zij ook de verantwoording voor de verschijningsvorm van de producten.

» Nu is het aardig om te horen op hoe verschillende wijze al die sprekers-fabrikanten hun these verdedigen. Sir Colin S. Anderson, commissaris van Anderson, Green & Co Ltd, en directeur van de Orient Line: « Toen ik belast werd met de verantwoordelijkheid van het schip « Orion », kreeg ik met mijn architect Brian O'Rorke « carte-blanche » van mijn hoofddirectie. Het is onmogelijk een succesrijke ontwerppolitiek te voeren, wanneer al onze ideeën gekeurd moeten worden door hoofddirectie of commissarissen. De verantwoordelijke « design-director » moet een positie bekleden, volkomen onafhankelijk van de aesthetische opinies van zijn mededirecteuren. Eén leidende persoonlijkheid is noodzakelijk en zijn onvermijdelijke fouten worden meermalen gecompenseerd door de waarde van zijn directe beslissingen zonder compromis. »

OPLAGE VAN DIT NUMMER :
*
2.600 exemplaren
(Nederlandse en Franse uitgaven).

Fig. 1.2. Report by Leon-Louis Sosset of the first International Design Congress in 1951 in London. Source: Sosset, Leon-Louis. "1951 Design Congress." *Kunstambachten en Kunstnijverheden. Maandelijks Inlichtingsblad der Provinciale Commissies voor Kunstambachten en Nijverheidsbedrijven*, no. 23 (1951).

After the war, Baugniet would — together with his friend and colleague Pierre-Louis Flouquet — continue to promote the social potential of the machine as a loyal advocate of industrial design.[16] For him, industrial design was the perfect domain for the artist to engage socially with life, one of the main concerns in his discourse.[17] He was a great adherent to Henry van de Velde's principles of modern esthetics, which he saw as the precursor of the industrial design doctrine. In that sense, it is not surprising that Fierens' short 1947 monograph on Baugniet argues that a talented "inventor of structures" and decorator is much more exceptional and valuable than a good painter.[18] In 1950, within the framework of his participation in the *Schoner Wonen/ Logis 50* exhibition, he stated: "We know that inexpensive furniture can only be obtained through one method: rationalization." In spite of his great advocacy for industrial design, Baugniet was quite critical of developments in the mid-1950s. He urged that, for it to fulfill its high expectations, intense and constant propaganda was needed both in schools and with regard to the general public.[19] Moreover, he commented on the new profession of industrial designer. For him, an industrial designer who did not consider the social function of a product was a *faux* artist.

Within the Formes Nouvelles organization, Baugniet was the only member that belonged to the old artistic generation of the interwar avant-garde to hold an executive position. Although mainly developed before the war, Baugniet's body of thought was apparently not outdated by the 1950s. Some of the younger Formes Nouvelles members shared a similar perspective. The president and art critic Léon-Louis Sosset (1913–1994) devoted several articles to the promising discipline of industrial design in the arts and crafts press.[20] The young architect Willy van der Meeren (1923–2002) was also an adherent to the social aspect of industrial production on a large scale.[21] Especially in the second half of the 1950s, Van der Meeren advocated industrial design through the press and his furniture and architectural creations (fig. 1.3).[22] His industrial aspirations became reality when the Belgian firm Tubax produced his furniture of metal frames and wood on a large scale. For critic Léon-Louis Sosset, this was the first action in Belgium concerning serially produced modernist furniture.[23] Throughout his career, Van der Meeren considered the industrial to be of paramount importance, albeit only in connection with the social.

Another figure that can be placed in the same line of thought is design critic Karel Horemans (1920–1993), known under his pseudonym

Fig. 1.3. Special issue on "Esthétique Industrielle" compiled by Willy van der Meeren for the periodical *Architecture* 55, no. 14, 1955. © Collection Flanders Architecture Institute - Collection Flemish Community.

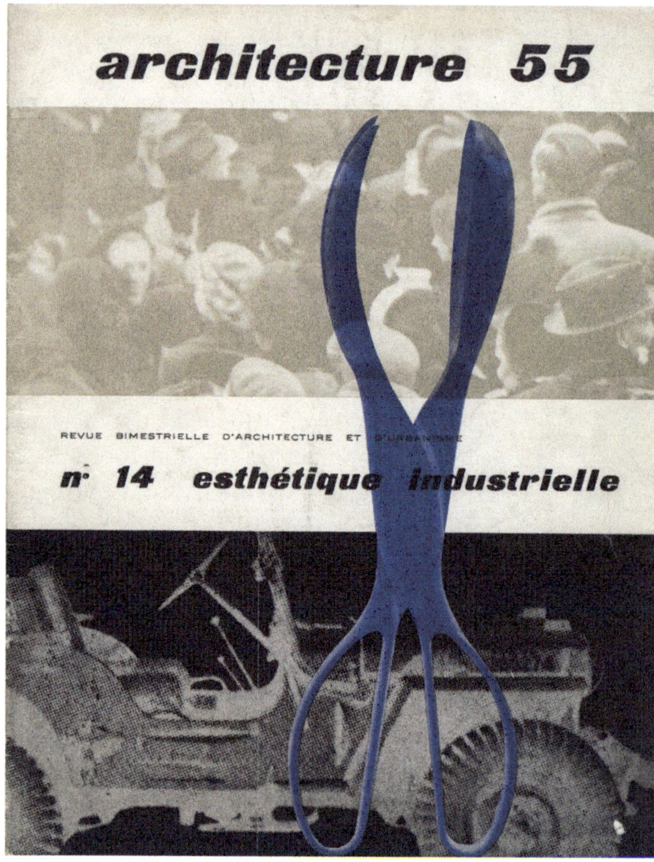

of K.N. Elno. He did not take part in the activities of the postwar organizations but followed them very closely in his capacity as critic. Design historian Fredie Floré elaborated on his intellectual journey and describes a movement from the realm of the fine arts to that of industrial design, architecture and interior architecture in the early 1950s.[24] As was the case for many artistic minds, the social possibilities of these domains appealed to him: the reconciliation of art and life. In his view, the true social responsibility of the artist was situated in the material fabric of everyday life. In view of the fact that useful objects designed by an artist function as "works of art" in series and through their daily use, art can become part of, and transform, everyday life.

2.2 Resuming Interwar Efforts to Unite the Arts and Industry

Besides Formes Nouvelles, a second organization moved by social and moral grounds took an interest in the promotion of industrial design in the early 1950s. This organization was Art et Travail (Art and Labor) which was created in 1953 in Brussels. The applied perspective of Art et Travail, however, was in contrast to that of Formes Nouvelles: instead of focusing on the domestic culture, it concentrated on the laborer's working environment. The movement was mainly based on the prewar efforts to unite the arts and industry that was then prevailing in La Cambre circles and the short-lived Art and Industry organization.

2.2.1 L'ART MONUMENTAL

The origins and initial impetus of the organization can be dated to 1952, when the Ministry of Public Education organized the exhibition *L'Art Monumental dans les établissements publics et industriels* (Monumental Art in Public and Industrial Buildings) at the Center for Fine Arts in Brussels (Palais des Beaux-Arts/Paleis voor Schone Kunsten) (fig. 1.4).[25] The primary aim of the exhibition was to illustrate the alliance between the arts and industry or aesthetics and the economy. Various art forms were put on show: sculpture, fresco, mural painting, ceramics, *dinanderie* and glass works. Some exhibits were, for example, Ossip Zadkine's bas-relief of a Brussels cinema (1932), Paul Delvaux's painting for the Casino in Ostend (1952), and Marc Mendelson's panels and textiles for the restaurant of a passenger ship (1950). In the catalog, Minister of Public Education Pierre Harmel (1911–2009) described the exhibition as inspiration for artists, managers and laborers.[26]

The critic Leon-Louis Sosset (1913–1994) reviewed the exhibition positively and described it as probably one of the first steps towards a true "synthesis of arts".[27] With this label, Sosset hinted at the interwar ideals of the *gesamstkunstwerk*. The connections with the interwar arts scene were indeed quite obvious. First of all, the exhibition was organized by the older Lucien Christophe (1891–1975), the General Director of the Fine Arts and Literature Department of the Ministry of Public Education, who was very well known for his 1936 report on the coordination between artistic and technical education and his plans to set up a service to unite art and industry. The exhibition seemed actually to restage his prewar interests on the unity of art and industry. Not surprisingly then, a large amout of the displayed art works was recycled from the 1937 Belgian participation at the international exposition in Paris. Moreover, many of the artists included were affiliated with the interwar organization Art and Industry that embodied these prewar networks and concerns.[28]

Critic Léon-Louis Sosset (1913–1994) applauded the exhibition and described it as probably one of the first steps towards a true "synthesis of the arts".[29] By using this label, Sosset hinted at the interwar ideals of the *gesamstkunstwerk*. The connections with the interwar arts scene

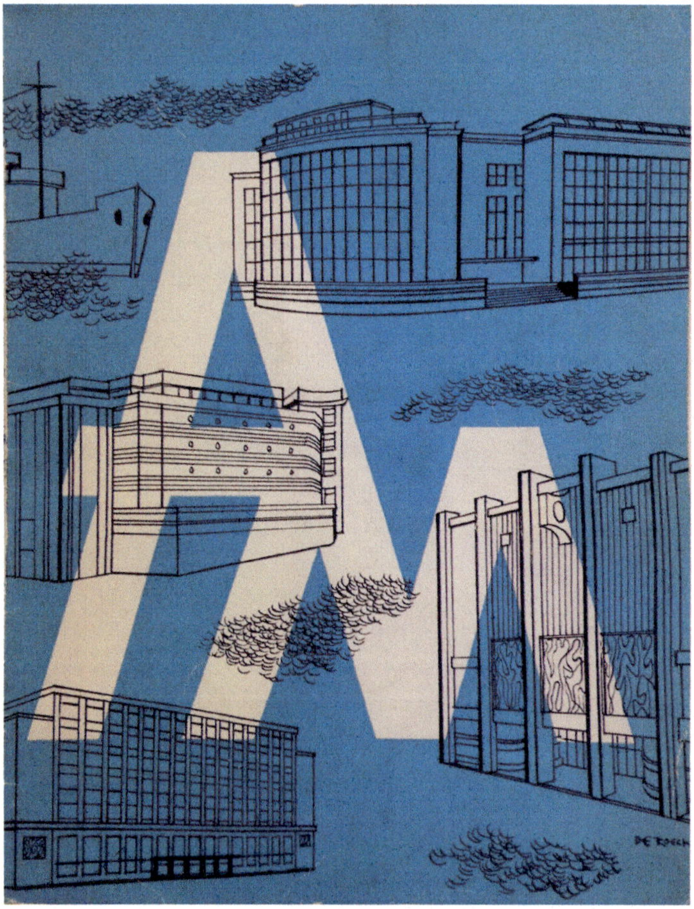

Fig. 1.4. Catalog cover designed by De Roeck for for the *"L'Art Monumental"* exhibition at the Palais des Beaux-Arts, Brussels, 1952. Source: Harmel, Pierre, and Lucien Christophe, "L'Art Monumental dans les établissements publics et industriels," edited by Palais des Beaux-Arts Bruxelles. Brussels: Editions de la Connaissance, 1952. © Lucien De Roeck, 2018, Fonds Lucien De Roeck www.lucienderoeck.be.

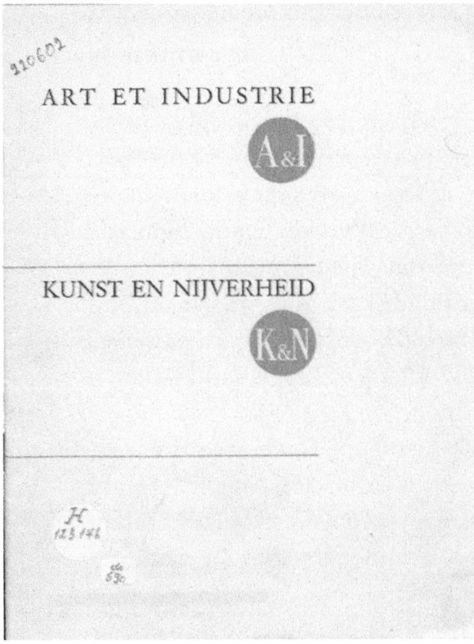

Fig. 1.5. Publication of the not-for-profit organization *'Art and Industry'*, Brussels, 1938. Source: H 123146, Collectie Stad Antwerpen, Erfgoedbibliotheek Hendrik Conscience.

were indeed quite obvious. First of all, the exhibition was organized by the older Lucien Christophe (1891–1975), the general director of the Fine Arts and Literature Department of the Ministry of Public Education, who was well known for his 1936 report on the coordination between artistic and technical education and his plans to set up a service to unite art and industry. The exhibition seemed actually to restage his prewar stakes on the unity of art and industry. Not surprisingly then, many of the artworks on show dated from the interwar period. A large number were recycled from the Belgian participation at the 1937 international exhibition in Paris. Moreover, many of the artists were affiliated with the interwar organization Art and Industry that embodied these prewar networks and concerns.[30]

The not-for-profit organization Art and Industry (Art et Industrie/ Kunst en Nijverheid) was set up in 1936 in Brussels following the example of the German *Werkbund* and the Swedish *Sloyd* (fig. 1.5).[31] The organization was composed of politicians, artists, craftsmen, industrialists, architects and men of letters.[32] The president, Louis Piérard (1886–1951), was a writer, journalist, art critic and socialist member of parliament. He was familiar with the Belgian avant-garde scene since he had relaunched, together with poet Pierre Bourgeois, the *Section d'Art* of the Belgian labor party.[33] The creation of this organization was in response to the 1936 report by Lucien Christophe that denounced the lack of coherence between technical and artistic education in Belgium.[34] Along with this reply to Lucien Christophe's insights, Louis Piérard also anticipated Art and Industry as a preparation for the Paris international exhibition entitled *Arts et Techniques dans la Vie Moderne* that would take place a year later, in 1937. Soon the organization was recognized by the Belgian government and was supported by Minister of Public Works and Unemployment Hendrik De Man, a socialist, and by Minister of Public Education François Bovesse.[35] Shortly after its establishment, Minister of Economic Affairs Philippe Van Isacker, a catholic, also supported the organization.

Many of Art and Industry's artistic members, some of whom were also included in the 1952 *L'Art Monumental* exhibition, taught at Henry van de Velde's School of Architecture and Applied Arts La Cambre/ Terkameren in Brussels (better known under its alias La Cambre), such as architect Victor Bourgeois, painter and decorator Charles Counhaye (1884–1971), graphic artist Joris Minne (1897–1988), architect Jean De Ligne (1890–1985) and textile artist Elisabeth de Saedeleer (1902–1972).[36] Not very surprisingly then, the principles of the organization were close to Van de Velde's principles applied at La Cambre to bridge the gap between artistic and technical education. In its 1938 publication, the young organization stated that technique and art needed to become intimately connected in the objects that shape our daily environment.[37] These objects had to be strong, practical and comfortable, and their shape had to correspond to their function. With the creation of this organization, Piérard wanted to wage war against "poor" Belgian products on the market: "The Belgians can no longer be the workhorse of Europe which brings half-finished and unfinished products to the market."[38] One of the proposed strategies of this organization to improve national production was to award a label of "good taste" to the best products of Belgian industry.

In contrast to its high ambitions, the actions of Art and Industry were rather limited. Its contribution to the Belgian participation in the 1937 international exhibition in Paris was restricted to the participation of its members in the artistic outcome and organizing committee. For example, Charles Counhaye designed the ceiling of the Belgian pavilion, and the Belgian display included a type design for the equipment of a library by Victor Bourgois and tapestries designed by Floris Jespers (1889–1965) and Rodolphe Strebelle (1880–1959). Art and

Industry was also represented in the Technical Committee under the direction of Henry van de Velde by several of its founding members such as: Jan-Albert Goris (1899–1984), Roger Avermaete (1893–1988) and its president, Louis Piérard.[39] However, it is difficult to estimate the precise influence of Art and Industry, given its infancy. At that point, the organization had existed for only one year. Shortly after the Paris exhibition, Art and Industry lost its government support because of a conflict between Louis Piérard and Henry van de Velde concerning the creation of a center for applied and decorative arts at the Heizel in Brussels.[40] The 1952 *L'Art Monumental* exhibition under the guidance of Lucien Christophe seemed to revive these same issues that were so pressing in the 1930s.

The real importance of this exhibition for the promotion of industrial design was manifested eight days before its opening by Queen Elisabeth, when Minister of Public Education Pierre Harmel gathered industrialists, artists and labor managers and urged them to unite their forces and work together. The Christian Democrat Harmel was at that time responsible for the organization of technical education. Accordingly, Harmel must have been familiar with the interwar ideas of his official Lucien Christophe and the problems that accompanied this issue. With this gathering on the eve of the exhibition, Harmel was aiming to bridge the historical gap between industry and art, as described in the catalog, by creating a favorable climate for developing a sense of taste and esthetics in the design of working places–as the movement's name "Art et Travail" indicates. The action was not only inspired by Christophe's plea to reconcile art and industry, but was also linked to Harmel's previous interest in bringing culture to the field of labor. At the close of the war he wrote a small booklet on how to overcome the gap between culture and profession in the prospect of the postwar restoration of Christian society.[41] The motives for this project then were grounded in a similar and rather paternalistic attitude towards the moral uplifting of the worker, as his speech of that evening illustrates:

> Labor must also be a source of beauty. Art must draw inspiration from the vigor and magnificence of labor and the laborer must open up to the beauty of art. Art must reveal to the laborer not only the dignity of his life, but also the disinterested aspects of culture. It is especially a question of the satisfaction of the workers, of their initiation to art and culture.[42]

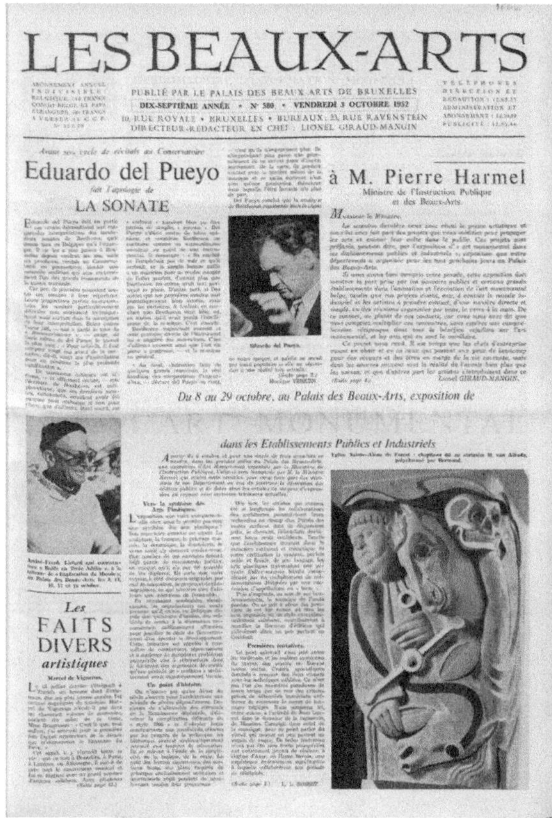

LES BEAUX-ARTS

PUBLIÉ PAR LE PALAIS DES BEAUX-ARTS DE BRUXELLES
DIX-SEPTIÈME ANNÉE • N° 580 • VENDREDI 3 OCTOBRE 1952
10, RUE ROYALE • BRUXELLES • BUREAUX: 23, RUE RAVENSTEIN
DIRECTEUR-RÉDACTEUR EN CHEF: LIONEL GIRAUD-MANGIN

Eduardo del Pueyo

LA SONATE

à M. Pierre Harmel

Ministre de l'Instruction Publique et des Beaux-Arts.

Du 8 au 29 octobre, au Palais des Beaux-Arts, exposition de

dans les Etablissements Publics et Industriels

Les FAITS DIVERS *artistiques*

Fig. 1.6. Open letter from Lionel Giraud-Manginto Minister Harmel in the periodical *Les Beaux-Arts*, no. 580, 1952. Source: H 70535, Collectie Stad Antwerpen, Erfgoedbibliotheek Hendrik Conscience.

Harmel's initial intention was to keep this movement informal as an encouragement for spontaneous collaborations and proposals that would organically inspire the future official action program of Art et Travail.[43]

Lionel Giraud-Mangin, director of the periodical *Les Beaux-Arts,* was invited to these gatherings and was very much enthused by Harmel's ideas.[44] He thought it would be a very big, advantageous and beautiful thing if indeed art did not begin at the museum's doors but was integrated into the construction of factories or the post office as something that goes without saying. Yet he argued in an open letter to the minister against this loose and informal character of the association in favor of a more solid shape (fig. 1.6). He was of the opinion that all the good intentions needed not only to be stimulated but also guided out of fear of mediocre or absurd results. He wanted, in his own words, to prevent the situation where "workplaces would become unappealing equivalents of memorials to the dead bound everywhere together with busts of King Albert that are an insult to the notable monarch."[45] Further, he advised the minister that such an organization should not be public but private, and that administrators from different backgrounds should be carefully chosen because, for him, this would influence the quality and value of the institution.

2.2.2 L'ART ET TRAVAIL

Giraud-Mangin's public letter seemed to have an impact, and by July 1953 the loose grouping was consolidated into a not-for-profit organization.[46] In this more rigid format, industrialists and officials connected to government services were members of the board of directors. In contrast to the older Formes Nouvelles, whose center of gravity was in artistic and creative circles, the Art et Travail movement became rooted in official and industrial circles. The initial aims still testified to an interwar inspiration as reflected by the *L'Art Monumental* exhibition: the design of workplaces and the search for beauty in manufactured objects.[47]

The most notable member of Art et Travail was its president, the industrialist and lawyer Jean del Marmol (1901–1971).[48] During World War II, he had won his spurs as a pivotal member of the anti-Communist Belgian Secret Army fighting the German occupier. He also played a major role in the organization of the 1935 world exhibition in Brussels, and worked together with Henry van de Velde for Belgian participation at the world exhibitions in Paris (1937) and New York (1939).[49] A deep friendship developed between the two men, and Jean del Marmol became a confidant of Van de Velde.[50] In 1950 Van de Velde even considered him as Teirlinck's successor as director of La Cambre.[51] Jean del Marmol saw Art et Travail as the continuation of these prewar initiatives, especially after his and Henry van de Velde's "failure" to continue the efforts of the 1937 Paris exhibition due to the departure of Minister Philippe van Isacker.[52]

At the time when Art and Travail was created, Jean del Marmol was the managing director of the Compagnie d'Afrique pour l'Industrie et la Finance[53] – a colonial company that was founded in 1951 to coordinate the colonial activities of the powerful Belgian concern of Léon Lambert.[54] In this organization, Jean del Marmol sat on the board of directors together with influential figures such as Léon Lambert

Léon-Louis Sosset even believed that of all the existing questions of the day, industrial design was the most important in 1954.[69] Press articles addressing the question of industrial design were burgeoning in specialized magazines for industry, trade[70] and the arts,[71] but also in daily newspapers. Architect Léon Stynen ascribes this attention to industrial design to the shock wave caused by the 1953 international design conference in Paris.[72] Under his directorship, in 1954 La Cambre opened the new Industrial Design Department, the first education program for designers in Belgium.[73] More importantly, however, 1954 was the year when industrial design was endorsed by official circles.

Minister of Public Education Pierre Harmel, who initiated the Art et Travail movement in 1952, began the year in March 1954 with a speech at the University of Liège on the subject of industrial design.[74] While Harmel in 1952 was still proclaiming a sort of paternalistic social concern to uplift labor through art, two years later with this speech he took a new radical turn in his discourse towards a more commercial approach. The tone of his discourse was colored by the writings of American designer Raymond Loewy, to whom Harmel owes the telling title of his speech: "La Laideur se vend mal", the French title of Loewy's straightforward book *Never Leave Well Enough Alone* from 1951. Raymond Loewy was one of the first-generation American designers that rose during the interwar years and gained celebrity status in postwar Europe. Most of his designs — from refrigerators to trains — made use of the popular American esthetic of streamlined curves, known as "styling." He positioned the role of design at the service of commerce and consumerism and was therefore known to define good design as upward sales curves. Although sharply criticized in Europe, his commercial approach was altogether of great influence — as illustrated by the reference made by Harmel.[75] In his speech, the Belgian minister regretted that Belgium was lagging behind with regard to industrial design and stimulated the start of a new movement in industrial circles to make up lost ground. This change of direction in Harmel's discourse — from esthetics to economics — was in line with the new direction of Art et Travail and marks the official attitude that was adopted in 1954 with regard to design promotion.

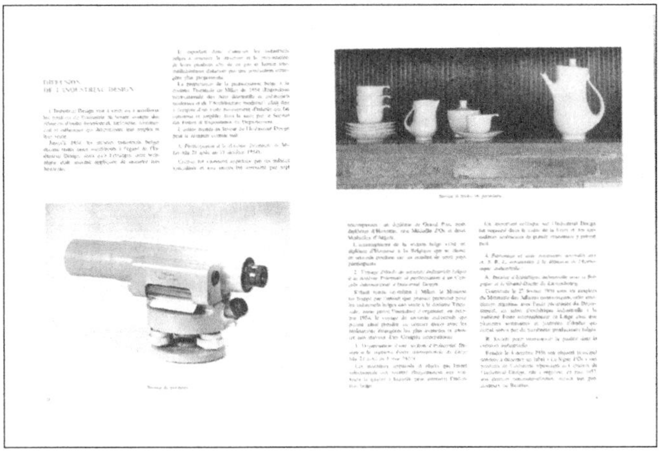

Fig. 1.7. The promotional brochure *Quatre Ans d'Action Économique* presented the results of four years of economic policy under Minister Jean Rey including the "Diffusion de l'Industrial Design." Source: "Quatre ans d'action économique." edited by the Ministère des Affaires Economiques. Brussels: Ministère des Affaires Economiques, 1958, 38-39.

3.1 Jean Rey's Policies

It was Minister of Economic Affairs Jean Rey (1902–1983), a young liberal, who was the first to actually include "industrial design" in his official policies (fig. 1.7).[76] Rey was a member of the first social-liberal cabinet under Prime Minister Achille van Acker that entered government in April 1954.[77] The ministry's justification for elaborating on the theme of industrial design made no bones about it:

> The objection will certainly be raised that *considerations of a commercial nature* are guiding the development of the "industrial design" movement; and that industrialists are resorting to this technique *to increase their sales figures and not out of love for art.*[78] (Italics added)

Rey's attitude toward industrial design was indeed largely inspired — just like that of the Christian Democrat, Harmel — by the commercial American approach. His 1954 speech on the dynamics of trade for an audience of Belgian industrialists in Liège is illustrative of his stand. In this talk, he drew special attention to the "creative"aspect which, next to production and sales, was growing in importance, but was in many cases misunderstood.[79] To make his point he gave the example of the exceptional boom in American industry that, according to him, was largely due to industrial design.

The Belgian government was not the only one to take an interest in industrial design at that time. Many European countries were investing in industrial design for similar commercial reasons. The initiatives were many. Take, for example, the creation of the Council of Industrial Design London[80] (1944), the Dutch Instituut voor Industriële Vormgeving (1950),[81] the German Rat für Formgebung (1951),[82] the French Institut d'Esthétique Industrielle (1951) and Beauté-France,[83] The Swiss Die Gute Form (1952), the Italian Associazione per il Disegno Industriale (1956), the Portuguese Art, Industrial Architecture and Industrial Design Nuclei of the National Institute of Industrial Research (1960).[84]

This larger movement has to be understood against the backdrop of the economic challenges raised after the war by the liberalization of world trade and the outlook of the European Common Market.[85] Instigated by the famous 1944 Bretton Woods Accords, which provided the economic and monetary structure for the postwar world and stimulated the liberalization of the global markets, and the 1947 American Marshall Plan, which had to stimulate European economic recovery, the European economies became modeled on liberal capitalism.[86] Europeanization only reinforced these structures. In 1951 the European Coal and Steel Community (ECSC) was founded by six countries: Belgium, France, Germany, the Grand Duchy of Luxembourg, the Netherlands and Italy. The same six countries in 1957 signed the Treaty of Rome, which stipulated free trade between the six member states within a timespan of twelve years and the subsequent creation of a European Common Market. All this resulted in the foundation of two new institutions: the European Economic Community (EEC) and the European Atomic Energy Community (Euratom).

Jean Rey's position on industrial design was a response not only to the intimidating prospect of the liberalization of the markets but also to internal issues. Right after the war, the jump-starting of the Belgian economy was internationally considered a true "Belgian miracle."[87] Because the Belgian industrial infrastructure was largely undamaged, Belgian industry could resume its prewar production quite rapidly. However, from 1950, Belgium lost its privileged position, since its well-preserved machinery was soon outdated. Very quickly other European economies caught up, partly with the help of Marshall dollars. Accordingly, Jean Rey during his term as minister faced a shrinking Belgian economy: industrial production was dwindling and foreign trade was diminishing. Besides, the internal market was barraged with foreign products. Add to this the important event of the 1958 World Exhibition in Brussels. Known as Expo 58, it was the first one organized after World War II and was a huge international event — especially in the light of growing tensions between the two Cold War superpowers, the United States and the Soviet Union. The organization of this extraordinary event was entrusted by the Belgian government to the Ministry of Economic Affairs under Jean Rey, since it was in charge of the Fairs and Exhibitions Department.[88] Unsurprisingly then, the government subscribed to an economic expansion policy to stimulate the production of consumer goods in Belgian industry, which was then still dominated by the production of semi-manufactured goods.[89]

This challenging economic context stimulated huge enthusiasm at the Ministry of Economic Affairs for industrial design as a factor of productivity and sales promotion. This resulted in the funding of the first major Belgian undertakings on this subject: the *Art et Travail Show* at the International Fair in Liège and Belgian participation in the X Milan Triennial. Each had a different significance for the promotion of industrial design. The first, the *Art et Travail Show* at the International Fair in Liège, was set up to make Belgian industrialists aware of the existence of industrial design and to convince them to implement this method in their companies. The second, the participation at the X Milan Triennial, aimed to convince the international public in Milan of the high quality of "Belgian design." Both events, however, demonstrate a clear shift: next to the social concern as a driving force for the promotion of industrial design, the commercial motives attracted more and more official attention.

3.2 The Art et Travail Show in Liège

On the occasion of the sixth edition of the Liège International Fair (Foire Internationale de Liège) an exhibition was set up that, for the very first time in Belgium, was dedicated entirely to "industrial design".[90] The exhibition was organized on the initiative of the Art et Travail organization with the help of the Ministry of Economic Affairs, the not-for-profit organization Grand Liège, the federation for metal industries, Fabrimetal, the Belgo-Luxembourg Steel Information Center (fig. 1.8). The display was christened the *Art et Travail Show*

Fig. 1.8. Minister of Economic Affairs Jean Rey visiting the *Art et Travail* exhibition at the sixth Liège International Fair in 1954. Picture by D. Daniel. Source: "Chronique." *L'Ossature Métallique* 19, no.5 (1954): 288. © Infosteel.

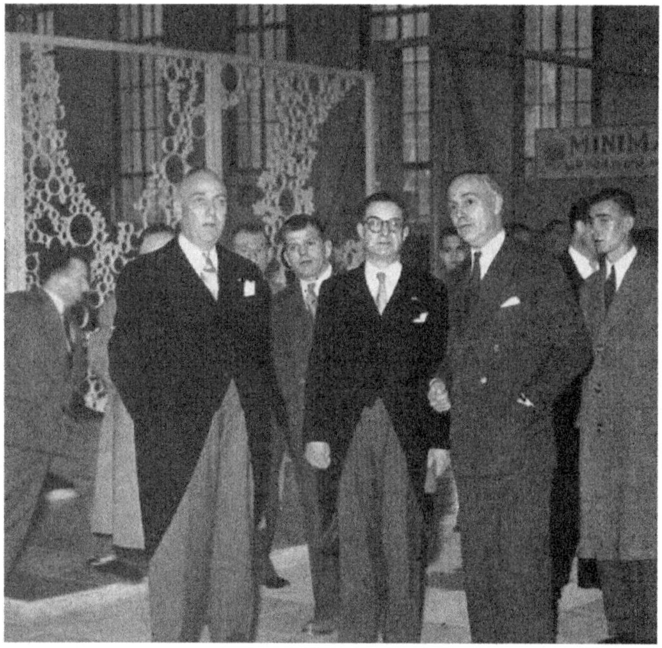

(*Salon Art et Travail*) after the young organization that took to heart the introduction of industrial design in Belgium.[91]

The *Art et Travail Show* was accordingly organized to make Belgian industrialists acquainted with "industrial design." The Liège International Fair was a well-chosen location for this, since it was an industrial, technical and rather specialist venue. The show was adjusted to its target group that was commonly seen as passive and traditional. Belgian industry was indeed often portrayed as suspicious of "esthetics."[92] Robert Bosmans of the Office of Foreign Trade, for example, stated that industrialists are convinced that the consumer's motive for buying a product does not depend on an esthetic appreciation conforming to the norms of an elite of "designers." Intending to convince Belgian industrialists, the *Art et Travail Show* defined design as something that was not added to make the object look better out of esthetic considerations, but as a technique applied from the conception of the object up to the completion of the prototype.[93]

The *Art et Travail Show* presented three separate exhibitions relating to industrial design. The first two were "imported" exhibitions:

The Beauty of Steel (*La Beauté de l'Acier*) by the German Paul Mahlberg and *Industrial Design in the USA* (*L'Industrial Design aux Etats-Unis*) by the New York Society of Industrial Design. The third was the *Industrial Design Show* (*Le Salon de l'Industrial Design*) by the duo made up of architect Constantin Brodzki and decorator Corneil Hannoset.

The desire to get these two foreign design exhibitions to Liège had already been expressed by Art et Travail at the 1953 Paris International Design conference. The exhibition of the New York Society of Industrial Design was only briefly mentioned in the official catalog and press.[94] This was probably due to its restricted documental scope. The exhibition was, after all, only based on documents sent by the Society on the "Problems of Industrial Design." Mahlberg's exhibition, by contrast, could count on much greater press coverage. The German Paul Mahlberg was a construction advisor for the German Steel Information Center, for which he conceived an exhibition on the topic of the beauty

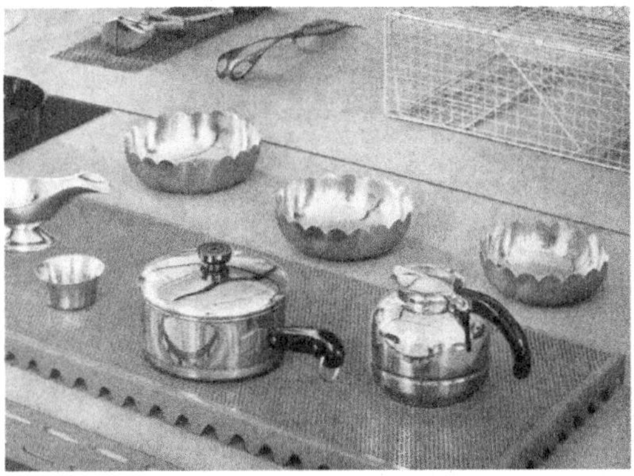

Fig. 1.9. Series of objects in stainless steel by Sartel (Liège) and Sidérur (Brussels) on display at Mahlberg's exhibition *The Beauty of Steel*. Source: Viénot, J. "Exposition 'La beauté de l'acier' à Liège." *L'Ossature Métallique* 19, no. 613-617 (1954). © Infosteel.

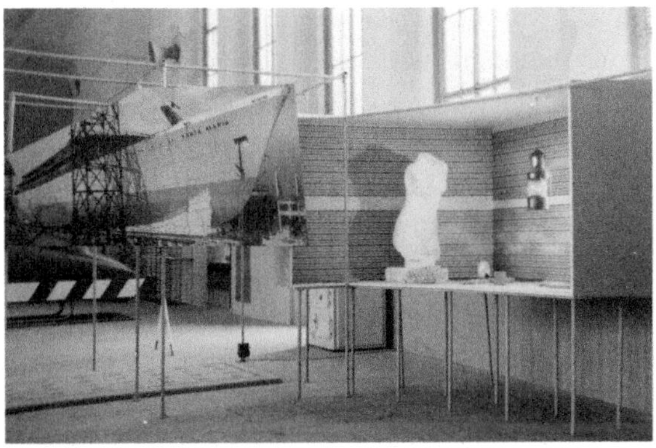

Fig. 1.10. View on Paul Mahlberg's exhibition *The Beauty of Steel* at the sixth Foire International in Liège in 1954. A picture of the Cockerill ship *Santa Maria* placed besides a Greek statue of Aphrodite. Source: Coutisse, Archives des Cressonnières.

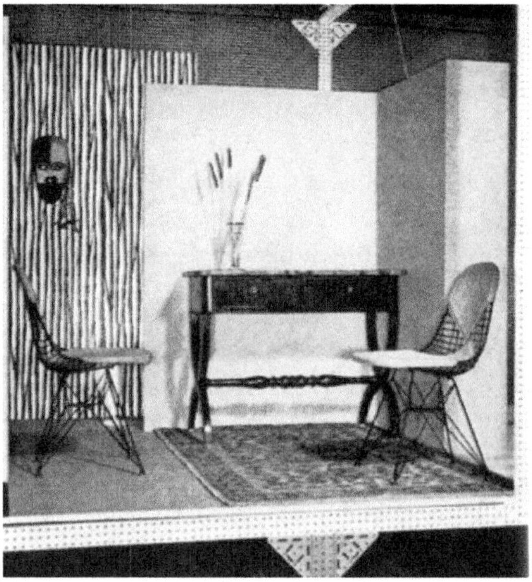

Fig. 1.11. Mahlberg presented domestic interiors where modern and old styles were mixed. Source: Viénot, J. "Exposition 'La beauté de l'acier' à Liège." *L'Ossature Métallique* 19, no. 613-617 (1954). © Infosteel

of steel in 1952.[95] Two years later, this exhibition was on show in Liège under the protection of the Belgo-Luxembourg Steel Information Center (CBLIA).[96] At the request of the CBLIA, the German exhibition was turned into an international one, although eventually most products on display were produced in Germany or Belgium.[97]

The spectrum of objects selected for this exhibition to illustrate the beauty of steel ranged from Miller furniture and a Melotte milking machine to a picture of Cockerill's huge *Santa Maria* ship and an artwork by German artist Norbert Kricke (figs. 1.9–10). Not only was the choice of objects Mahlberg's responsibility, but also the scenography. Press reports described, for example, an impressive piece of décor which consisted of a monumental frame made of a thousand sections of tubes of different diameters.[98] Mahlberg also recreated domestic interiors blending modern and old styles. Modern Miller chairs were, for example, combined with a primitive mask and an antique sideboard (fig. 1.11). According to art connoisseur and professor of Design Theory at La Cambre, Robert Delevoy, the virtue of the exhibition was its attempt to humanize the material of steel.[99]

Along with Mahlberg's show *The Beauty of Steel*, a third exhibition within the framework of the *Art et Travail Show* introduced industrial design to the public in Liège. This exhibition was entitled the *Industrial Design Show*.[100] Architect Constantin Brodzki (b. 1924) and decorator Corneille Hannoset (1926–1996) were appointed to set up the exhibition with the help of the young Italian artist Enrico Castellani and photographer Serge Vandercam. At the entrance, a large wall was devoted to the question "Qu'est-ce que l'Industrial Design?". Several definitions were suggested which were especially attuned to the interests of industrialists. Further on, a large diagram indicated the place

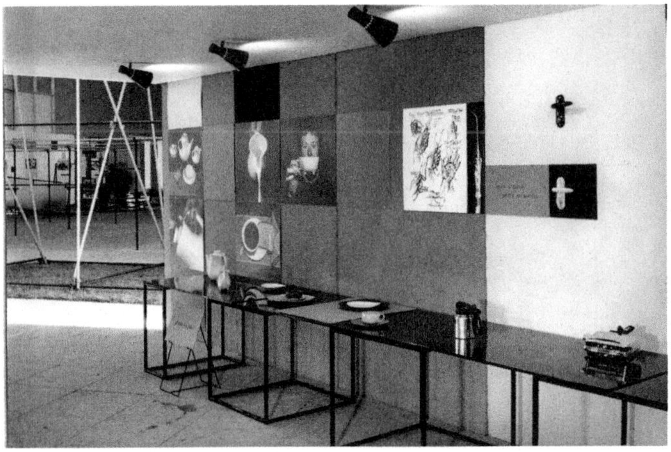

Fig. 1.12.View of the *Industrial Design Show* at the sixth Liège International Fair in 1954. In the front is Olivier Strebelle's coffee set for Cerabel. Source: Fonds Hannoset © AML (Archives et Musée de la Littérature).

of industrial design within the company. It was positioned at the same level as the traditional departments such as technical, manufacturing and sales. However, it was more intimately linked to the latter two. All this was done very consciously to stimulate industrialists to appeal to the process of industrial design.

The setting included objects as well as pictures to make this point clear (fig. 1.12). Pictures were especially used to represent larger heavy products, such as airplanes, trains, electric turbines and factory equipment. Small signs mounted on a separate support guided the attention of the visitors to specific aspects of small groups of objects. For example, the coffee set by Olivier Strebelle for Cerabel together with some other ceramics were accompanied by the words "a pleasure for the eyes" and Knoll furniture was juxtaposed with the sign "curiosity for forms" and so on (figs. 1.13–14). Starting with rather formal aspects, the guidelines on the small signs evolved, as the visitor progressed through the exhibition, towards the more technical aspects drawing attention to modes of production and the choice of materials. Belgian and foreign objects were mixed in this setting.

The three exhibitions were complemented by a one-day international conference on industrial design chaired by the director of the Usines Tubes de la Meuse, Léon Fraigneux, and the president of Art et Travail, Jean del Marmol.[101] The conference included foreign and Belgian speakers such as Kalff (director of Philips in Eindhoven),

Fig. 1.13. The coffee set by Olivier Strebelle for Cerabel together with some other ceramics was accompanied by a small sign saying "pleasure for the eyes". Source: Fonds Hannoset © AML (Archives et Musée de la Littérature).

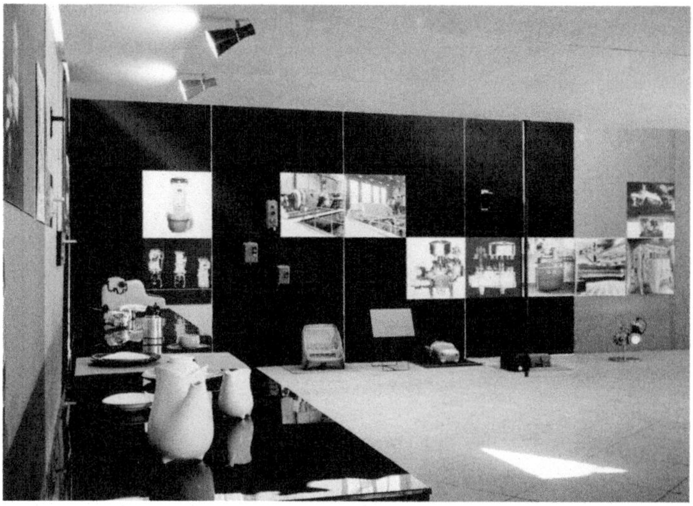

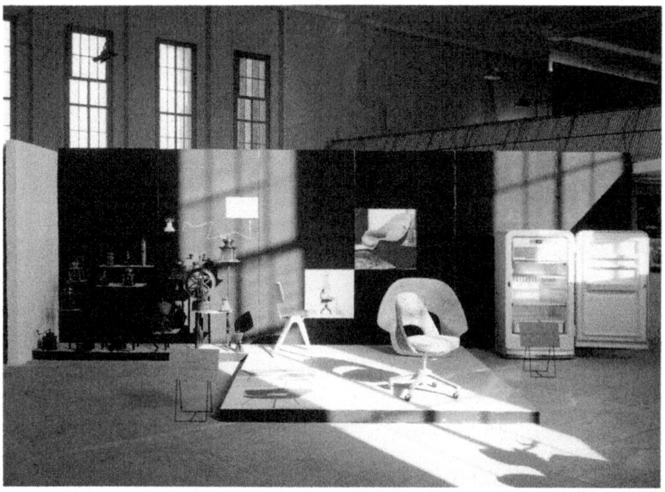

Fig. 1.14. In the middle Knoll furniture is placed next to the sign "curiosity for forms". On the left different models of coffee grinders from antique fanciful models to sober logical modern solutions illustrate an evolutionary process. Source: Fonds Hannoset © AML (Archives et Musée de la Littérature).

Sanders (director of the Institute for Industrial Design in Amsterdam), Mahlberg (an advisor of the German Steel Information Center in Düsseldorf) and Stynen[102] (director of l'Ecole nationale supérieure d'Architecture et des Arts décoratifs de la Cambre).

The Liège event hit the mark. In the early autumn of that year, a few months after this first and large show of industrial design, the first gathering of the "Commission Technique des Industrial Designers" was organized, reuniting the participants in the Liège conference.[103] At this meeting, unanimous agreement was reached on the need to promote industrial design for the development of the national economy. Subsequently, it was decided to found a national movement in order to disseminate this "new doctrine". Less than one month later the commission was invited by Jean Rey to visit the second pivotal event that established Belgian industrial design for the first time in the international arena, the Milan Triennial.

3.3 Belgian Design in Milan

The Milan Triennials — described by Paul Betts as the Olympics of good taste — have a history as international events for design, fine arts and urban planning that dates back to the 1920s.[104] Belgium was a loyal participant from the beginning.[105] The previous Belgian entry, for the ninth edition of 1951, organized by the Ministry of Public Education and Fine Arts, was strongly criticized for its chaotic and tasteless display. The organization of the Belgian section for the following edition in 1954, the first in fact to be dominated by industrial design, was entrusted to the Ministry of Economic Affairs.[106] In the eyes of the Belgo-Luxembourg Steel Information Center's magazine, this meant a clean break with the artistic craft tradition:

> Breaking with a tradition that had led it to confine itself to the realm of crafts (and we know how mediocre the experiences have been at that level), Belgium has made, for the first time, an effort to highlight the value, not only technical but also aesthetic, of our products.[107]

The X Milan Triennial in 1954, whose theme was "The honor of mankind and his work", was focused on industrial design, which was also notable in the Belgian section.[108] In contrast to its 1951 participation, Belgium showed a sober arrangement by architect André De Poerck, with mainly highly technical products on display, among which were microscopes, cameras, guns, a theater projector, electricity transformers and a portable device for industrial radiography (fig. 1.15).[109] This technical presence was completed with some products for domestic use and a few crafted products: glasses by Cristalleries du Val Saint-Lambert, a chair by architect Roger Thirion, a glass table by the designers René Baucher and Sylvie Feron and a coffee set by sculptor Olivier Strebelle for Cerabel. This "clean break" with the crafts — as the Belgian steel industry put it — was well received at the Triennial: it was rewarded with three honorary diplomas, two gold medals and a silver.[110] Overall, the Belgian press thought it was a praiseworthy at-

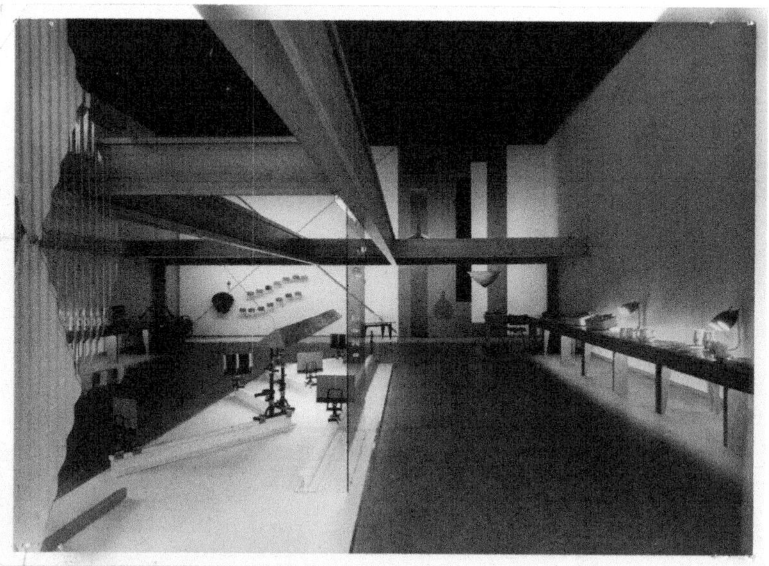

Fig. 1.15. View of the Belgian section at the X Milan Triennial. Picture by Serge Vandercam © SABAM Belgium 2018. Source: Archivio Fotografico © Triennale di Milano.

tempt, given the infancy of Belgian design.[111] Many critics did note, however, that the section was not a faithful representation of the Belgian situation but, in Sosset's words, "[...] concerned in fact only sporadic contributions, lost in everyday routine production."[112]

To bolster the impact of this Milan event, Jean Rey took the initiative to travel with a delegation of sixty Belgian industrialists (among whom were six members of the Liège "Commission Technique des Industrial Designers") to visit the X Milan Triennial and to participate in the corresponding international conference on industrial design

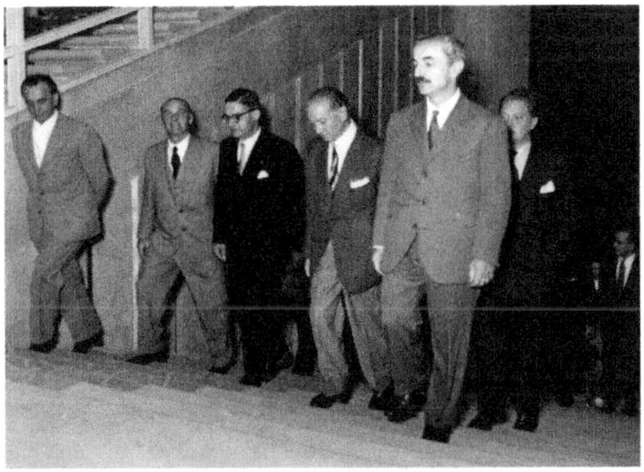

Fig. 1.16. The visit of Minister Jean Rey (the man in the middle wearing glasses) to the X Milan Triennial with Tommaso Ferraris, the Secretary of the Triennial (left) and the architect Carlo di Carli (right). Source: Archivio Fotografico © Triennale di Milano.

(fig. 1.16).[113] This three-day conference was an impressive gathering of many notable design personalities such as Gillo Dorfles, Alberto Rosselli, Jacques Viénot, Walter Dorwin Teague, Paul Reilly and Max Bill.[114]

4. THE FIRST INSTITUTIONS FOR INDUSTRIAL DESIGN

Rey's strenuous action in favor of industrial design would eventually result in more durable initiatives. Especially the industrial bastion in Liège proved to be a fruitful ground for the industrial design movement. The idea of uniting all those involved into a sort of organization after the by then well-known foreign examples thrived from 1954. But in Brussels too, a group of artists and policymakers around the Center for Fine Arts took an interest in promoting "good" industrial design.

4.1 The Takeoff Run in Liège

In Liège, the enthusiasm triggered by the significant design events of 1954 was translated into the foundation of two institutions: the Belgian Association of Industrial Designers (1954) and the Industrial

Design Institute (1955). The first was established by the "Commission Technique des Industrial Designers" in November 1954. Particularly stimulated by the Milan trip with Minister Jean Rey, this Liège-based "commission" would swing into action and establish, with the moral support of the industry, this new Belgian Association of Industrial Designers (Association Belge des Industrial Designers).[115]

The primary aim of the Association was to unite everyone who was active as a *styliste industriel*. This resulted in a mash-up of engineers, technicians, commercial artists, industrial draftsmen, publicity agents, decorators and architects. Furthermore, the association was concerned with setting up a center of documentation, information and orientation at the service of Belgian industry. The center had to stimulate the realization of "models with an aesthetic value combined with a better economy of means and greater technicity."[116] The bottom line, however, was the introduction of industrial design to increase the sales of Belgian products.

The older Charles Graffart (1893–1967) became its president.[117] He was at that time head of creation at the Cristalleries du Val-Saint Lambert, and could look back on a rich career as a glass designer.[118] His work was very well received at the Paris Exposition des Arts Décoratifs in 1925 and his modern designs won a gold medal at the XI Milan Triennial in 1957. Around 1950 he embarked on a long study journey to the United States. Although we lack tangible evidence, Graffart probably encountered the phenomenon of industrial design on this journey. Three vice-presidents were also appointed. Each had to cover a different city, respectively Liège, Antwerp and Brussels. The Brussels vice-president was the architect Henvaux. He was soon replaced by Jean-Jacques Thiry, the deputy director of the Belgo-Luxembourg Steel Information Center (CBLIA), previously affiliated with the Brussels-based Formes Nouvelles.[119] The Antwerp vice-president was Jan Lauwers, director of the Nationaal Hoger Instituut voor Bouwkunsten en Stedebouw (NHIBS). The third vice-president responsible for the Liège area was Maurice Paquay, head of advertising at the Usines Tubes à la Meuse in Liège and closely involved in the 1954 *Art et Travail Show* as "aesthetic adviser" for the Liège Fair. Two secretaries completed the board: the French-speaking Louis Desamory and the Dutch-speaking Johan Fleerackers. It was, in particular, the architect Louis Desamory who would take on an important role in establishing industrial design in Belgium. As a professor at the Institut supérieur d'Architecture de Liège, he was charged by its director Joseph Moutschen to prepare a department on "Esthétique Industrielle."

Although its name contained the English designation "industrial designers", the association stressed that the founding members were not designers. On the contrary:[120]

> The founding members do not claim to call themselves designers. It is too early in Belgium, except for some rare exceptions, for anyone to claim that title and to pride themselves on a practice and sufficient results. However, the founding members represent, in their respective capacities, the different sectors of Belgian industry and the different regions of the country.[121]

Yet, in spite of this statement, three founding members would call themselves "designers" in the association's articles, using the English designation. These were its Antwerp vice-president Jan Lauwers, its Liège vice-president Maurice Paquay, and René Houben, who was affiliated to the Ateliers de Constructions Electriques de Charleroi (ACEC). Along with these three "designers", the founding members were all personalities with diverse backgrounds such as decorator Marcel Baugniet from Formes Nouvelles, archaeologist Adelbert Van de Walle of the Museum for Decorative Arts in Ghent, and business manager Louis van Veen, commercial director of General Motors Continental in Antwerp.

From the beginning, the Belgian Association for Industrial Designers fostered very close contacts with foreign counterparts. It claimed to maintain correspondence with its American, British, French and Dutch colleagues. In December 1954, one month after its foundation, the association invited the head of the French Institute of Industrial Design, Jacques Viénot, to Liège.[122] In front of more than five hundred people, Viénot gave a lecture on industrial design as the main factor for productivity. A few months later, it was the turn of Dutchman Karel Sanders, the director of the Dutch Institute for Industrial Design. He gave a lecture in the Antwerp buildings of General Motors in front of three hundred people. In addition, the young organization planned visits to its foreign counterparts, such as Paul Mahlberg's "Industrieform" in Essen and the Dutch Institute for Industrial Design in Amsterdam. Close relations were also maintained with architect and designer Henry van de Velde in Switzerland, who agreed to become its adviser and honorary president in 1956.[123]

In March 1955, a second institution saw the light of day in these Liège circles: the Industrial Design Institute (L'Institut Industrial Design).[124] Its president was Louis Fraigneux, the director of Usines à Tubes de la Meuse in Liège, who was very active from the beginning in this Liège movement around industrial design. The new institute was not devoted to the profession of industrial designer like the Association for Industrial Designers did, but wanted to group industrialists and businessmen to spread the commercial benefits of industrial design for the national industry.

By the next edition of the Liège International Fair in April 1955, two industrial design institutions — one for "designers" and the other for businessmen — thrived in the Liège stronghold. No wonder the fair once more reiterated the topical subject of "industrial design" for its seventh edition.[125] This time it was categorized under "sales promotion" next to "Design in packing material" and "Industrial publicity."[126] The "Industrial Design" section included two exhibitions: the American *United States Industrial Design*, and the Belgian *Section d'Industrial Design*. As it did the year before, the Liège International Fair had invited the New York Society of Industrial Designers to participate. This time, its exhibition was far more ambitious and was set up with the help of the US Department of Commerce.[127] On an area of 360 square meters, sixty-six products were displayed, spanning office equipment, toys, industrial products, ceramics, kitchens, furniture, optics and radios. The icing on the cake was a four-color illustrated catalog called *What Is a Product*, which visitors could take home.

Next to this American design manifestation, the Belgian Ministry of Economic Affairs organized an even greater representation of Belgian industrial products. The show was actually a rerun of the successful Belgian exhibition at the 1954 Milan Triennial showing similar highly technical and industrial objects such as guns, motorcycles, milking machines, turning lathes, optical instruments, gravy boats and cutlery (fig. 1.17). However, the architecture of the exhibition by André De Poerck was adapted. The Liège version was far more theatrical, thanks to the use of large black and silver surfaces.

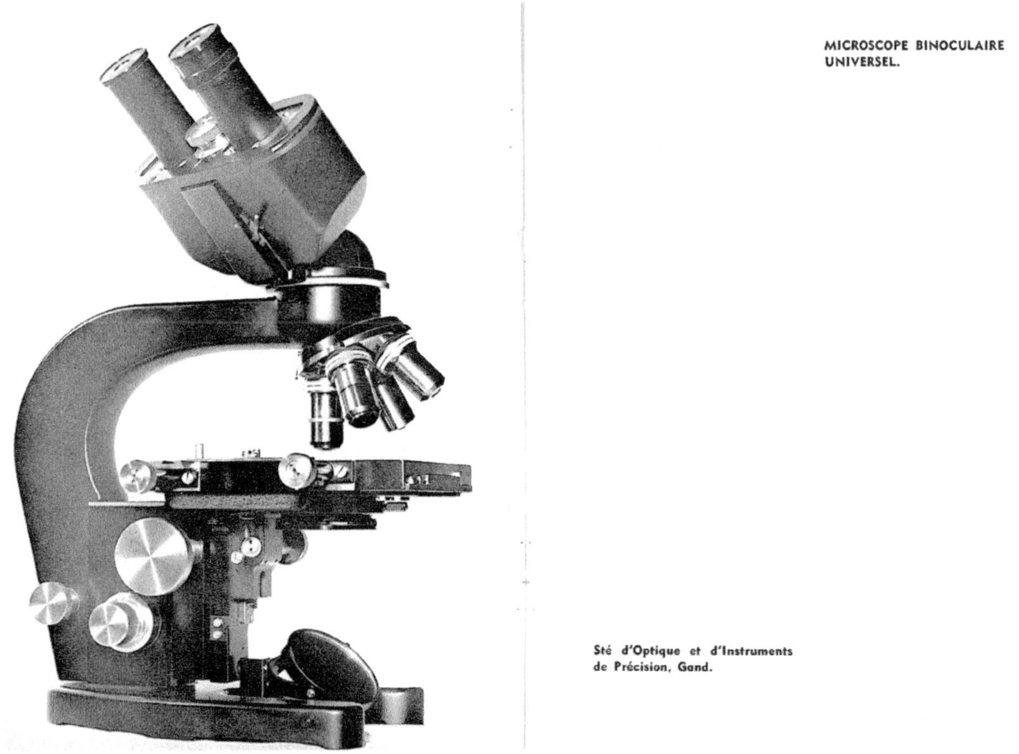

Fig. 1.17. The binocular microscope of Societé d'Optique et d'Instruments de Précision (Ghent) in the catalog for the *Second Industrial Design Show* at the seventh Liège International Fair, 1955. Source: "L'Esthétique Industrielle. La 7me Foire Internationale de Liège," ed. Ministère des Affaires Economiques (Ministère des Affaires Economiques, 1955).

4.2 The Institute for Industrial Design for Belgium and the Grand Duchy of Luxembourg (1956)

Both the Belgian Association for Industrial Designers and the Industrial Design Institute were granted patronage by Minister of Economic Affairs Jean Rey.[128] The Industrial Design Institute under Louis Fraigneux had already obtained official recognition right after it was created in March 1955. The Belgian Association of Industrial Designers under Charles Graffart had to wait a little longer, but official recognition finally came in February 1956. Probably on the initiative of the Ministry of Economic Affairs, which by then patronized two Liège-based institutes, the two organizations merged in April 1956 under the pretext: "This is no time for confined and isolated initiatives [...] one has to move fast, one has to think big."[129] This new institution, which was the result of the two-year takeoff run in the industrial Liège milieu, was christened the Institute for Industrial Design for Belgium and the Grand Duchy of Luxembourg (Institut d'Esthétique Industrielle pour la Belgique et le Grand-Duché de Luxembourg/Instituut voor industriële vormgeving voor België en G.-H. Luxemburg).[130] The Institute took up the previous Liège objectives: to disseminate industrial design in Belgian industry in order to boost the sales of Belgian products on internal and external markets.

The board of directors of the Institute for Industrial Design for Belgium and the Grand Duchy of Luxembourg was marked by its genesis. The members primarily had a background in the industrial and technical realm, and only some were architects or had a background in education.[131] The former Industrial Design Institute provided the president (Louis Fraigneux), while the previous Belgian Association for Industrial Designers delivered the rest of the board. Its two vice-presidents, Jan Lauwers and Jean-Jacques Thiry, together with its two secretaries, Louis Desamory and Johan Fleerackers, and its honorary president, Henry van de Velde, were all reclaimed. Its former president became the new Liège vice-president and Maurice Paquay was promoted to director.

4.3 The Signe d'Or Award (1956)[132]

While Liège seemed to belch out one industrial design initiative after the other, there was also something brewing in the cultural scene in Brussels. In October 1956, a new institution was set up in Brussels with the help of the Ministry of Economic Affairs: the Foundation for the Promotion of Quality in Industrial Creation (Société pour promouvoir la qualité dans la création industrielle/Vereniging ter bevordering der kwaliteit in de Nijverheidscreaties).[133] As the Liège milieu was obviously excited by industrial design because of its potential to increase productivity, the Brussels cultural scene was especially interested from an educational perspective. The new organization hoped to preach "good design" to the industry and public:

> The association aims to promote [...] the manufacture and commercial distribution of functional objects, produced on a large or small

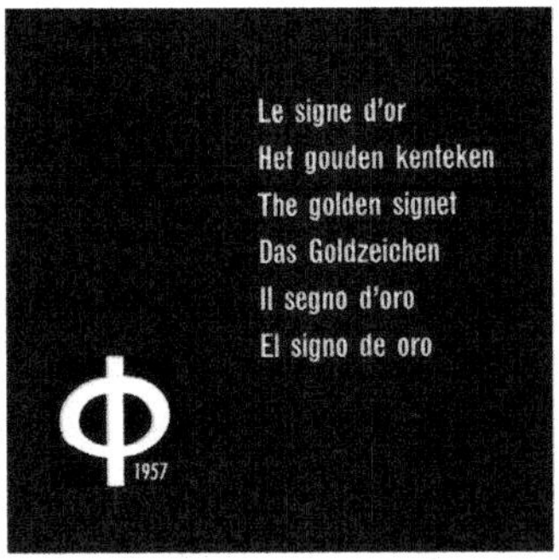

Fig. 1.18. The logo of the Signe d'Or designed by Corneille Hannoset. Source: *Signe D'or. Het Gouden Kenteken. The Golden Signet. Das Goldzeichen. Il Segno D'oro. El Signo De Oro.* Brussels: Published with the co-operation of the Belgian Office of Foreign Trade, 1957.

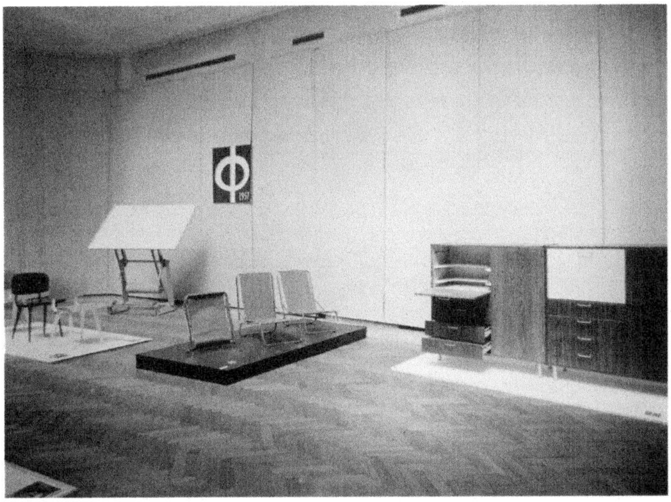

Fig. 1.19. View of the first Signe d'Or exhibition held at the Center for Fine Arts in Brussels, 1957. Source: Coutisse, Archives des Cressonnières.

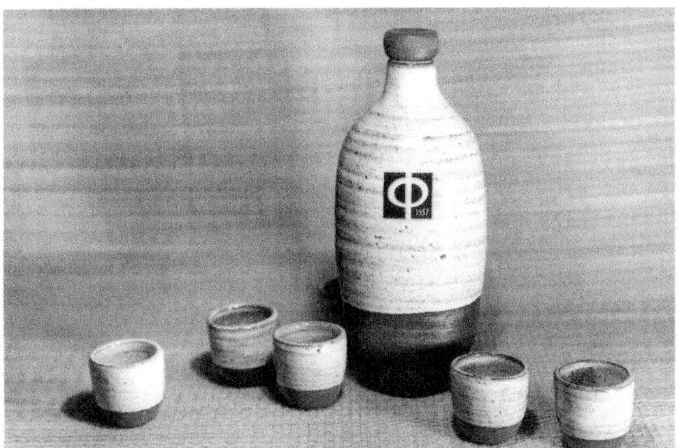

Fig. 1.20. Antoine de Vinck (1924-1992), stoneware liqueur set, laureate of the Golden Signet, 1957. Source: © Antoine de Vinck

many countries to observe new design trends that she developed her natural *goût esthétique*.[140] She attributes to this position her knowledge of Belgian and foreign markets, the problems of distribution and creation, and of course: the introduction of industrial design. A couple of months before she joined the board, Des Cressonnières had also been involved in the Institute for Industrial Design as one of its founding members.

Although most of its founding members had won their spurs in the interwar artistic and cultural world, the Signe d'Or would, however, not take up interwar discourses on social art such as Formes Nouvelles or the early Art et Travail. Instead, it would adopt the discussions on "good" design, following the lead of its foreign counterparts, such as the Swiss Die Gute Form/La Forme Utile (1952), the French Beauté-France (1953), the Italian Compasso d'Oro (1954) and Kaufmann's exhibitions at MoMA (1950–1955). In particular, Josine des Cressonnières would take the lead and aim to install industrial design "as a prestigious discipline in the international stream of taste."[141]

5. CONCLUSION: A CONGLOMERATE OF ART, INDUSTRY AND POLITICS

On the rare occasions that Belgian design institutionalization is mentioned, it is stated that it started in 1956 when the Institute of Industrial Design and the Signe d'Or were established with support from the Ministry of Economic Affairs. Indeed, 1956 is a watershed in the history of Belgian design promotion, yet an important historical chapter precedes this date. The first postwar signals of the international "industrial design" phenomenon were picked up in the early 1950s by people previously involved in the interwar artistic circles. In particular, the painter, decorator and critic Marcel Baugniet, who had previously developed a strongly committed discourse on the machine in the pages of *7 Arts*, engaged himself in the postwar bodies for the promotion of design. He was, among others, the vice-president of the organization Formes Nouvelles, where he found a solid base for promoting the rationalization of domestic furniture and continuing his crusade for the social role of the machine. Also pivotal was the Art et Travail movement instigated in 1952 by the Ministry of Public Education under Pierre Harmel, which reached back to the interwar concerns about uniting art and industry that surrounded the 1937 Paris exhibition.

Along with the continuation of these prewar traditions, national economic policies played a major role in the institutionalization of industrial design in Belgium. That the Ministry of Economic Affairs, under the leadership of none other than the liberal Jean Rey, included "industrial design" in 1954 in its policy indicates the extent to which industrial design was seen as a close ally in facing the new postwar economic order. The main economic challenge was the outlook of the European Common Market that was under construction. The national economy was considered under threat by the feared foreign competition on the increasingly liberalized markets. Moreover, after its miracle years after the war, the Belgian economy was suddenly reducing, mainly due to an outdated infrastructure. The introduction of the modern technique of industrial design had to provide Belgium with a modern liberal face on the international markets and increase the productivity of Belgian industry, and thus increase national welfare.

While the subject of industrial design in Belgium made few waves in 1950, by 1954 it was all the rage in artistic, cultural, industrial and official circles. In 1956 this chaotic medley between artistic and industrial traditions, fueled by postwar economic challenges, merged into two government-endorsed organizations: the Institute for Industrial Design in Liège, and the Signe d'Or in Brussels. As the next chapter will show, this multifaceted character and the underlying conflicts of this first upsurge of industrial design in Belgium would haunt the heyday of Belgian design institutionalization. This is best exemplified in the ambivalent mobilization of Henry van de Velde's legacy.

CHAPTER II

Reconfiguring Henry van de Velde (1956–1964)

1. INTRODUCTION: HENRY VAN DE VELDE AT THE INDUSTRIAL DESIGN SHOW IN LIÈGE

In 1956 the Institute for Industrial Design organized its first activity with the support of the Ministry of Economic Affairs: the third edition of the *Industrial Design Show* (Salon d'Industrial Design) at the Liège International Fair (fig. 2.1).[1] From 1954 the government consistently supported the annual organization of the industrial design shows at the International Fair within the framework of Jean Rey's new design policy.

The exhibition scenography was assigned to two founding members of the new Institute for Industrial Design: architect duo Lucien Kroll (b. 1927) and Charles Vandenhove (b. 1927).[2] In an area of 1,000

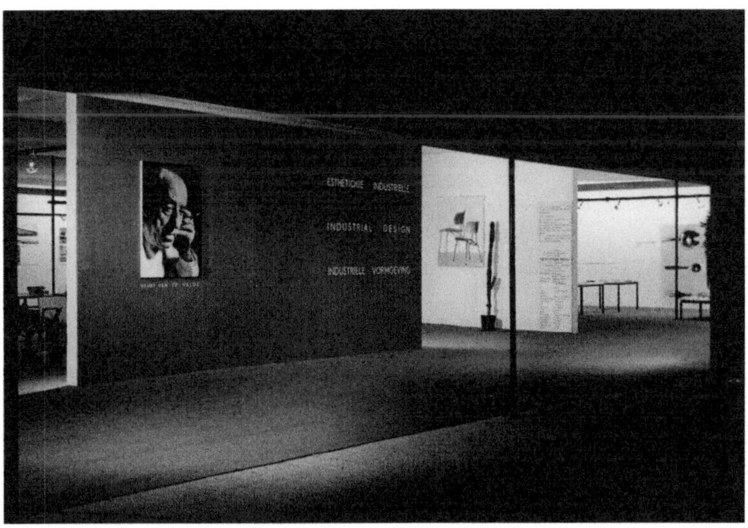

Fig. 2.1. View of the *Industrial Design Show* at the Liège International Fair, 1956. Source: Coutisse, Archives des Cressonnières.

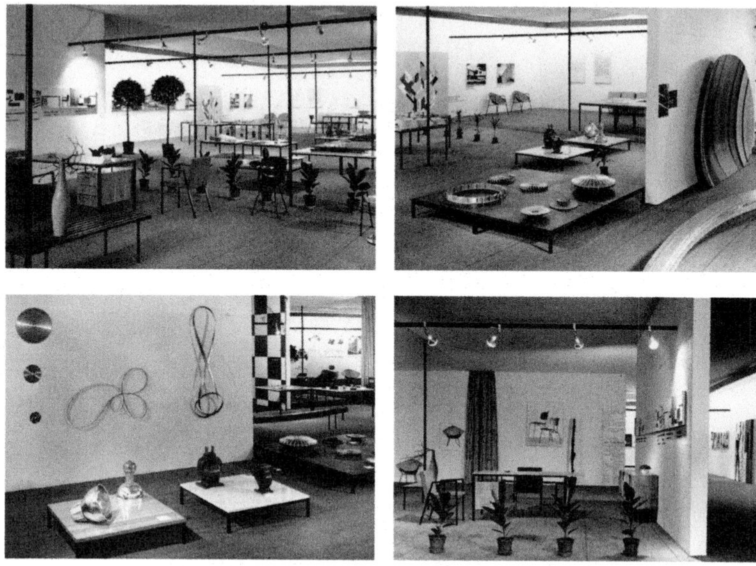

Figs. 2.2–5. View of the *Industrial Design Show* at the Liège International Fair, 1956. Source: Coutisse, Archives des Cressonnières.

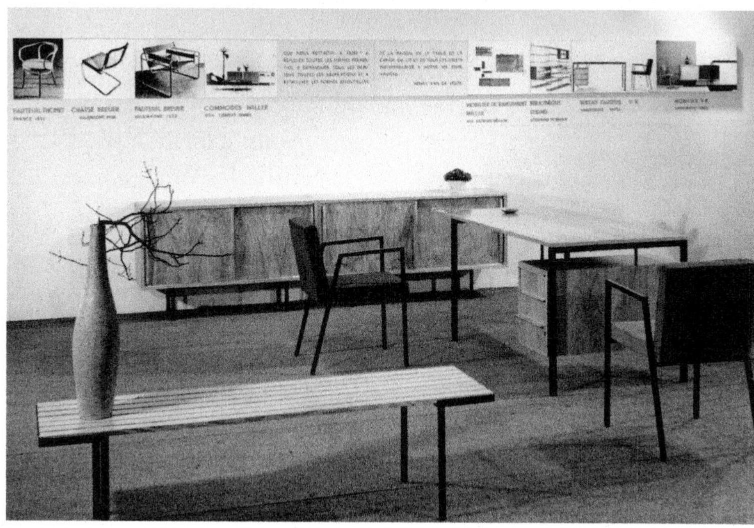

Fig. 2.6. View of the *Industrial Design Show* at the eighth Liège International Fair, 1956. The architect duo Kroll and Vandenhove included didactic and instructive photographic reproductions on the walls in the lay-out of the *Industrial Design Show* in Liège, 1956. Source: Coutisse, Archives des Cressonnières.

square meters more than two hundred Belgian objects selected by an international jury were displayed (figs. 2.2–5).[3] The architects were responsible for the theme, the texts, the images and the display devices. The main materials used in the layout were white marble, polished slate and plastic panels in red, white, grey and black. Kroll and Vandenhove arranged the objects in glass showcases and on square tables of different heights. While the objects displayed had to testify to "good" industrial design in Belgium, the didactic photo frescos situated them in international developments, referencing leading design figures such as Max Bill, Walter Gropius, Henry van de Velde, Le Corbusier and Theo van Doesburg, or model examples such as the Italian Olivetti and refined Japanese design (fig. 2.6).

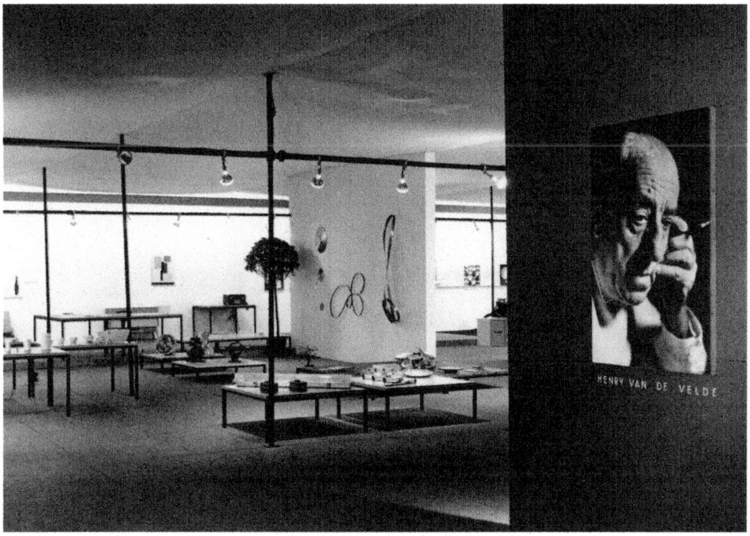

Fig. 2.7. A large portrait of Henry van de Velde as the honorary president was featured on a red wall at the entrance to welcome visitors to the *Industrial Design Show* in Liège, 1956. Source: Coutisse, Archives des Cressonnières.

At the entrance, a large portrait of Henry van de Velde filled a red wall (fig. 2.7). [4] He was internationally considered as one of the founding fathers of industrial design. Appealing to his reputation, the Institute for Industrial Design made him its honorary president, and accordingly his image warmly welcomed the visitors to the exhibition. His portrait was not only an attempt to inscribe industrial design in the Belgian tradition or to connect with the modern movement, but was more importantly mobilized by the two architects — in Kroll's words — as an "invocation".[5] Van de Velde's "humanism" was called upon to avert the commercialization and scientific rationalization of design. Putting their heads in the lion's mouth, the architects criticized the commercial approach that, as from 1954, was very much in vogue with the Belgian government that funded this exhibition and with the industrialists who were the main target group of the exhibition.

Thus, the image of Henry van de Velde did not, in this case, have purely promotional value, but was used by two members of the

Institute to criticize developments in the field. Within the framework of this official promotion at the Liège Fair, the architect duo subtly integrated their own agenda, distinct from the Ministry's intentions or even the Institute's official program. The Belgian design bodies were, as a result, no smooth emulsion of artistic and commercial interests, nor did they seem to fit the intentions of the government that funded them. Instead it was a polyphonic history, of which this incident only raises a corner of the veil.

As the first chapter described, the founding of the Institute for Industrial Design and the Signe d'Or award scheme was a blend of the interwar legacy — characterized by ideals around the social function of art and design — and the commercial enthusiasm in the Liège industrial milieu. This cocktail, however, was not very well mixed. At the time of its funding, the Signe d'Or housed mainly Brussels cultural policymakers, while the Institute represented primarily the industrial milieu of Liège. Both were, however, used by the Ministry of Economic Affairs to tackle international competition prompted by the prospect of the liberalization of the European Market. As the Kroll-Vandenhove case shows, this cocktail of industrialists and aesthetes started to mix immediately. The dichotomy between artistic and commercial approaches divided the ranks of the design bodies.

While chapter one explored the emergence of "industrial design" in Belgium, the second chapter will look at how the past was recruited to root this new phenomenon in a Belgian tradition. In particular, Henry van de Velde and his pre-war design legacy proved to be a timely and welcome icon for the Belgian movement. Van de Velde enjoys great fame in the historiography of the modern movement, yet his role in the heyday of industrial design in Belgium remains largely underplayed.[6] This chapter explores then this forgotten title role of Van de Velde in the design institutionalization in Belgium. As the Kroll-Vandenhove case illustrates, this Belgian design pioneer was a crucial point of reference for different interest groups to appropriate the new domain of industrial design within the Belgian context. In this chapter, we are not so much interested in his biography or oeuvre as in the exploitation of his legacy within the context of the Belgian postwar design institutionalization. Why is his image used in debates to express criticism? What visions are projected onto this figure? Which interests and tensions are grafted onto the image of this design pioneer? By exploring the use of Henry van de Velde by diverse actors to represent, legitimize and advance competing opinions on the nature of design in Belgium, this chapter touches upon different attitudes toward the concept of design, Belgian design traditions, American design, European art traditions and the internationalization of design.

2. THE HENRY VAN DE VELDE "MYTH"

2.1 The Doyen of Belgian Industrial Design

Henry van de Velde is one of the few internationally renowned Belgian "designers." He is considered one of the founding fathers of the functionalist esthetics that shaped the modern movement of the twentieth century. The editor of the Italian magazine *Casabella*, Ernesto N. Rogers, underlined his singular influence three years after his death: "In his long life the most important artistic movements of a century mold him, imbue his spirit, affect him, derive from him, or are devised by him: Naturalism, Symbolism, Impressionism, Expressionism, Rationalism, and maybe even a little Surrealism."[7] His impressive and exceptionally long career, from the late nineteenth to the latter half of the twentieth century, however, is difficult to restrict to one professional domain. He was trained as a painter in his home town, Antwerp, but soon embarked upon various branches of visual arts. Because of the multifaceted character of his work, Henry van de Velde is miscellaneously categorized as: painter, craftsman, furniture designer, graphic designer, architect, interior designer, pedagog, industrial designer – and some more professions are probably ascribed to him too. It was especially during the 1950s that his capacity as "industrial designer" was emphasized in debates and the press in Belgium, not just as a random industrial designer, but as one of the pioneers of industrial design. This all had to do with the recent developments. In the context of the postwar reality (outlook on international competition, regeneration of the national economy and the Marshall Plan) industrial design became part of Belgium's national economic policies. In the wake of the creation of the Institute for Industrial Design and the first events of industrial design described in chapter one, Van de Velde's legacy, as one of the design pioneers, was revisited.[8] Henry van de Velde was then almost 92 years old and residing as a voluntary exile in Switzerland.

The Institute of Industrial Design appropriated the figure of Van de Velde as the symbol of its new national movement. The organization asked him to become its honorary president and to provide advice

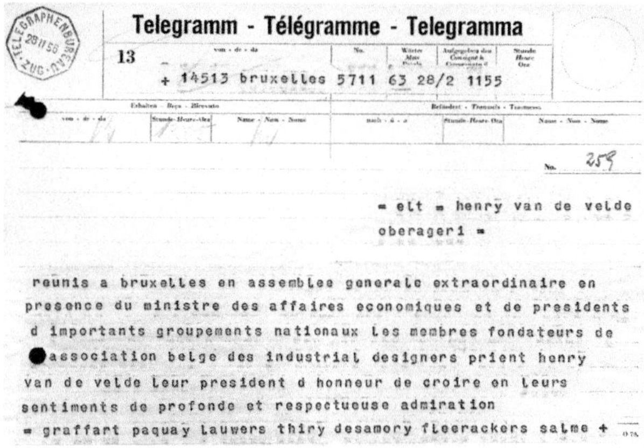
Telegramm - Télégramme - Telegramma

13

+ 14513 bruxelles 5711 63 28/2 1155

No. 259

= elt = henry van de velde
oberageri =

reunis a bruxelles en assemblee generale extraordinaire en presence du ministre des affaires economiques et de presidents d importants groupements nationaux les membres fondateurs de association belge des industrial designers prient henry van de velde leur president d honneur de croire en leurs sentiments de profonde et respectueuse admiration
= graffart paquay lauwers thiry desamory fleerackers salme +

Fig. 2.8. Telegram from the Belgian Association of Industrial Designers to its honorary president, Henry van de Velde, in Switzerland, February 28, 1955. Source: Fonds Henry van de Velde © AML (Archives et Musée de la Littérature)

on its first programs and actions.[9] He accepted this invitation after close contacts with the founders of the Institute (fig. 2.8).[10] To validate his symbolic role, a few member designers, among them Kroll and Vandenhove, went to Switzerland after the official establishment of the Institute to show respect to "le maître" as their honorary president.[11] In particular, two achievements in Van de Velde's career made him very appropriate to act as a symbol of this Belgian movement. First, with his 1894 text *Déblaiement d'art*, a plea for art at the service of society, he was internationally considered one of the founding fathers of industrial design. Authorities like Nikolaus Pevsner mentioned him as one of the pioneers of modern design.[12] Second, he is often portrayed — not least by himself — as the spiritual father of the Bauhaus. This owing to the fact that the Bauhaus sprouted from two schools in Weimar of which one, the Grand Ducal Saxon School of Arts and Crafts (*Kunstgewerbschüle*), was created by Van de Velde in 1907.[13] In 1915, he recommended the younger Walter Gropius (1883–1969) among two others as his successor. Gropius then eventually opened the notable Weimar Bauhaus after the First World War in 1919 in the buildings previously designed by Van de Velde.

The new Institute drew extensively on the image of Van de Velde to represent itself in its first years. The front page of the first issue of the Institute's magazine featured, for example, a large portrait of the latter with the caption "Henry VAN DE VELDE. Honorary President of the Institute for Industrial Design" (fig. 2.9).[14] The portrait was made

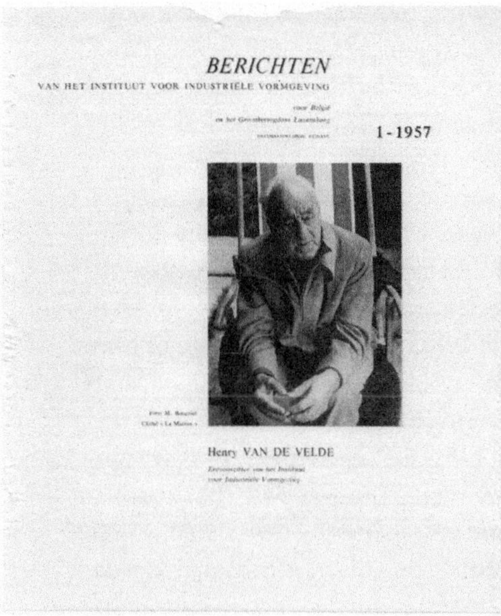

Fig. 2.9. The front page of the first issue of the institute's magazine featured a large photographic portrait of Henry van de Velde taken by Marcel Baugniet. Source: Berichten van het Instituut voor Industriële Vormgeving, no. 1 (1957).

by one of its founding members, Marcel Baugniet, and showed a seated Van de Velde from a frontal viewpoint. Most telling was the use of Van de Velde's image at the Institute's first action, the third *Industrial Design Show* at the Liège International Fair in 1956. As mentioned, architects Lucien Kroll and Charles Vandenhove placed a large photographic portrait of the honorary president by Willy Kessels on a red wall at the entrance to welcome visitors. Despite the additional meaning given by the architects, the image here served as a signboard of the new Institute and of Belgian design.

2.2 His Home Country, Belgium

The postwar design institutions in Belgium gratefully drew on Henry van de Velde's bold reputation in the international modernist arena. From a political perspective, however, this was not obvious. In the words of Kathleen James-Chakraborty, "Van de Velde had an extraordinary aptitude for political ineptness."[15] After all, Van de Velde was not always seen as a hero in his home country. On the contrary, on many occasions he was *persona non grata*. In the aftermath of both world wars, he was accused of collaboration with the Germans and, although never proved or even refuted, this unsound political reputation dogged him. In addition, his cosmopolitan, international and European attitude fell on barren ground with the nationalist camp in Brussels. His reception in his homeland is best described as an unstable love-hate affair.

After the First World War, Henry van de Velde was even portrayed as "anti-Belgian" by the nationalistic press.[16] His stay in Germany during the First World War offended many and led him to be called a "Boche" and "Mof", derogatory terms in French and Dutch, respectively, used to refer to Germans during the First World War.[17] This image of him as a political traitor was widespread in the Belgian nationalistic, conservative and catholic press. Van de Velde himself said that a part of the Belgian press was stirred up against him and "used the most cowardly and hateful imputations."[18] This was indeed the case with the Brussels-based, French-language nationalistic newspaper *La Gazette*, in which his work was ridiculed.[19] Roger Avermaete similarly recalls that Van de Velde's name was dragged through the mire by the "hyper-nationalistic press."[20] The Belgian politician Camille Huysmans, a good friend of Van de Velde's, contends that it was jealous rivals who made him the center of controversy and portrayed him in the context of the First World War as a supporter of Germany.[21] In reaction to the nationalistic slant of the assaults, the target of these attacks called himself an "irreducible pacifist and internationalist."[22] Partly because of these attacks, Van de Velde did not return to his home country immediately after the war, but resided for the first few decades in Switzerland and the Netherlands.[23]

This besmirching of Henry van de Velde's reputation did not stop the younger generation Brussels avant-garde from reintroducing him onto the Belgian interwar art scene as an inspiration for the younger generations of architects and artists.[24] In 1923, the Brussels publisher *L'Equerre,* set up by Georges Rens and the Bourgeois brothers, issued the revised version of his *Les formules de la beauté architectonique moderne* (1916–1917).[25] For this occasion, the *Lanterne Sourde*, the literary and artistic student circle of the Université libre de Bruxelles, invited Van de Velde to give a lecture on modern esthetics at their university.[26] This relaunch of Van de Velde in Belgium as an important international figure of the modernist movement could not counter the latent prejudice against him as "anti-Belgian." When his name was put forward as commissioner general for the 1925 *Exposition des arts décoratifs* in Paris, not only did the nationalistic press voice its opposition, but so did artists. Architect Victor Horta (1861–1947) drafted an incriminating note that was signed by several artists to protest against the assignment.[27] Indeed, Van de Velde came away empty-handed, while Horta obtained the assignment for the Belgian pavilion in Paris.

Gradually, however, Van de Velde could count on more and more official support in Belgium. On the initiative of the socialist Camille Huysmans, in his capacity as Minister of Fine Arts and Education, Henry van de Velde was finally invited back to Belgium in 1925 to set up and direct at the age of 62 the La Cambre School of Architecture and Applied Arts.[28] After ten years, in 1936, Van de Velde resigned as director and received many important official commissions, among others, the organization of the Belgian participation in the world exhibitions of Paris (1937) and New York (1939).[29]

After World War II, Van de Velde was again accused of collaboration.[30] This time the accusations were based on his work as aesthetic adviser to the Commission for the Reconstruction of the Country under the German occupier.[31] Although the charges were dropped in 1946, he left Belgium and retreated to Switzerland where he lived for the rest of his life in voluntary exile. He felt like a *persona non grata* in his home country. Disillusioned, he declared bitterly that Belgium was a country where nothing finds any response.[32]

In the mid-1950s, however, the hate mail again turned into a love affair when Van de Velde came back into favor in the discussions of art and architecture magazines and official design bodies as a prophet of industrial design and a pioneer of a new style. Advanced by their biggest advocates, such as Marcel Baugniet,[33] Pierre-Louis Flouquet,[34] Léon Stijnen,[35] Roger Avermaete[36] and Léon-Louis Sosset,[37] the first official design institutions gratefully embraced Van de Velde's legacy to gain a firm footing for industrial design in Belgium. So, just before his death at the age of 92, Van de Velde turned from an undesirable person to the official doyen of Belgian industrial design.

3. GLOBAL FREELANCER

3.1 Between Reason and Intuition

> The words [Van de Velde] pronounced in 1892 retain in 1923 a sense of immediate truth: could they be any less topical in 1952?[38] (Victor Bourgeois, 1923)

Architect Victor Bourgeois' suggestive question in 1923 seems to become a self-fulfilling prophecy. Van de Velde's relevance did not seem to fade in the 1950s. On the contrary, many references in postwar design and architectural debates were made to his architectural work and his theories. He even became the official mascot of the newly founded Institute for Industrial Design in 1956. His "words", however, were characterized by ambiguity and were open to a range of interpretations. In the words of architect and historian Pierre Puttemans, "No personality could be stranger or more contradictory than that of Henry van de Velde."[39] Already in 1923, Victor Bourgeois describes this contradictory quality in terms of *intuition* versus *reason*:

> Van de Velde probably ignores the duality of instinct and intelligence: for him these two forces are natural allies, and what his intuition proposes was secretly desired by his reason.[40]

Bourgeois very elegantly describes Henry van de Velde's intrinsic and complex synthesis between a rational and an intuitive conception. Design historian Elie Haddad indeed demonstrates that these conflicting terms drove the development of his esthetic theories.[41] She describes his stance as "a synthesis of the two opposite poles of rational conception associated with machine production and an 'empathic' conception of art articulated around a particular understanding of the ornamental function."

In many interpretations, however, this complexity is reduced to a single dimension — either artistic intuition or rational logic. The rationalist interpretation is triggered by Van de Velde's praise of machines and the engineer's logic. In particular, in his texts *Déblaiement de l'Art* (1894), *Amo* (1909) and *Voie Sacrée* (1933), his tributes to reason and machines are very pronounced. In the latter text, for example, Van de Velde stated that "the most adequate means to fight ugliness is the return to reason, to the strict logic of the shapes of all things."[42] Maurice Casteels' 1932 monograph depicts him as the savior of "la Raison" because of his passionate reasoning.[43] Advancing on this focus, Nikolaus Pevsner's authoritative study *The Pioneers of Modern Design,* first written in 1936, considers the Belgian as one of the five architects who were the first "to admire the machine, and to understand its essential character and its consequences in the relation of architecture and design to ornamentation [...]."[44] He recalls Van de Velde's claim that engineers would become the future architects. Pevsner's lineage of pioneers was highly influential and became internationally widespread, so that this "rationalist" or "engineer" image of van de Velde was widely disseminated.

At the same time, however, Van de Velde was not only the "defender of reason", but functioned also as a symbol for artistic freedom and personal intuition.[45] It was especially his reaction against Muthesius' Werkbund theses which earned him the title of the "artist" designer *par excellence* and made his image very attractive to use in discussions on individual and artistic approaches towards design — as did Kroll and Vandenhove when they evoked Van de Velde's "humanism" in Liège. At the controversial "Werkbund" conference in Cologne in 1914, Van de Velde after all reacted passionately to Muthesius' proposition to develop a universally valid taste through standardization.[46] Van de Velde saw his suggestions as restrictive of the artistic mind:

> By his innermost essence the artist is a burning idealist, a free spontaneous creator. Of his own free will he will never subordinate himself to a discipline that imposes upon him a type, a canon.[47]

This confrontation was the occasion for the famous caricature by the German Karl Arnold that was widely circulated and is still known today by many design enthusiasts.[48] Arnold's drawing is a kind of triptych. The first section depicts Van de Velde with a pair of compasses and a triangle leaning on a decorative Jugendstil chair with the caption "van de Velde designed the *individual* chair." Next to it is a portrait of Muthesius with a chair of pure simplicity accompanied by the words "Muthesius designed the chair-*type*." Both personalities are flanked by an unknown carpenter who rests his arm on a normal chair "to sit on" — ridiculing the whole debate between the individual and standardization.

Besides embodying opposing concepts of design (reason versus intuition), the ambiguity of Van de Velde's architectural and theoretical oeuvre is heightened by several factors. Design critic K.N Elno tried to put his finger on the problem on the 100th anniversary of the designer's birthday.[49] He observed that Van de Velde's work is obscure and multifaceted, and partly ascribes it to the difficulties in summarizing this figure's credo in a simple line. Architect and historian Pierre Puttemans is more systematic and gives us three reasons why it is so hard to categorize him. The first difficulty lies in the equivocal vocabulary he used. Second, his position was obscured by the half-and-half attitude he adopted towards mechanization, industrialization and rationalization. Third, his theoretical and architectural oeuvre did not evolve in an analogous manner.

His Janus-faced legacy makes him not only a complex figure to pin down into one clear category, but also gave him an ideal profile as global freelancer in postwar Belgium: adherents to different movements and even opponents could all subscribe to his honorable endeavor to improve the industrial products in our daily life, strategically highlighting the most suitable angle for their opinions.

3.2 Janus-Faced Exploitation

The versatile character of Van de Velde's oeuvre made it fit easily into different categories, from artistic and intuitive to rationalist and technological. As such it could be used in a variety of — even opposing — design discourses. In 1950s Belgium, the Henry van de Velde "myth" was used to underpin two different design visions, cherry-picking at will the most suitable elements from his oeuvre. His "rationalist" face was exploited by officials, industrialists and rationalists to prove the commercial benefits of design. His "empathic" face was mobilized by artists, architects and art critics to protect the humanist conception of design. Even within the young Institute for Industrial Design, this duality was present. Although Van de Velde himself during his life opposed the use of symbols in his design theories, both movements would use his image as a symbol to represent, legitimize and advance its opinions on the nature of design in Belgium.

3.2.1 VAN DE VELDE'S LEGACY IN TERMS OF THE NATIONAL ECONOMY

Especially in official and industrial circles, Henry van de Velde was presented in the 1950s as a rationalist whose principles generated great economic success in foreign countries, the United States being an outstanding example. Belgian industry was thus incited to embrace the American design approach — which was presented as the implementation of Van de Velde's principles.

The first to sketch such an image of the "economically relevant" Belgian was rationalist architect Léon Stynen. In his capacity as director of La Cambre and vice-president of Art et Travail, he was invited to speak at the design conference of the 1954 Liège International Fair. Stynen was enthusiastic about the recent attention devoted in Belgium to the economic and commercial possibilities of industrial design. At the same time, however, Stynen expressed his astonishment at the late endorsement of industrial design. His compatriot and predecessor as director of La Cambre, Henry van de Velde, had, after all, already expressed these ideas almost 50 years before. He portrayed Van de Velde for this audience of industrialists as the progressive detractor of the "beaux métiers artisanaux" and the advocate of the nascent industrial powers.[50] With this depiction, Stynen was in fact distorting Van de Velde's ideas, which were much more opaque and ambiguous, so as to win over the minds of industrialists to the case of industrial design. He continued that it was quite paradoxical that, not Belgium, but the United States had implemented the experiences developed by Van de Velde.[51] He concluded that Belgium now imported masses of American products that were consumed in preference to Belgian equivalents because of their high quality, which actually originated in Van de Velde's principles.

The narrative that the success of American design sprouted from Van de Velde's doctrine was adopted in the Institute's first documents. It was claimed, in a similar vein to Stynen's speech, that American industrialists had in fact hijacked Van de Velde's doctrine.[52] According to the Institute, his principles circulated widely in the United States thanks to the American "exodus:"

> [...] the Nazi accession to power resulted in the closure of the Bauhaus. Gropius, Breuer, Mies van der Rohe and some others left Germany to work and teach in America. Provided with exceptional financial means, it only took Americans a few years to expand their production in an extraordinary manner. Today, thanks to the system of imports, we buy an important quantity of products from the US owing to techniques resulting from the doctrine of Van de Velde, and that in preference to Belgian products.[53]

This simple narrative by the Institute echoes the clichéd story that prevailed at the time around the Bauhaus legacy. As architectural historians Paul Betts and Kathleen James-Chakraborty have argued, the widespread story about the American Bauhaus success was in fact a

Cold War fable to underline American preeminence as fertile soil for the European modern movement.[54] However, the Institute used this story, not to prove American excellence, but rather to trace the cradle of the modern movement down to Van de Velde's principles. Thus, through a very simplified and reduced genealogical line, the whole Werkbund-Bauhaus legacy — including its American sequel — was somehow identified with the compatriot Henry van de Velde. This was only strengthened by the statements of Van de Velde himself.

In 1955, the Ulm School of Design (HfG) opened its doors under the directorship of Swiss artist and former Bauhaus student Max Bill (1908–1994).[55] This new school was seen as the legitimate West German heir to the Bauhaus after World War II and was therefore often referred to as the "New Bauhaus". In a similar vein, Van de Velde christened the HfG the "fourth citadel of modernism" following its prewar predecessors (fig. 2.10).[56] He considered the Arts and Crafts school in Weimar (1907) to be the first citadel of modernism, followed by the notable Bauhaus (1919) and La Cambre (1926). With these designations, he actually emphasized that these four design schools were spiritually rooted in his principles developed in Weimar.[57] It is true that from 1950 Max Bill sought Van de Velde's advice on the Ulm issue. He included him in preparatory activities and invited him to become a member of the HfG *Ehrencollegium*. However, Bill did not honor him as the father of the Bauhaus.[58] In his 1953 speech at the university of São Paulo, he especially lauded Van de Velde for the social and moral character of his architecture and did not even mention his educational innovations.[59] As was the case in most postwar narratives, it was the one-generation younger Walter Gropius that got all the credit from Bill as the founder of the Bauhaus.[60]

The most explicit articulation by Van de Velde himself of the claim to this mythological "fatherhood" of the modern style is found in his posthumously published memoirs *Geschichte meines Lebens*. This narrative is quite similar to that circulating in Belgium and first articulated by Léon Stynen in 1954:

> I noticed that among the adepts of a new, rational architecture and of the tendency for a pure, primary design, which originated in Europe and which besides had also appeared in the United States, I was seen as the "father" of the new style.[61]

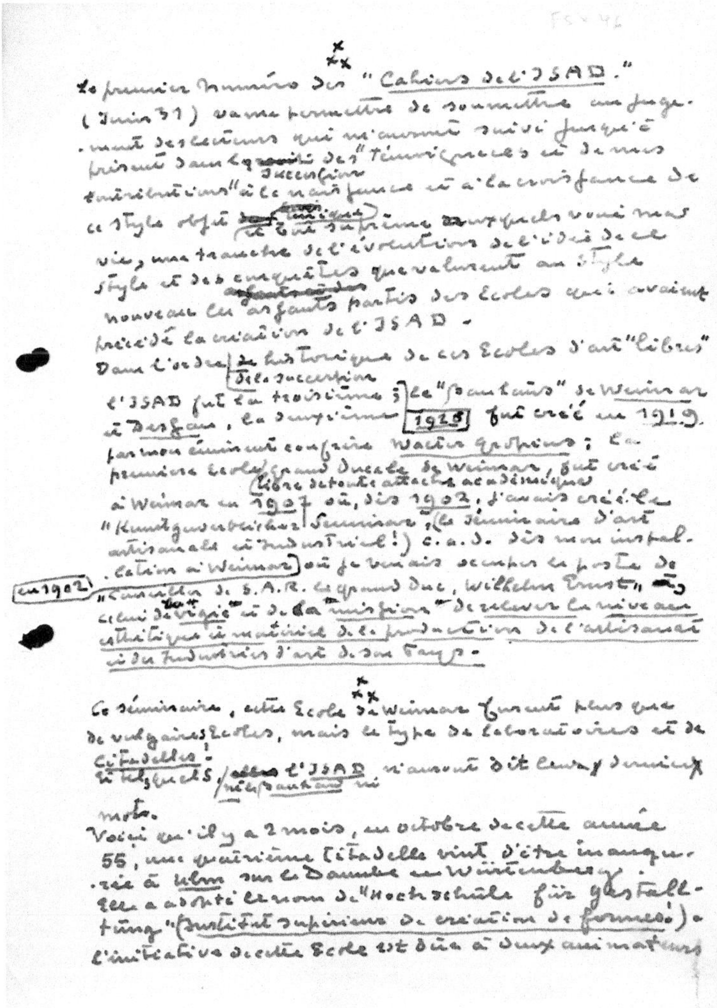

Fig. 2.10. Henry van de Velde's manuscript *Les Citadelles* on the four citadels of modernism, 1955. Source: Fonds Henry van de Velde, © AML (Archives et Musée de la Littérature)

This Atlantic Henry van de Velde narrative — full of clichés, simplifications, and distortions — circulated widely in Belgian design circles in the 1950s. According to most of the design critics and advocates of Van de Velde, it was indeed quite paradoxical that industrial design was (re)introduced into Belgium after the example of the United States followed by Italy, Switzerland and Great Britain, while the ideas themselves originated in Belgium.[62] Their national hero Henry van de Velde was after all — as the myth lingered — the founder of industrial design.

However, the deployment of this myth in the context of postwar design promotion can be understood as a double strategy. On one hand, Van de Velde's legacy was used to explain the success of American design. This has to be partly understood in the wake of the Marshall Plan. Initiated after World War II, this program launched a productivity campaign in Europe to boost economic recovery and protect Europe from the communist threat. Through different media the American way of life was promoted and disseminated in Europe through what is now called a "soft power" approach, in order to contain communism in the context of the Cold War. Conducted within the framework of this cultural war, these actions generated the dissemination of American design goods and opinions.[63]

Although Belgium could only benefit to a small degree from the Marshall Plan because its economy was reasonably healthy after the war — France and Great Britain were entitled to the bulk of the money — the Marshall Plan had an impact.[64] In Belgian official circles, industrial design indeed became included in economic policies after the example of the American economic miracle. Both Pierre Harmel and Jean Rey — the first ministers who drew attention to the phenomenon — referred to the American practise as a positive argument to introduce industrial design in Belgium too and subsequently to support the establishment of the design bodies.[65] The hauling-home of Henry van de Velde was probably a strategy to appropriate the exotic new discipline and to inscribe it in its own history, since the practise of industrial design did not have any remarkable tradition in Belgium. Van de Velde was thus a grateful and timely asset to give industrial design a notable pedigree rooted in the national soil.

Second, this myth was a well-structured argument to convince Belgian industry to embrace Van de Velde's principles since he reaped such great success abroad. The whole story was depicted as a lost opportunity for Belgium to validate the principles of its compatriot. Emphasis was put in particular on the fact that foreign industries, such as the American one, reaped the harvest of his activities, instead of the Belgian one.[66] In general, it was accepted that, given the "massive passivity" and "narrow-mindedness" of the industry and its strong ties to tradition, Van de Velde's ideas could not flourish in Belgium and he had to move to Germany and Switzerland to further develop his practise and educational programs.[67] It was thus the "blind" Belgian industry itself that was to blame for this situation. The problematic and politically sensitive relation between Van de Velde and his home country was never mentioned.

3.2.2 VAN DE VELDE'S LEGACY AS ANTIDOTE FOR COMMERCIALIZATION

As the opening anecdote of this chapter shows, the Institute not only used the "rational" face of Henry van de Velde to give industrial design its commercial credentials, but also deployed his "emphatic" face for quite the opposite. At the Institute's first event, the 1956 *Industrial Design Show* in Liège, a large portrait of its honorary president featured in the entrance, as described above. The exhibition's architects, Kroll and Vandenhove, used this mythical figure to draw attention to their own concerns in the design field at that moment. Kroll explained in hindsight that Van de Velde's image had to serve on the one hand as an antidote for the commercialization of industrial design, and on the other hand as the godfather of the artistic design tradition of the Bauhaus:

> His humanism could protect us from a professionalization that was too commercial and also humanize the "Bauhaus" models which were the most topical but, confusingly, the most abstract.[68]

Paradoxically, although Kroll claimed that Van de Velde's message was ignored, hidden and deformed, he himself adjusted Van de Velde's image to underwrite his own vision of that moment.[69]

Like many Western European intellectuals and architects of his day, Lucien Kroll pondered the question of the individual and his personal freedom in a mechanical and standardized architectural system.[70] He used Van de Velde's theory to deal with his concerns. He visited the voluntary exile in Switzerland with a group of other founding members to seal the foundation of the Institute for Industrial Design and to celebrate their honorary president. That occasion gave rise to correspondence between Kroll and Van de Velde on the issue of mechanization. He was of the opinion that:

> The continuation of [Van de Velde's] effort towards beauty would have to focus this time, not on mechanization, which had become too brutal, but on respect towards the person and his right to be

Fig. 2.11. Message from the new director, Pol Provost, to the members of the Institute for Industrial Design. Source: Provost, Pol. "Message." Bulletin d'Information. Institut D'Esthétique Industrielle, no. 4 (1959): 1.

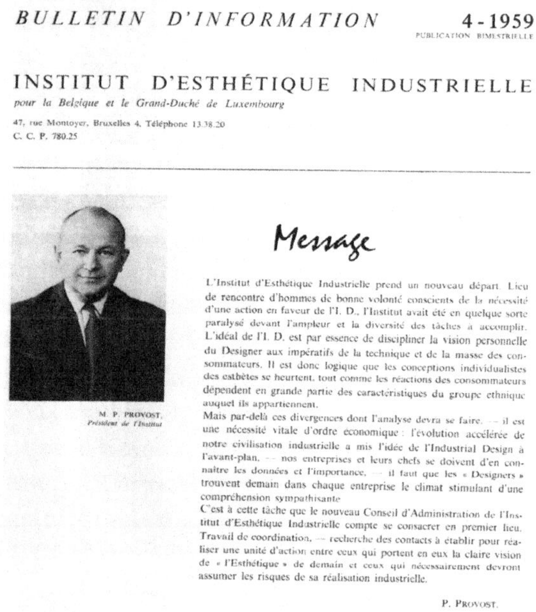

BULLETIN D'INFORMATION **4 - 1959**
PUBLICATION BIMESTRIELLE

INSTITUT D'ESTHÉTIQUE INDUSTRIELLE
pour la Belgique et le Grand-Duché de Luxembourg
47, rue Montoyer, Bruxelles 4, Téléphone 13.38.20
C. C. P. 780.25

M. P. PROVOST,
Président de l'Institut

Message

L'Institut d'Esthétique Industrielle prend un nouveau départ. Lieu de rencontre d'hommes de bonne volonté conscients de la nécessité d'une action en faveur de l'I. D., l'Institut avait été en quelque sorte paralysé devant l'ampleur et la diversité des tâches à accomplir. L'idéal de l'I. D. est par essence de discipliner la vision personnelle du Designer aux impératifs de la technique et de la masse des consommateurs. Il est donc logique que les conceptions individualistes des esthètes se heurtent, tout comme les réactions des consommateurs dépendent en grande partie des caractéristiques du groupe ethnique auquel ils appartiennent.
Mais par-delà ces divergences dont l'analyse devra se faire, — il est une nécessité vitale d'ordre économique : l'évolution accélérée de notre civilisation industrielle a mis l'idée de l'Industrial Design à l'avant-plan, — nos entreprises et leurs chefs se doivent d'en connaître les données et l'importance, — il faut que les « Designers » trouvent demain dans chaque entreprise le climat stimulant d'une compréhension sympathisante
C'est à cette tâche que le nouveau Conseil d'Administration de l'Institut d'Esthétique Industrielle compte se consacrer en premier lieu. Travail de coordination, — recherche des contacts à établir pour réaliser une unité d'action entre ceux qui portent en eux la claire vision de « l'Esthétique » de demain et ceux qui nécessairement devront assumer les risques de sa réalisation industrielle.

P. PROVOST.

serving the renewed aims of the Belgian design bodies, namely the promotion of design to benefit industry.

4.1 "A Purely Industrial Society"

Conflicts between "industrialists and aesthetes" seemed to plague the Institute in its heyday. It was eventually reorganized in 1959 under the presidency of prominent businessman Pol Provost (1907–1990), then director of the firm De Coene.[76] In his message to the members as new president, Provost announced a brand-new start by building a bridge between the quarrelling groups that formed the Institute (fig. 2.11). Despite his good intentions, the reformed Institute was chiefly focused on the Belgian industry.[77] When Provost described the Institute's tasks in 1959, he acknowledged that industrial design is composed of many different elements and domains.[78] However, he emphasized that the economic questions were the most "urgent" since they were currently posed by the Ministry of Economic Affairs and Belgian industry. In this vein, Provost completely renewed the executive board with members coming mostly from the industrial or commercial world. As such, the new board no longer suffered from its earlier heterogeneity but targeted solely the Belgian industry.[79] Josine des Cressonnières, the secretary of the Signe d'Or, reported enthusiastically on this change in a letter she wrote to the London-based design Misha Black, then the Vice-President of the International Design Council for Societies of Industrial Design:

> The truth is that the "old" Institute is dead, as I had told you. The actual one only keeps the name and the "façade", but it is entirely renewed, with a new Board of Directeurs [sic], and new company articles. All the wrong people are washed out! and this is going to be a purely industrial society. All the Directors (elected 10 days ago) are industrialists or representatives of industrial groupements. [...] The new president, Mr. Provost, is a friend of Mr. Giron, my president. He is an energetic, clever, and dynamic caracter [sic]. It is a very happy event that he became president.[80]

The new start of the Institute was symbolized by a new visual image. By 1961 the new house style was implemented in all its communications (fig. 2.12). The most visible aspect was the new logo — a sober

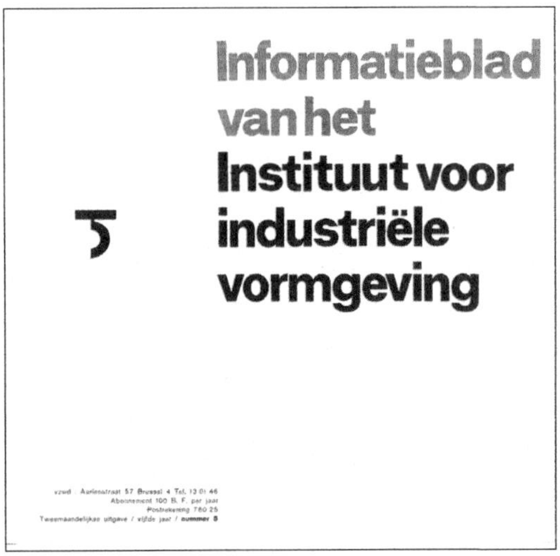

Fig. 2.12. Around 1961 the new house style of the Institute for Industrial Design was implemented. The logo was designed by Michel Olyff and the magazine *Informatieblad* was designed by Corneille Hannoset. Source: Informatieblad van het Instituut voor Industriële Vormgeving, no. 5 (1962).

composition of a thick straight line and a thick curved line — designed by graphic designer Michel Olyff.[81] The new logo had to symbolize the essence of design: one can design anything — from a spoon to a city, as the modern adage goes — with the help of a straight and a curved line.

Thanks to this new élan and the reorganization of the Institute, the Ministry of Economic Affairs, then headed by the liberal Minister Jacques Van der Schueren, designated the Institute as the coordinating body for industrial design in Belgium.[82] It was tasked with managing the credits that had to be divided between the initiatives in favor of design. The Ministry of Economic Affairs also demanded the merger of the activities of the Institute of Industrial Design and of the Signe d'Or.[83] The two institutions consequently shared the same office and a secretariat headed by Josine des Cressonnières. However, they both remained separate legal entities.[84]

This centralization in the womb of the Institute enabled a stronger standardized policy. However, it failed to bridge the difference between the industrialists and the aesthetes. The Institute's clear position in service to the industry could count on some criticism. The Institute was accused of completely ignoring the social and cultural aspects of design.[85] Moreover, cultural and artistic actors felt left out in the cold. Since the new Institute for Industrial Design was now in charge of the credits for industrial design promotion, the funding for this cultural perspective dried up.

4.2 Connecting with International Developments

With the new direction of the Institute, the "artist" industrial designer Henry van de Velde was swapped for a more hardcore industrial conviction of design, one that could stimulate production and industry. The new discourse was no longer marked by national heroes but connected with new international tendencies such as the Ulm philosophy.

It was, in particular, through the impressive network of Josine des Cressonnières that the Belgian design scene became intimately linked with international developments. In her capacity as secretary of the Signe d'Or, Des Cressonnières had, from 1957 onwards, invited key design personalities to sit on the Signe d'Or award jury, among them the notable Eliot Noyes and Raymond Loewy (figs. 2.13–14). Many of the invited jury members were involved in the International Council of Societies of Industrial Design (ICSID), which had been newly founded in Paris in 1957.[86] For example, its president, Peter Müller-Munk, sat on the first Signe d'Or jury in 1957.[87] He was followed by the vice-presidents Enrico Peresutti and Micha Black in 1958, the secretary Mia Seeger and the treasurer Pierre Vago in 1959, and the vice president Sigvard Bernadotte in 1960. Because of the good relations

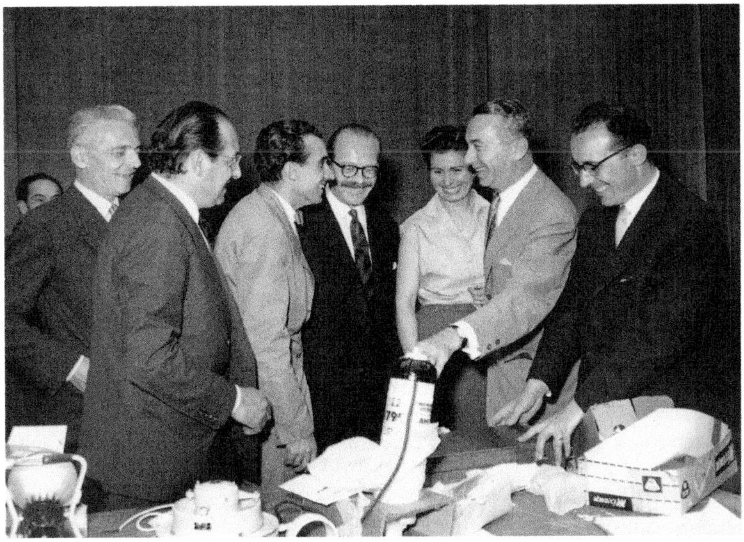

Fig. 2.13. The first Signe d'Or jury in 1957. From left to right: M. Stoffel, Robert Giron, Alberto Rosselli, Jaap Penraat, Josine des Cressonnières, Peter Müller-Munk and Louis Desamory. Source: Coutisse, Archives des Cressonnières.

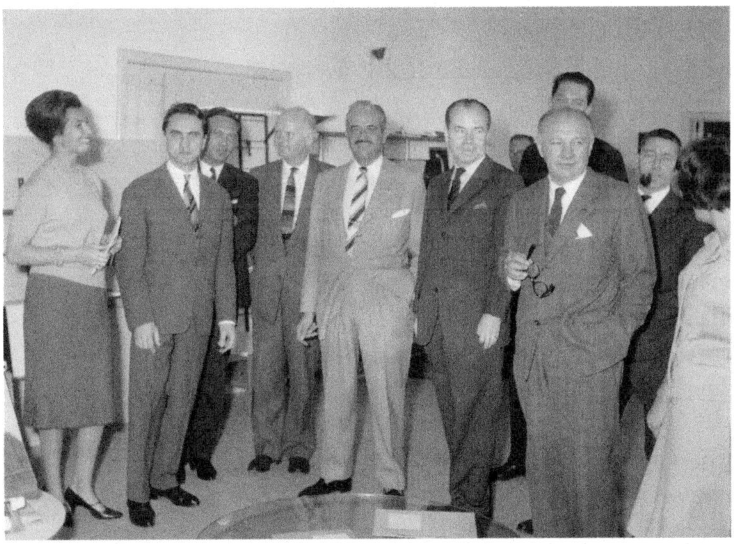

Fig. 2.14. The Signe d'Or jury in 1960. From left to right: Josine des Cressonnières, André De Poerck, P. Blijweert, Paul Reilly, Raymond Loewy, Sigvard Bernadotte, Jules Wabbes, Marcel Mabille, and Nanny Still. Source: Coutisse, Archives des Cressonnières.

promotion and "rationalization." Moreover, he argued that the monolithic understanding of design should be replaced by a pluralistic understanding because there was no single reality of industrial design, but rather many realities.[98]

This successful lecture at the Federation of Belgian Industries was the immediate reason for organizing a trip in May 1962 with the Institute's Commission of Education to the Ulm School of Design.[99] Maldonado welcomed the Belgian Commission and guided them through the school. The aim of the trip was to observe the pedagogical methods applied in Ulm and to exchange views with the corps of professors. A few months later, in November 1962, Maldonado was again invited to speak in Belgium,[100] this time about his vision on design in the context of the Enlarged European Market at the symposium in Brussels of the Liaison Committee for Industrial Design in the Common Market.

It was also in Belgium that Maldonado developed a definition of design that would become the international reference standard. In 1964, Josine des Cressonnières organized the first ICSID Seminar on Industrial Design Education in Bruges supported by UNESCO. On this occasion, Maldonado proposed the following working definition for industrial design:

> Industrial design is a creative activity whose aim is to determine the formal qualities of objects produced by industry. These formal qualities include the external features but are principally those structural and functional relationships which convert a system to a coherent unity both from the point of view of the producer and the user. Industrial design extends to embrace all aspects of human environment which are conditioned by industrial production.[101]

That same year, this definition was immediately adopted by the Belgian design institutes, and would, in 1969, also become the official ICSID definition.[102]

5. CONCLUSION: THE "MYTH" DISMANTLED

While Henry van de Velde was hailed as mascot for the Belgian industrial design movement in the late 1950s, he had lost his mediating role as a "design pioneer" by the early 1960s. After his death, Van de

Velde's image was no longer apt to communicate various design visions or to represent the Belgian design movement. Instead, his myth was dismantled and became reduced to a piece of the puzzle in the history of design. Maldonado's vision of the designer as an active partner of industry blended better with the commercial ambitions of the state-sponsored institutions than Van de Velde's ambiguous discourse.

Van de Velde's "words" of the 1890s, which had been fished up in the 1920s by the interwar avant-garde and again in the 1950s by design adepts, seemed after all to have lost their relevance by the 1960s. At Van de Velde's centenary, in 1963, six years after his death, design critic K.N. Elno reflected on his legacy. While Elno was an ardent supporter of the Belgian's theories and their importance for developments in Belgium in the early 1950s, he changed his tone in 1963:

> His attacks on the tastelessness and suffocating bad taste of a bourgeois and slightly cowardly nineteenth century are formulated today with more caution and obviously in a more detached manner. His odes to technique and industry now hardly seem less romantic than those of Henri de Saint-Simon and it is not excluded that they sound awful to many of our contemporaries for whom the industrial technology is an obvious and appreciated reality [...][103]

CHAPTER III

Visions and Fantasies of "European" Design in the Early Cold War Years (1961–1970)

1. INTRODUCTION: THE GALLIC 16

In 1957 Roger Tallon (1929-2011), affiliated to the Paris-based design agency *Technès*, designed a turning lathe for the Belgian company *La Mondiale*.[1] The new mechanical utensil was christened the *Gallic 16* (fig. 3.1). The term "Gallic" was a historical reference to the pre-Christian territory of Gaul, which coincided with modern Western Europe and was populated at the time by the Celts, who were known for their extraordinary bravery and strength, characteristics that suited the desired image of *La Mondiale*'s machines. In this product range of *La Mondiale* with its fearless and bold names, the *Gallic 16* was the first simplified and clean-looking design in conformity with the modernist credo.

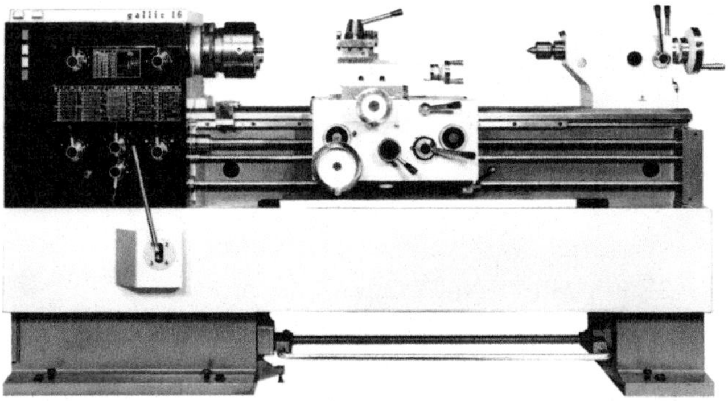

Fig. 3.1. The *Gallic 16* turning lathe designed by Roger Tallon, affiliated to the Paris-based design agency *Technès*, for the Belgian firm *La Mondiale*, 1957. Source: Paris, Musée des Arts Décoratifs (Archives Roger Tallon, don 2008) © distr. MAD, Paris/ADAGP, Paris (Tour Gallic Tallon pour la Mondiale, 1957)

Roger Tallon's functionalist design apparently also broke with precedents outside the company, given its large resonance in the field of mechanical utensils and industrial design. In advertisements, exhibitions and press accounts, the *Gallic 16* was lauded as a quantum leap for the design of mechanical equipment.[2] Many even made the comparison with the Citroën *DS 19*. Like this French design goddess, the joints between the iron compartments were not puttied up, but instead the connections between the iron sheeting were left open. One other asset of the lathe was the application of ergonomics — the new fashionable word of that time. Part of this was the user-friendly graphic design of the diagrams and the use of colors. The moving parts of the lathe were lacquered in bright yellow, which stood out in sharp contrast with the other static parts painted black, grey and white. But more importantly, soon after its launch on the market, the *Gallic 16* was seen as the materialization of the Europeanization that was then gaining much traction.

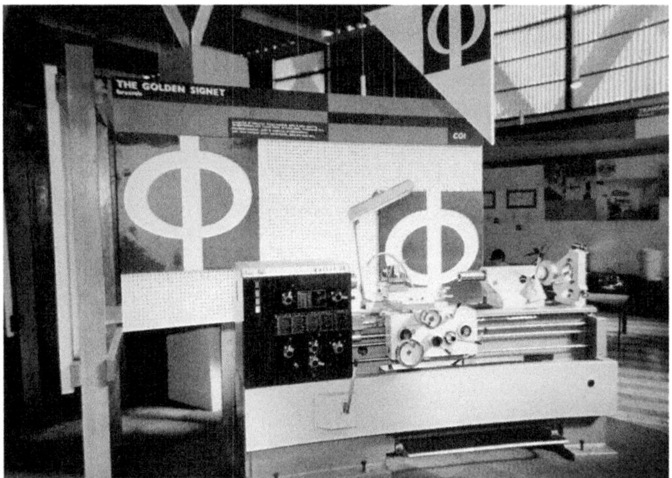

Fig. 3.2. The *Gallic 16* was awarded the Signe d'Or in 1960. Here on display at the Belgian pavilion in Johannesburg (South Africa), 1961. Source: Coutisse, Archives des Cressonnières. Picture by Jacques Dolphyn.

In 1959 the French magazine *Esthétique Industrielle* described the *Gallic 16* as a shining example of "Europeanization."[3] Indeed, the lathe emanated from a transnational collaboration between a French-based design studio and a Belgian company. According to the magazine's editorial board, "European" design was creditable if it was inscribed in the logic of present-day production, and especially if one could *not* attribute a specific nationality to it other than "European." The *Gallic 16* was thus considered exemplary, foremost because its nationality was unidentifiable — French, Belgian, German? — rendering it more "European." Not only did the French magazine consider the *Gallic 16* as a promising specimen of "European" design, but Belgian design bodies also adhered to this European dimension. In 1960 the *Gallic 16* won the Benelux Signe d'Or award (fig. 3.2).[4] Three years later, in 1963, it was selected for Signe d'Or's European sequel: the *European Selection of Industrial Design.*[5]

This selection — although never fully realized — was one of the first actions of the Belgian initiative: the Liaison Committee for Industrial Design in the Common Market. This committee was established at the initiative of the Institute for Industrial Design for Belgium and Luxembourg and grouped together all European — Belgian, Luxembourg, Dutch, Italian, French and German — design bodies in the scope of the new extended European market. Accordingly, its main *raison d'être* was to coordinate and protect the common interests regarding industrial design of the member countries to European officials.[6] Secondly, the committee aimed to bring the "European" product to the public and to create a panorama of European design — here is where the *European Selection of Industrial Design* and the *Gallic 16* fit in.[7] The discourse of the Liaison Committee emphasized the important economic and cultural position of Europe between the two Cold War superpowers — the United States and the Soviet Union - that required a European answer of equal value. A simplified and idealized vision of the rich European traditions was proposed as a basis for this European answer — or better design.

Although the Liaison Committee for Industrial Design achieved hardly any actual results and its *European Industrial Design Selection* with the *Gallic 16* as one of the nominees was never carried through, this chapter will take it as a lead to explore the desires and fantasies projected onto "European" design by the Belgian design institutions. What was this "European" design? How was it understood and envisioned? And what were the Belgian motives in setting up an initiative for its promotion?

While chapter two focused on the ambiguous exploitation of the design icon Henry van de Velde in the heyday of Belgian design institutionalization, this chapter highlights the unique supranational character of design promotion in this same period. Similar themes are touched upon — such as the import of American design ideas, European design consciousness and the Ulm philosophy of Tomás Maldonado — but are now elucidated from the perspective of "European" design. This chapter first of all contextualizes the set-up of the Liaison Committee for the Common Market in the supranational tradition of the Belgian design policies and contrasts it with the national design actions. Furthermore, it analyses the discourse that accompanied the "European design" of the Liaison Committee and frames it in the larger geopolitical tensions of the day. Lastly, it looks at the persistence of the nation-state as the dominant category for organizing design in the postwar period, even when talking about or exhibiting this alluring "European" or "international" design.

2. THE EARLY STAGES OF DESIGN INSTITUTIONALIZATION IN BELGIUM: BELGIAN OR SUPRANATIONAL?

The Belgian initiative to set up the Liaison Committee in the Common Market and pursue the promotion of European design has to be framed in the unique supranational tradition of Belgian design promotion, an aspect that is left largely unexplored in the literature.[8] While the importance of a national context cannot be denied in the development of the postwar design cultures, the European perspective and the supranational consciousness were, however, quite prevalent in the Belgian case. In contrast to many other Western European countries, Belgium's official design promotion was not just concerned with reinforcing the national image, but was also interested in the construction of a Benelux, European and international design.

2.1 The Supranational Scope

The institutionalization of industrial design in Western Europe dates from the decades after World War II, when national design bodies proliferated throughout Western-minded countries: from Canada and Australia to Greece and Japan.[9] Much of this movement was endorsed by the various national governments at the onset of international liberalization. In Belgium, this trend emerged under the first socialist-liberal government elected in 1954.[10] After the short takeoff run described in chapter one, two institutions were founded in 1956 under the patronage of the Ministry of Economic Affairs: the Institute for Industrial Design for Belgium and Luxembourg, and the Benelux Signe d'Or award. Curiously, although the Belgian government explicitly considered industrial design as a tool to stimulate the dynamics of the national economy, Belgian design institutionalization was marked by an international and supranational scope.[11] The blueprint of both organizations founded in 1956 reflected the latest political agreements that would eventually lead toward European unification: the Belgium-Luxembourg Economic Union (1922) and the Benelux (1948).

The design bodies in Belgium, although backed by the national government, were not restricted to Belgian territory. The Institute — as its full name indicates — encompassed the territory of Belgium and the Grand Duchy of Luxembourg, reflecting the Belgo-Luxembourg Economic Union (BLEU), founded in 1922 and seen as the forerunner of the Benelux. The Signe d'Or, on the other hand, was a Benelux prize, rewarding products from Belgium, the Grand Duchy of Luxembourg and the Netherlands. The Benelux was an economic union between these three neighboring countries that came into operation in 1948, and is seen as the predecessor and great example of the European Economic Community. The same holds for the Brussels Design Centre, a not-for-profit organization established some years later in 1962 under the patronage of the Ministry of Economic Affairs and the Belgian Foreign Trade Office. Although its articles of association state that one of the main goals was "to promote the prestige of Belgian industry, particularly on foreign markets",[12] the center deliberately outlined a policy aimed at both Belgium and Luxembourg.[13]

This supranational trend in the institutionalization of Belgian design was unique compared with its foreign counterparts. As design historian Penny Sparke observes, most postwar national design policies were intended to create new national identities or reinforce existing ones for cultural and commercial purposes.[14] There is also the famous Scandinavian example, where a regional design identity was constructed in favor of a national one.[15] The difference here is that the different national institutions endorsed this notion of Scandinavian design, while in the case of Belgium the design institutions pursued alone a policy towards the Benelux or the BLEU. This leaning towards supranationalism can probably be explained by Belgium's position as one of the driving forces behind European unification. It was indeed from the prospect of the international competition in the Common Market as stipulated in the 1957 Treaty of Rome that industrial design annual show contained the maximum dollar amount for the items displayed, e.g. *Useful Objects of Design Under $10.* Its postwar successor was the "Good Design" program (1950–1954) and was the brainchild of Edgar J. Kaufmann Jr., director of MoMA's department of industrial design.[25] The "Good Design" exhibitions were a close collaboration between the Museum and America's largest wholesale marketer to select "good" household goods and exhibit them in commercial stores as well as at the museum. These two actions, where commercial and cultural aspirations concerning design merged, proved to be a guiding example for Josine des Cressonnières. Indeed, she expressed the wish that similar initiatives could be made in European department stores with the help of design bodies. This tentative proposal by des Cressonnières is a clear prefiguration of the European design selection.

2.2.2 THE INTERNATIONAL DESIGN EXHIBITION AT THE AU BON MARCHÉ IN BRUSSELS

A first attempt to reach this European sweep was the 1960 Signe d'Or exhibition at the *Au Bon Marché* department store in Brussels.[26] On the occasion of its 100th birthday, *Au Bon Marché* invited the Signe d'Or to organize a prestigious industrial design exhibition. The collaboration probably came about thanks to the experiences of Josine des Cressonnières in Belgian distribution, which was indeed her home base with regard to design. She started in 1953 at *À l'Innovation — Au Bon Marché*'s rival — as the first Belgian "stylist." She quit the job to devote herself completely to the foundation of the Signe d'Or, but in 1960 she returned to try out the Signe d'Or's new orientation.

The original idea of the anniversary exhibition was to display *European* industrial design. However, when the International Council of Societies of Industrial Design (ICSID) was asked to be one of the patrons, the scope shifted toward *international* industrial design.[27] That the authority of the ICSID was invoked was no surprise, since the ICSID and the Signe d'Or were, after all, no strangers to each other.[28] Their relationship was even a well-established one given that many of the leading ICSID figures — Peter Müller-Munk, Alberto Rosselli, Enrico Peresutti, Misha Black, Pierre Vago, Mia Seeger and Sigvard Bernadotte — had been invited by des Cressonnières to sit on one of the international juries of the Benelux Signe d'Or award organized each year from 1957 to 1960.[29] The ICSID was willing to grant its protection to the exhibition only if it could ensure the high quality itself.[30] As such, the vice-president of ICSID, Misha Black, was appointed as the foreign counselor of the exhibition.[31] The great involvement concerning the exhibition's content — switching it from European to international design — and the precision applied in the preparation by the ICSID with regard to this "Belgian" exhibition can be explained by the fact that it was a "rookie." The ICSID was created only in 1957 and the first executive board still had to prove the validity of the notion of *international* design. Moreover, this *Au Bon Marché* exhibition was the first that they endorsed. Accordingly, it was advised not only to exhibit European design, but to turn it into an *international* Industrial Design exhibition — including American and Canadian design organizations. This broader scope also met the aim of the first president, Peter Müller-Munk, who considered the ICSID as the start of a transatlantic design community.[32]

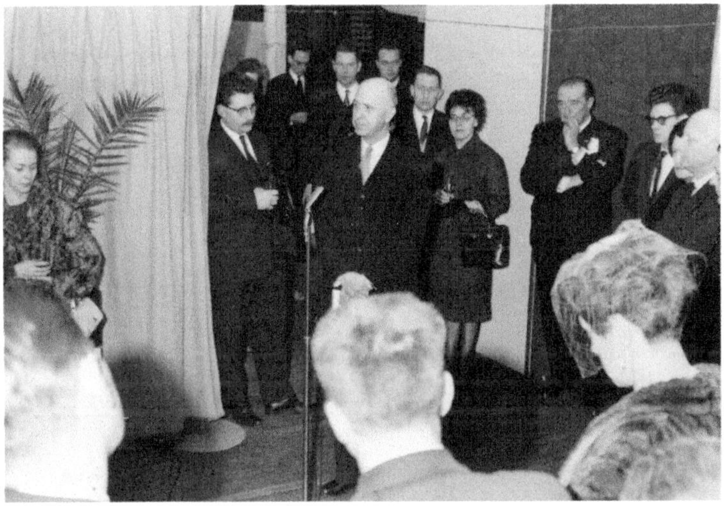

Fig. 3.4. Belgium's Minister of Cultural Affairs, Pierre Harmel, opens the *Au Bon Marché* anniversary exhibition by the Signe d'Or, entitled *International Design Exhibition,* February 1960. Source: Brussels, Archives et Bibliothèques de l'Université libre de Bruxelles, Fonds GIB, 25Z.

Again, as was the case for the Benelux Signe d'Or, this "international" scope was similarly turned into a national lucrative action. The exposure of "good design" from 15 foreign countries where the discipline of industrial design had already gained ground was understood on this occasion to be an opportunity for the Belgian public — industrialists and consumers — to familiarize themselves with the idea of "industrial design." The *Au Bon Marché* anniversary exhibition by the position as it was sandwiched between the two large design regimes of the Cold War protagonists: communism and consumerism.[51]

3.1.1 ECONOMIC DETERMINISM

What is immediately striking is that the Committee understood industrial design here as "a cultural assimilation of scientific and technological progress."[52] The Committee deemed that each economic type produced its own kind of product and consumer. This economic design determinism — the type of economy defines the genre of its production — was probably also a way to play up to the European government. Indeed, the economy was central to the European integration as formulated by the Belgian Paul-Henri Spaak and other "founding fathers" of Europe,[53] as well as in most postwar reconstructions, where many equated economic prosperity with peace and security.[54]

This economic deterministic view resulted in a simplified bipolar global design landscape. The American economy was seen as plethoric and based on competition, directing industrial design to the principle of obsolescence and "styling" — or the fast and superficial renewal of products. The Soviet economy, on the contrary, was directed by the government, resulting in design "rid of the stimulus or danger of competition." Between the two, the committee believed that Europe had to orient itself and plot European design as a third way. It was claimed that Europe had the advantage of having the oldest tradition, which could act as a source of inspiration, and the best developed industrial structure. So, for the committee, Europe had to find a well-balanced formula integrating its cultural heritage with the demands of a technological civilization.

This standard story of Europe's rich culture is similarly used to emphasize the unique character of the "European" consumer. In April 1961, the president of the Institute for Industrial Design, Pol Provost, was invited to give a talk at a conference on industrial design organized at Unesco in Paris. His talk was entitled: "How to imagine a product with a large sales potential within the framework of the extended market (the Common Market)." On this occasion he elaborated on the character of the European consumer. He claimed that his taste was shaped by the thousand-year-old European tradition. In contrast to his communist and capitalist counterparts, the European consumer was portrayed as "choosy" and "individualistic." "Choosy" because

— as Provost's line of reasoning went — he was not forced to acquire particular products, as was the case in the Soviet Union, but could choose freely from a large range of products. "Individualistic" because, in comparison with the American consumer, he was considered by Provost not to be "price-minded" since his purchasing behavior was an emotional matter guided by his subconscious.[55]

3.1.2 SANDWICHED BETWEEN EAST AND WEST DESIGN REGIMES

In this bipolar design landscape, the Liaison Committee wanted to conquer a third equal and autonomous position. However, its discourse and origins betray clear influences from both sides and have biased attitudes toward certain design practices from East and West. For example, the American influence is undeniable in the set-up of this European design venture — the European design selection was after all directly infused by the MoMA example of the "Good Design" and "Useful Objects" exhibitions. At the same time, however, a negative attitude toward American design surfaces in many statements of the Committee: arguments are advanced to drive back American influence, and instead the positive aspects of communist mechanisms of collective consumption are examined. As many design historians have already noted, the reception of American design was ambivalent in most European nations. In general, American "styling" was completely condemned by the European modernist design scene, while it embraced without reservation the modernist MoMA selection of "international" design with the Knoll and Miller furniture of its most famous exponents.[56]

What is peculiar in the case of the Liaison Committee is that this aversion to American styling is argued by means of a politically charged example in the context of European integration. The committee chose to warn about the import of US industrial models and the danger of "styling" by pointing out the decline of the British radio and television industry. The Committee claimed that British producers, stimulated by the annual radio and television fair, were forced to present new models each year. This practise prohibited new fundamental research and only allowed superficial novelties. Once they entered the international market, British manufacturers were outdone by foreign producers who focused not only on formal renewal but also on structural and functional renewal. To understand the politically delicate nature of picking out a British example, one has to take into account the political context of the day. In 1961, Harold Macmillan, the British prime minister, applied for EEC membership.[57] Yet, British emphasis on its special relationship with Washington was alarming, especially to the French, who feared US influence on European politics through the outpost of Britain. Analogous with this political situation, the committee saw British industry as a potential "American Trojan Horse" — in the words of French President de Gaulle as he portrayed Britain back then — for the European design practice that it wanted to protect from this Atlantic threat.[58]

Tomás Maldonado proposed several alternatives to the capitalist American model during his talk at the 1962 symposium in Brussels. As rector of the Ulm School of Design, he was one of the prominent design authorities of that time. In his vision we can detect a rather closer affinity with communist production than with a capitalist consumer model. At the symposium, he advised large companies not to turn to the dangerous American method of "styling," but to take up the creative tradition of the former industry, which was based on the integral renewal of the product, to recapture new competition.[59] He also noted that products for individual consumption are overly developed in comparison with products for collective consumption. In his view, this new attention for the collective demand, together with long-term research, would result in industrial design that would not stimulate "excessive waste of energy and materials, nor meet the demand of prestige and flamboyance," but would contribute to a better distribution of energy and materials and thus assist the cultural and social progress of the people.[60]

3.2 European Design: A Collage of National Design Identities?

Whatever the Committee may have envisioned with its actions, the concept of "European" design looked in reality rather fragmented. This is especially perceptible when taking a closer look at the practical organization of the *European Selection of Industrial Design*. It was carried out in two phases. First, a selection of 50 products was made at a national level by the different design bodies. Second, an international jury traveled from city to city to select the best products of these national selections in order to establish a "European" compilation. Remarkably, to address the "European" spirit or style, they first had to go through the national level. European design was, in contrast to the

the selection's accompanying discourse, understood as a sum of products, each with its own nationality. It seems that the committee was aware of this shortcoming since it foresaw the unification of national selection criteria and harmonized European production, implying that this unity was not yet in place. [61]

This fragmented understanding of European design seems to be quite prevalent also outside the Liaison Committee. On the European consumer market, European design manifested itself as a collage of national particles. Many scholars have indicated that the national characteristics of products gained a lot of value in postwar global trading where nations used design to secure their position on the extended markets. As Penny Sparke states, "as the global context threatened their demise, the meaning of 'local' and 'national' were intensified."[62] The president of the ICSID, the American designer Peter Muller-Munk, almost foresaw this strong postwar emphasis on national design identities in his short 1959 account of industrial design in the Common Market.[63] He claimed that growing standardization did not necessarily imply that the nationality of products would become unascertainable. On the contrary, increasing competition in the Common Market would only make the national characteristics of each product clearer.

The fragmented conception of "European design" arose especially in the postwar period as can be seen in its descriptions and depictions in magazines and shops. One example is the "European kitchen."[64] The Belgian architectural magazine *La Maison* reported on the "European kitchen" in its issue of February 1961 — around the same time the Liaison Committee was established.[65] The kitchen was equipped with labor-saving precision appliances from Switzerland, gadgets from the Netherlands, as well as practical and elegant plastic furniture from Italy. All this was complemented with techniques and taste from Germany, and highly qualified craftsmanship and creativity from France. European was thus here understood as a collection of national images. In one of *La Maison's* French counterparts, the magazine *Esthétique Industrielle,* a similar listing was made while describing the "European" dinner table that was a compilation of Swedish dinner plates, French cutlery and German glasses.[66] One other telling example is a huge billboard that covered the façade of the *Galeries Lafayette* department store in Paris.[67] "Ici l'Europe" was written in huge capital letters, while the background was filled with logos from different European countries, including Eastern and Northern ones. The logos were designed as if each nation were a brand.

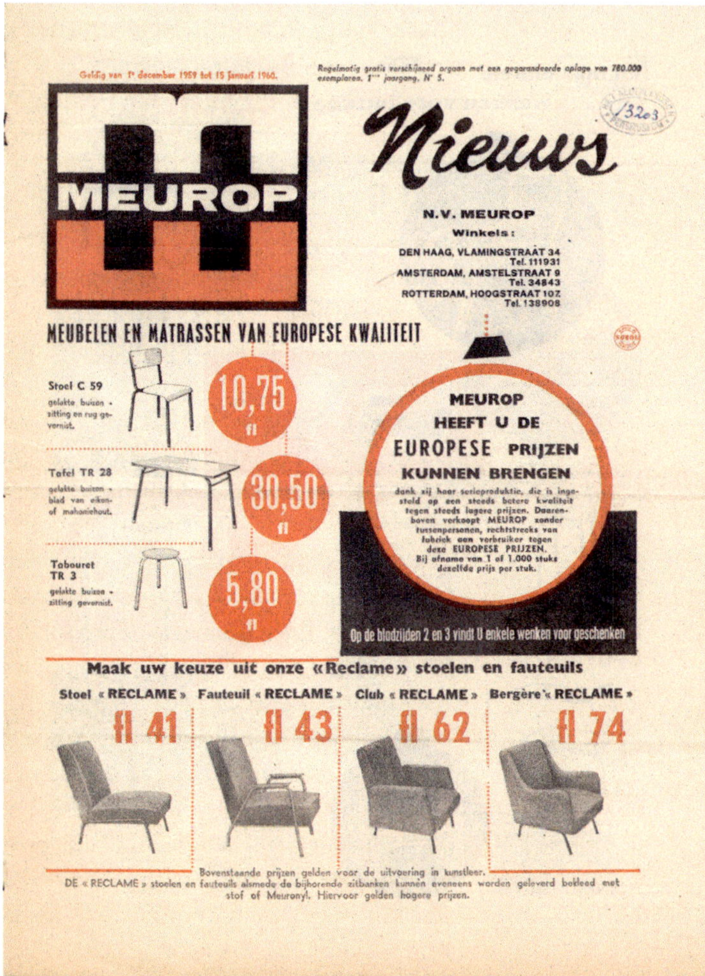

Fig. 3.10. "Meurop brought you the European prices." Source: *Meurop Nieuws*, 1 (1959) 5. Collection: International Institute of Social History (Amsterdam).

This "European" design as an exotic patchwork of national brands seems to have become embedded in the Belgian consumer culture of the 1950s and the early 1960s, where the term "Europe" became part of a commercial vocabulary. The postwar consumer vocabularies seem to become expanded by the code word "European" as a synonym for economic and advantageous, alluding to the promising prospects for the Common Market for EU consumers. Names of shops and products were inspired by the European integration that was associated with economic prosperity.[68] Some companies even stipulated a new policy oriented toward a European future. This was, for example, the case for the Belgian furniture company *Meurop*. The company was founded in 1958 in the light of the new extended market and established as one of the first European distribution networks, with shops in northwestern Europe.[69] Its name gives away its European ambitions: "Meurop" is a contraction of the French word *meubles*, meaning furniture, and *Europe*. In its advertisements for 1959 and 1960, Europe was often mentioned, mostly to emphasize the economic advantages for consumers, but also to refer to a certain modern style and industrial quality (fig. 3.10).

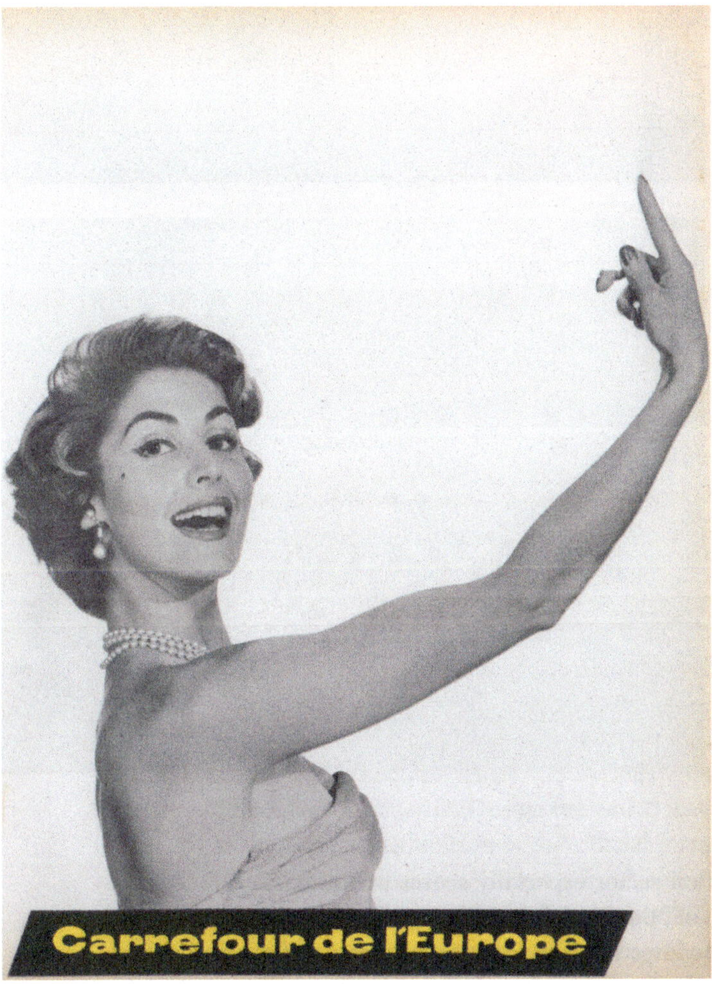

Fig. 3.11. Advertisement for the *Carrefour d'Europe* exhibition organized by *Au Bon Marché* in 1956. Source: Brussels, Archives et Bibliothèques de l'Université libre de Bruxelles, Fonds GIB, 25Z.

linked to national economic interests. As the account illustrates, both aspects are just two different sides of the same coin. For example, the supranational character of the first Belgian design bodies — all politically inspired by recent political agreements of the Belgian government — had its limits: the organization, funding and administration were purely Belgian, and the foreign products were only seen as stimulators for Belgian industry. The establishment in 1961 of the Liaison Committee for Industrial Design in the Common Market, on the contrary, can be seen as a true endeavor to initiate supranational design promotion because it was a transnational collaboration between the design bodies of the six EEC member-states. The Liaison Committee hoped to promote "European" design as a third way along with capitalist US design and communist Soviet models. However, although envisioned as a harmonized whole, this "European design" seemed in reality to be trapped in the national. It manifested itself indeed rather as the sum of national products instead of as a true synergy surpassing the national. Although aspirations to raise design promotion to a supranational level ebb away in Belgian design promotion after the mid-1960s, the European, international and national continued to flirt with each other in the case of the Brussels Design Centre. The changing conditions of production in light of European economic liberalization prompted the national institution to question the sustainability of national labels of products and revealed the malleability of the national category.

CHAPTER IV

Design Exchanges between Brussels and Moscow in the Early 1970s

1. INTRODUCTION: LEAPING OVER THE IRON CURTAIN

In March 1973 Prince Albert of Belgium stood in front of a small scale-model of a huge Russian construction crane on pebbles in the exhibition space of the Brussels Design Centre (fig. II.1).[1] He had been invited by the Design Centre and the Belgian Foreign Trade Office to open the first show in the West of the Russian All-Union Research Institute of Technical Aesthetics (known by its Russian acronym VNIITE), the Soviet counterpart of the Brussels Design Centre. Flanked by a female translator and accompanied by the Russian ambassador and the Belgian minister of foreign trade, the Belgian prince took considerable interest in the designs conceived on the other side of the Iron Curtain.[2] The products selected for the show in Brussels had to illustrate the wide range of Soviet production: from heavy industrial machines and medical products to crafted domestic goods and sports equipment. One year later, in 1974, the roles were reversed and VNIITE hosted the Brussels Design Centre in Moscow showing a panorama of modern Belgian designs.

These design exchanges across the Iron Curtain, between the heart of Europe and that of the Soviet Union, in the midst of the Cold War, may seem striking. Most historical studies on European Cold War design focus indeed rather on the "Americanization" of European design cultures.[3] In a sense, rightly so: this period, from the 1940s to the mid-1970s, is indeed called the "classic American century" since America was at the height of its political and economic power.[4] In studies on postwar Belgian design too, elaborate attention is given to the inflow of American design culture and the ambivalent reactions it provoked.[5] Much more obscure in these European Cold War narratives, however, are the design exchanges with the second Cold War superpower: the Soviet Union. This article sheds light on this lesser-known aspect and explores the institutional and personal connections between Belgian and Soviet design cultures in the postwar era.

2. THE DESIGN CENTER AS TRANSNATIONAL FORMAT

The partition of the world into two opposing geopolitical camps – the capitalist American "free world" and the communist Soviet society, which was so widely depicted and carried out, did not, as one might have assumed, preclude West-East design exchanges.[6] As Susan Reid and Rosalind Blakesley point out, "The Iron Curtain may have affected the transmission and reception of cultural emanations from both East and West, but it never created the separate, hermetically sealed cultures and societies that could be implied by Churchill's famous phrase."[7] This is confirmed when looking at the far-flung transnational network of national bodies for design promotion.

The institutionalization of industrial design in Western Europe dates from the decades after World War II and was largely directed by government-backed manifestations.[8] Modern design was endorsed by nation-states for various reasons including the creation of a national design identity, Cold War propaganda, the improvement of the national economy and its position in global markets. One of the most popular formats used to promote national industrial production was the design center. Through the promotion of "design", such a center aimed to increase national prestige and consequently the national export and welfare. Accordingly, design centers are mostly seen as transmitters of national design cultures. However, looking at the flow of goods, ideas and people that crossed the boundaries of the nation-state, one can actually account for the hybrid face of these bodies. Indeed, such centers not only had a national activity radius, but shaped, and were shaped by, the transnational design community in the postwar era.

The American designer Arthur Pulos situates the "start" of such a global network of designers and design bodies at the 1946 *Britain Can Make It* and 1951 *Festival of Britain* exhibitions.[9] The first International Design Conference was indeed organized by the British Council of Industrial Design (CoID) within its framework in 1951. The second International Design conference was consequently organized in Paris in 1953, and this international gathering led directly to the foundation of the International Council of Societies of Industrial Design (ICSID).[10] In the light of the emergence of this global design scene, Arthur Pulos also situates the proliferation of the design center format in virtually every "country — from Argentina to the Soviet Union — that could lay claim to being industrialized."[11] Most of these were modeled on CoID's famous Design Centre in London, established in 1956 and allegedly the first.[12] Adopting this British format, most importantly, implied becoming part of an international community of design centers and sharing experiences through letters, exhibitions and meetings. In 1967, these promotional bodies even became officially recognized as important nodes of the transnational design community when they were allowed to become members of the ICSID.[13] In that sense, these national design centers incorporated a remarkable paradox: while they connected on a global level driven by the modernist belief that design was something universal, these bodies simultaneously gave credibility to design as the marker of the nation. It is this double play that characterizes the exchanges between the Brussels Design Centre and VNIITE.

Both the Belgian and Russian organizations were established in 1962 and insinuated themselves into the international tradition of design promotion, stimulated by the debates within the ICSID community and especially by contacts with the director of the British CoID and its Design Centre, Paul Reilly. In 1957 Reilly visited Moscow where he met with Russian designer Yuri Soloviev (1920–2013) whom he invited to London for a couple of weeks to acquaint himself with industrial design practises in Britain.[14] These London connections encouraged the creation of a Russian design institute (VNIITE) a couple of years later under Soloviev's direction.[15] It was a gigantic organization working under the umbrella of the State Committee of Science and Engineering of the Council of Ministers of the USSR and included not only design promotion, but also design research and practise. At its height, VNIITE employed around 2,000 people and encompassed fourteen departments from 'Exhibition and Multimedia' to 'Transportation Design and Ergonomics', managing ten regional branches from Moscow.

The Belgian Design Centre was equally indebted to the British example. The good connections between the first ICSID executive board and Josine des Cressonnières, the secretary of the Belgian design award, the Signe d'Or, were decisive in the adoption of the

"British formula". As discussed in chapter three, many of the leading ICSID figures had been invited to sit on the international juries of the Signe d'Or and in 1960 both organizations joined forces to mount the *International Design Exhibition* at the *Au Bon Marché* department store in Brussels. At the suggestion of the ICSID President Peter Muller-Munk, Josine des Cressonnières put forward a proposal to take the Signe d'Or closer to the British example and set up a Design Centre.[16] This included a permanent exhibition (which changed every three months and which displayed objects selected by a jury), regular thematic exhibitions, a design and designer index, and a library. The difference from the British situation was that the Brussels Design Centre was not connected to a Council of Industrial Design, but operated as an individual force in the Belgian design scene.

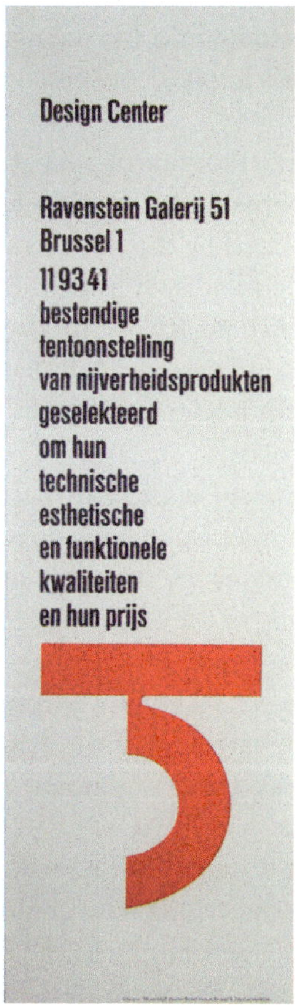

Fig. 4.1. The first poster of the Design Centre still mentioned its original American spelling "Design *Center*". In 1963 British spelling was adopted. Source: © Michel Olyff. Ittre, Private archive Michel Olyff.

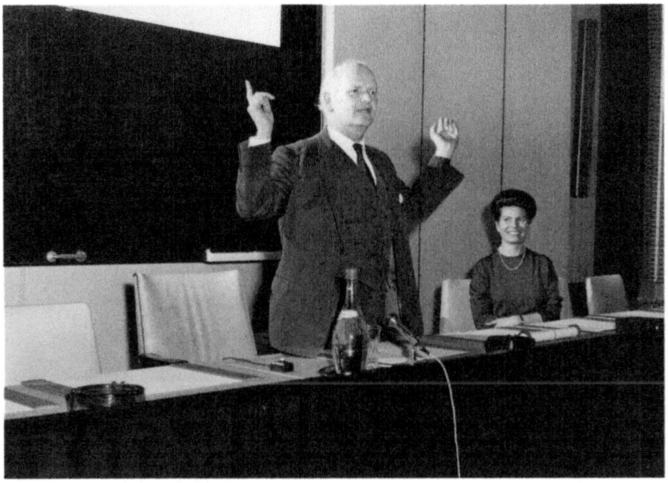

Fig. 4.2. Paul Reilly's lecture in Brussels within the framework of the preparations for a Belgian Design Center, May 1963. Source: Coutisse, Archives des Cressonnières.

The British influence was also noticeable in the name change in 1963, swapping its original American spelling — Design *Center* — for the British spelling — Design *Centre* (fig. 4.1).[17] This change was likely part of the positioning of the new design organization and reflected a move away from the American design approach that was very influential in the formative years of design institutionalization.[18] The early success of the American design scene had of course much to do with the Marshall Plan that initiated a productivity campaign in Europe since the late 1940s and which facilitated the easy inflow of American practices. As many design historians have described, American design was, however, received with ambivalence by European nations.[19] In general, the European modernist design scene embraced without reservation the modernist Museum of Modern Art (MoMA) selection of "International" design with Knoll and Miller furniture as its most famous exponents, while it completely condemned the so-called American practise of "styling" — or the fast and superficial renewal of products. Also in Belgium, the American practice of boosting sales curves by giving existing products a new exterior without changing much of the technology inside was strongly criticized by the first design institutes. The rejection of the American spelling was probably a way to avoid association with those kinds of practises.

The official reason for this change was that similar centers used British spelling, revealing its aim to join the international design community. Even the Minister of Foreign Affairs, Maurice Brasseur (1909–1996), explained during the Belgian Design Centre's announcement that the British name "Design Centre" — and not a Flemish or French version — was consciously chosen because of its international renown and in order to follow its foreign counterparts.[20] However, the visit of the CoID's director Reilly to Brussels a couple of months earlier was probably one of the main incentives to alter the name (fig. 4.2).[21] He was invited as advisor and shared his experience of design centers with the representatives of the main interest groups that funded the Belgian version, such as the general director of the Belgian Foreign Trade Office, Georges Burnay, the director of the "Fairs and Expositions" office, Louis Loncin, and the new board of directors.

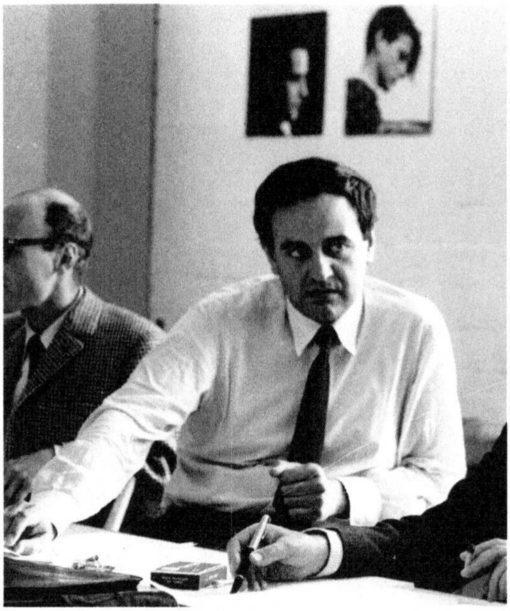

Fig. 4.3. Tomás Maldonado at the Ulm School of Design (HfG), 1965. Source: Ulm, HfG-Archiv Ulm.

Along with the appeal of this British model for its blueprint, the Belgian Design Centre was also heavily inspired by the design philosophy prevailing at the Ulm School of Design (HfG) of which the Argentinian Tomás Maldonado (1922–) was one of the most renowned exponents (fig. 4.3).[22] Maldonado was known for his innovative and controversial theories on design. He rejected the artistic

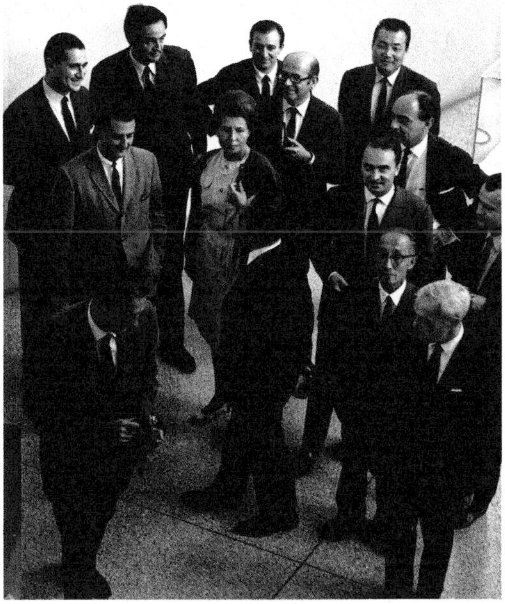

Fig. 4.4. Josine des Cressonnières at the Second ICSID Seminar on The Teaching of Industrial Design in Ulm, together with Arthur Pulos (US), Roger Tallon (France), André De Poerck (Belgium), Misha Black (UK), Zvonimir Radic (Yugoslavia), Tomàs Maldonado (West Germany), Shinji Koioke (Japan), Nathan Shapira (US), Basilio Urbine (Argentina), Gino Valle (Italy), Manuel Villazón Vázquez (Mexico) and Alexandre Wollner (Brazil), 1965. Source: Ulm, HfG-Archiv Ulm.

Bauhaus pedagogy as well as the American commercial approach as outdated conceptions. He proposed a third way: a sociological and science-based product management and system analysis.[23] The adoption of Maldonado's views was hardly surprising since — as described in chapter two — close contacts were established between the Belgian design institutes and the Argentinian at the ICSID conference in Venice in 1961, which resulted in several lectures by Maldonado in Belgium and a Belgian delegation visiting the HfG in Ulm. Moreover, the Ulm philosophy was very much in vogue in international design circles and Maldonado's definition became the official ICSID definition in 1969.

The Russian and Belgian design centers each had a prominent position within the transnational design community. Key in both cases were the powerful network and personal skills of the director. Despite the Cold War circumstances VNIITE was an outward-looking organization. According to former VNIITE designer Dmitry Azrikan, it was "probably the most globally open Soviet institution behind the Iron

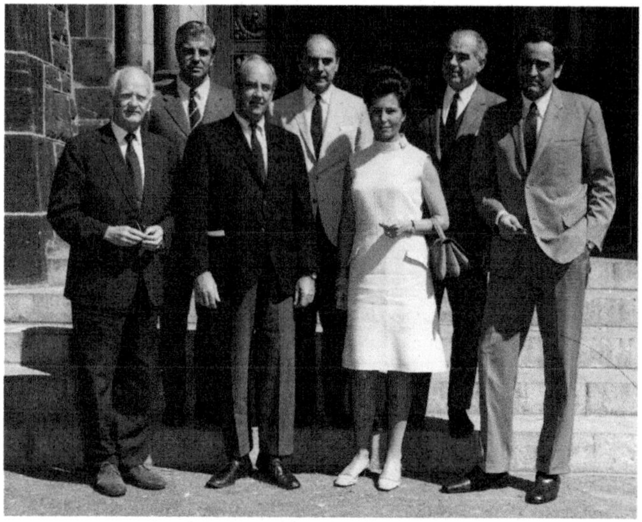

Fig. 4.5. The ICSID board in Montreal in 1967. From left to right: Paul Reilly, André Ricard, Richard Latham, unknown, Josine des Cressonnières, Jacques Guillon and Tomás Maldonado. Source: Brighton, University of Brighton Design Archives, ICSID Archive (2-3-7).

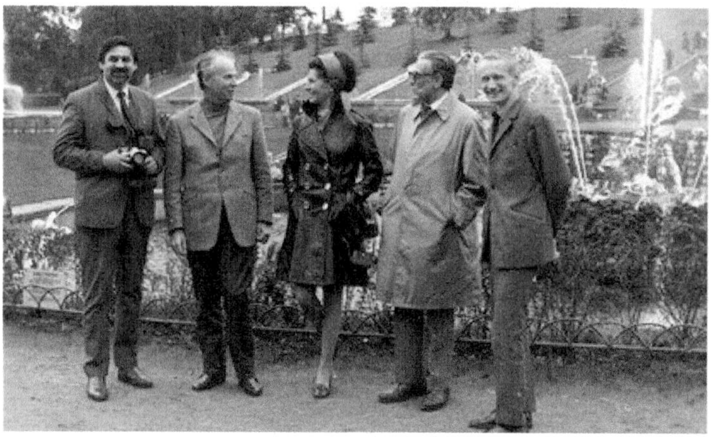

Fig. 4.6. Board Meeting in Leningrad in June 1970. From left to right: John Reid, Yuri Soloviev, Josine des Cressonnières, Edgar Kaufman and Andrjez Pavlowski. Source: Brighton, University of Brighton Design Archives, ICSID Archive.

Curtain, except for the Bolshoy (sic) Ballet."[24] From 1965 VNIITE was involved in ICSID activities as one of the only non-Western members. The high point of VNIITE's international orientation was the invitation of designer Raymond Loewy, the ultimate American design icon, to Moscow in 1973. Azrikan called this international orientation a true VNIITE enigma and attributed it to the singular personality of Yuri Soloviev, whom he portrayed as a talented designer with a love for fashion who worked hard to convince the communist authorities of the importance of design or, in Russian, "*Teknicheskoy Estetiki*" (technical aesthetics). A decisive factor was Soloviev's powerful network and his close connections with the top state administration.[25] Apparently, he was an expert in striking the right diplomatic attitude: "Like many people in the Soviet Union, he played two contradictory roles: the bright, Western-oriented intellectual and artist on the one hand, and the 'disciplined Party member' on the other."[26]

The Brussels Design Centre profiled itself in the ICSID community as the "older brother" beside the British parent institution, advising many younger ones in their setup.[27] It was, for example, involved in the organization of design promotion in Colombia and Brazil.[28] This prominent position of the Brussels Design Centre was largely due to the pivotal role of its director: Josine des Cressonnières was for 16 years (1961 to 1977) also general secretary of ICSID (figs. 4.4–8). Des Cressonnières had a background as an art historian and came into

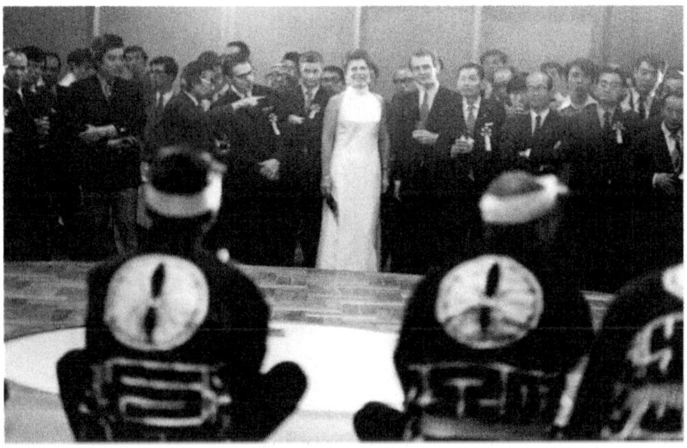

Fig. 4.7. Josine des Cressonnières at the 8th ICSID Congress in Kyoto, 1973. Source: Brighton, University of Brighton Design Archives, ICSID Archive (8-3-3).

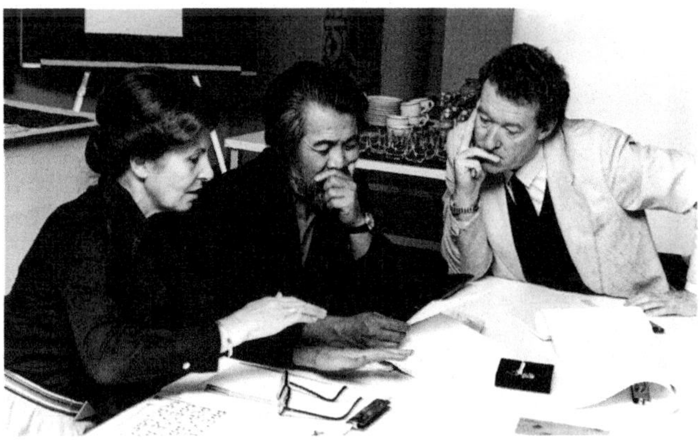

Fig. 4.8. Josine des Cressonnières in discussion with her good friends Kenji Ekuan and Loek van de Sande, 1980. Source: Brighton, University of Brighton Design Archives, ICSID Archive (10-3-2).

contact with industrial design in the early 1950s through her job as the stylist of the Belgian department store *À l'Innovation* in Brussels.[29] Soon after, she became involved in the first design organizations in Belgium and steadily occupied a central position in both the Belgian and international design scenes.

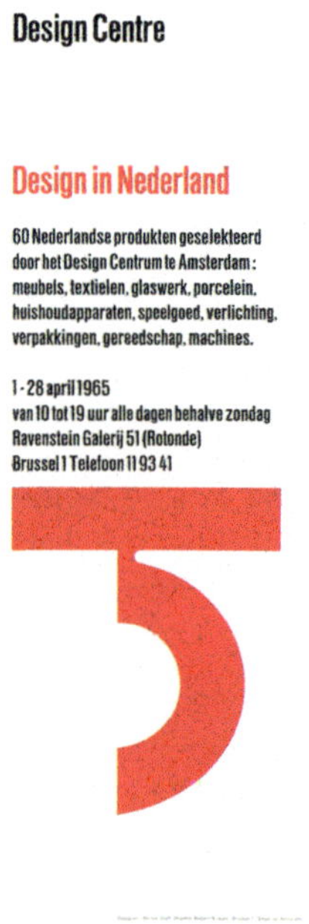

Fig. 4.9. "Design in the Netherlands" (1965), poster designed by Michel Olyff. Source: © Michel Olyff. Ittre, Private archive Michel Olyff.

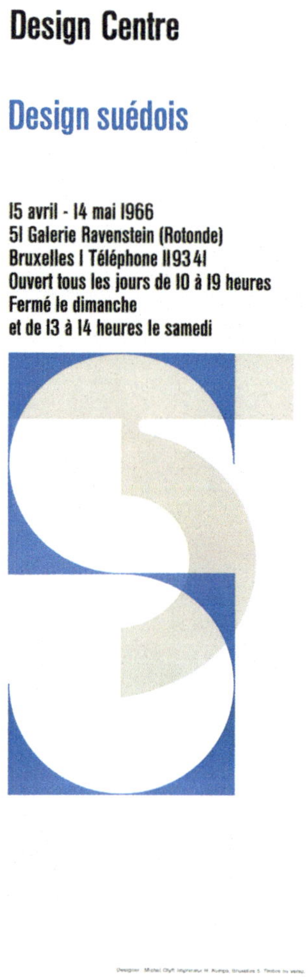

Fig. 4.10. "Design Suédois" (1966), poster designed by Michel Olyff. Source: © Michel Olyff. Ittre, Private archive Michel Olyff.

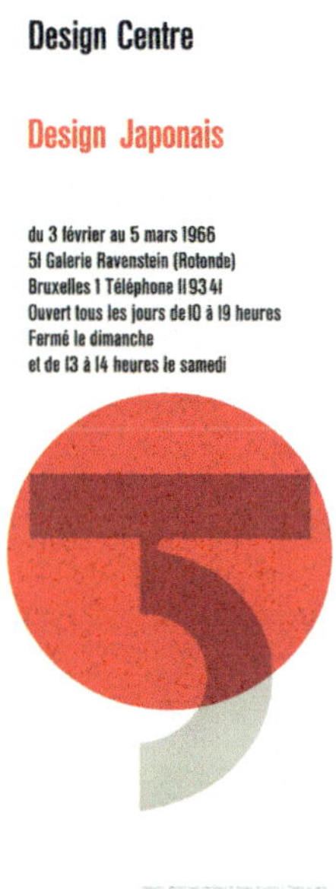

Fig. 4.11. "Design Japonais" (1966), poster designed by Michel Olyff. Source: © Michel Olyff. Ittre, Private archive Michel Olyff.

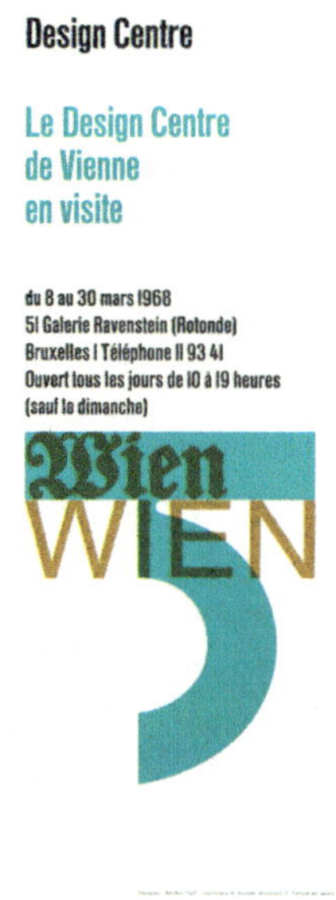

Fig. 4.12. "Le Design Centre de Vienne en visite" (1968), poster designed by Michel Olyff. Source: © Michel Olyff. Ittre, Private archive Michel Olyff.

2.1 Exchanging Exhibitions

The display of Soviet design at the Brussels Design Centre was not an isolated incident. It had become common practice to exchange exhibitions between these kinds of design institutions. The Brussels Design Centre had been one of the driving forces behind this idea of exchange. Even its first 1962 articles of association stated that the Brussels Design Centre had to create a flow of exchange with its foreign equivalents.[30] Accordingly, its inaugural exhibition displayed twenty-one foreign objects selected by six of its foreign counterparts.[31] This welcoming attitude dovetailed nicely with the Design Centre's hybrid setup, which not only related to an "altruistic" design attitude but also corresponded to national concerns, that is, to the focus on increasing trade and exports. The underlying premise was the economic liberalization in light of the European Economic Community (EEC) or Common Market. The display of foreign objects within the safe walls of the Brussels Design Centre was a somewhat artificial simulation of the fierce competition in the world beyond.

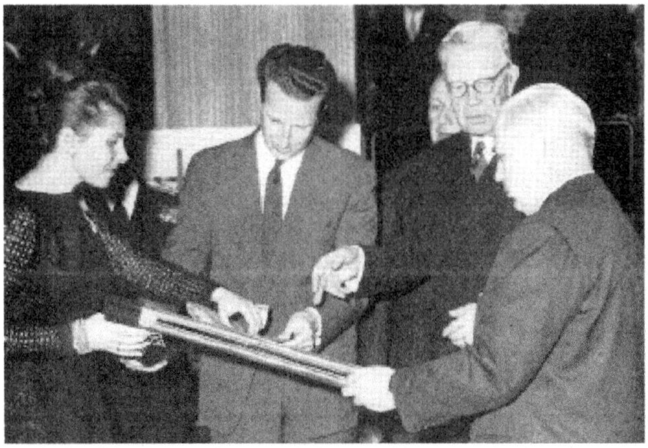

Fig. 4.13. Opening of the Swedish design exhibition at the Design Centre, 1966. From left to right: Josine des Cressonnières, King Baudouin of Belgium and King Gustav VI of Sweden. Source: des Cressonnières, J. (1969). Le Design Centre a cinq ans. *Industrie*, 356-380.

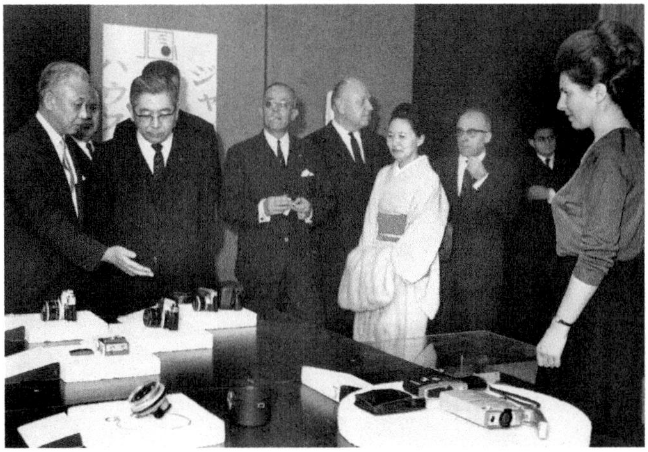

Fig. 4.14. Opening of the Japanese design exhibition at the Brussels Design Centre, 1966. Source: Coutisse, Archives des Cressonnières.

Once the Brussels Design Centre had opened its doors in 1964, design shows were exchanged with its counterparts in Japan, Finland, the Netherlands, Germany and beyond (figs. 4.9–14). In the case of VNIITE, the exchanges were slightly more intricate in the context of the Cold War. VNIITE had hosted design shows of its Western counterparts such as Great Britain, Italy and Germany.[32] Many of these international design exchanges between these national design bodies were paired with commercial, diplomatic or economic missions within the framework of the Cold War.[33] Hence, kings, queens, ambassadors, ministers and other officials were mobilized to formally inaugurate these politically motivated design events. These exchanges thus have to be seen as diplomatic encounters where state authorities frequently manipulated their country's image by presenting carefully selected material, and thus offered a biased representation of the nation, even while pretending to be comprehensive.

3. BELGIAN-SOVIET DESIGN EXCHANGES

While it was commonplace to swap exhibitions within the network of national design bodies, the reeling in of a Soviet exhibition by the Brussels Design Centre was considered an exceptional success. It was, after all, the first VNIITE show outside communist territory. Hosting this première was a matter of honor within the ICSID design community, especially since the Belgian Design Centre had to compete with the British one to get its first showing. Paul Reilly, who had maintained good relations with Yuri Soloviev, was equally eager to receive the Soviet exhibition and initial plans had been made.[34] However, this was thwarted by the political events of 1971 at the Soviet Embassy in London. In 1972, he wrote to des Cressonnières:

Congratulations on your answer from VNIITE. You are clearly going to win the race. Had our Government not deported 101 Russian spies last year, we might have been the first Design Centre to stage Yuri's exhibition. We will all make a point of coming to see it if you do really pull it off![35]

The competition between London and Brussels was not only colored by the prestige of getting the first VNIITE exhibition, but was probably also informed by recent conflicts within the ICSID. Around that time des Cressonnières's powerful position at the core of the ICSID was under attack. The President, Henri Viénot, viciously challenged her role as ICSID's figurehead and stated that he was "against the pursuance of any direct influence of Josine des Cressonnières on the Secretariat."[36] He even suggested moving the secretariat to another city to reduce her influence on the Secretariat.[37] Briton John Reid, the previous ICSID president, took up this suggestion and proposed to move it to the Nash house in London.[38] Against the background of these strained relations and power struggles within the ICSID, the arrival of the VNIITE show in Brussels instead of London must have been an important move for Josine des Cressonnières, underlining the importance of the Brussels Design Centre and her political capacities.

Notwithstanding the political tensions between the Soviet Union and Great Britain and the frictions within ICSID, one other decisive reason was the symbolic value of Brussels as the capital of Europe and the location of the NATO headquarters. In 1958, Brussels became the temporary seat for the European Economic Community (EEC) and the European Atomic Energy Community (Euratom) and host of the first postwar world fair, the iconic Expo 58.[39] Consequently, Brussels became the diplomatic and symbolic heart of Europe, positioned between the Cold War protagonists: the United States and the Soviet Union. In addition, Belgium had traditionally good relations with Eastern Europe and, in the 1960s, played a remarkable role in bridging the East-West gap.[40] After World War II, the Belgian government under socialist Prime Minister Paul-Henri Spaak, endorsed a neutral position between the two superpowers, presented as a "third way". At this time, Spaak even supported pro-Soviet attitudes in the hope of achieving good relations with both blocs. By the early 1950s, however, Belgium's political position changed. Under pressure of the increasingly tense geopolitical situation, heightened by the recent setup of the Marshall Plan, NATO and European integration, Belgium became a loyal ally of the US and adopted an Atlanticist policy. Although from then part of the Western bloc, Belgian diplomats continued with attempts to bridge the East-West gap. Belgian politicians Spaak, Paul van Zeeland and Pierre Harmel became respected protagonists in advocating East-West détente, which resulted in the famous Harmel doctrine that aimed to bring about conciliation between European states and the East. Although Belgium was considered only a "small country" in the Cold War, it played a significant role in Cold War diplomacy as a "go-between" for the two hostile superpowers.

Next to Belgium's reputation as an intermediary in the East-West dialogue, the design exchange is probably best framed within the Belgian-Soviet trade agreements. In 1971 the Soviet government signed a trade treaty with the Benelux countries to expand the business transactions between the two parties.[41] The Belgian Office for Foreign Trade underlined the importance of trade with communist countries, partly because they expected new needs in the field of consumption goods resulting from the gradual increase in the standard of living.[42] In addition, preliminary negotiations for the Helsinki Accords of 1975 — one of the few actual successes in East-West détente — started in 1973, the same year Brussels hosted the Soviet exhibit.[43] One of the three spearheads of the negotiations was that the West and the communist bloc both promised fuller economic, scientific and technological cooperation. Anticipating these favorable trade prospects, this Belgian-Soviet design exchange had to forge new commercial relations between Brussels and Moscow. The Belgian exhibition in Moscow in 1974 was framed as a Belgian commercial mission led by Prince Albert, the figurehead of Belgium's Foreign Trade Office.

Beyond this détente context, a more personal reason to hold the first "Western" VNIITE exhibition in Brussels was the friendship between Soloviev and des Cressonnières. Soloviev even began his memoirs with a reference to their close relationship: "Josine des Cressonnières was an extraordinary woman. She was not only beautiful and intelligent but also had excellent managerial abilities. [...] We became intimate friends."[44] They met in 1967 during the fifth ICSID Conference in Montreal, where they spent time at Canadian designer Jacques Guillon's country house. In the following years their paths often crossed. In 1968 the two were members of the small international jury for the 3rd Biennial of Industrial Design (BIO) in Ljubljana. The same year VNIITE organized the UNESCO Symposium in Tbilisi, which des Cressonnières also attended. In 1969 Soloviev was elected Vice-President and thus joined des Cressonnières on the ICSID executive board. Moreover, the 1973 VNIITE exhibition in Brussels coincided with the start of the preparations for the IX ICSID Conference and General Assembly planned in Moscow in 1975.[45] At the time of the exhibition's opening, des Cressonnières arranged an ICSID meeting so the full board could see this prestigious premiere: the Japanese Kenji Ekuan, the Italian Rodolfo Bonetto, the Viennese Carl Auböck and the Frenchman Henri Viénot were all present on this occasion.

3.1 Glancing behind the Iron Curtain

The first VNIITE exhibition in the West was carefully organized by its department of Exhibition and Multimedia.[46] A scale model of the whole was made, followed by a full-size mock-up before the exhibition traveled to Brussels. In its final arrangement, the exhibition comprised three sections: heavy industry, consumer goods, and crafts. The first part included many scale models of large industrial products, such as a model of a 110-ton truck, a 10-ton crane and a boat.[47] The second part showed consumer products, including domestic furniture, cameras and a mini scooter. The third part featured a range of traditionally crafted products, for example: a silver wine set, linen carpets and matryoshkas. Many of the products were small-scale prototypes that yet had to be produced or further developed. A slide show added a bit of boldness to the models by showing large-scale images of industrial design accomplishments (fig. 4.15). Large panels with wall texts aimed to make Belgian visitors aware of Soviet efforts in the field of design. One panel, for example, informed visitors about the 2,200,000 living spaces built in 1973 where "individual preferences were to be reconciled with mass production." Other panels underlined the high level of scientific design research, especially with regard to ergonomics, one of VNIITE's strengths.

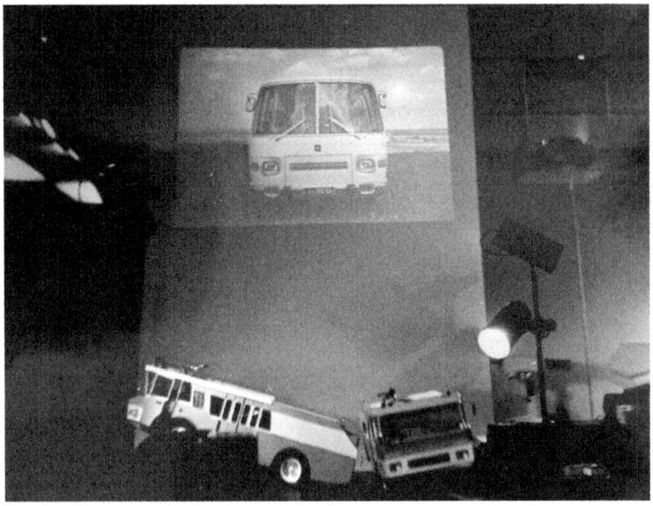

Fig. 4.15. A projected image of the PAZ Tourist-lux bus by Marc Demidovtshev and two prototypes of a fire-engine by Lev Kutsmitchev, Vladimir Ariamov and Alexander Olshanetski. Source: VNIITE Archive, report #1001, 1971, Moscow Technological University.

The political tensions of the Cold War increased the appeal and colored Belgian reactions to the exhibition. The exoticism of this hard-to-reach Cold War superpower gave the event an exclusive character. The press release explicitly announced it as the first Soviet design exhibition outside the communist bloc during the Cold War. It was indeed the first VNIITE show in the West, but certainly not the first one showing Soviet design. The Soviet Union organized several industrial exhibitions in the West and non-aligned countries from the late 1950s and participated in the Brussels Expo 58.[48] However, most of the press articles rephrased this scoop, calling it an historic event, and some even dedicated their headlines to it: "World première in Brussels" or "First Soviet design show on exhibit in Brussels". Accordingly, the prevailing Cold War stereotypes surfaced between the lines in many reviews. In general, it was positively reviewed as a window giving visitors a view of how Soviet life was designed, like peeking behind the Iron Curtain. Soviet design was perceived as "different" from Belgian or European industrial design in that it revealed, according to the press, clear marks of a planned economy considering the emphasis on heavy machinery instead of consumer goods.[49] According to des Cressonnières, the singularity of Soviet design was explained by the divergent criteria that were prioritized in the design process.[50] The function, technical quality and price/quality ratio were of special importance in the Soviet Union, while in Europe, des Cressonnières argued, much more attention was given to imagination and esthetics.

There was a tendency to emphasize the popular image of Soviet design as progressive in the field of heavy industry while backward in the area of domestic and consumer goods.[51] The review of the Flemish liberal newspaper *Het Laatste Nieuws*, for example, criticized the ponderous appearance of a television set.[52] Not all reactions to the exhibition espoused these stereotypes about the backwardness of Soviet quality.[53] The same review also described the "surprising freshness" of a Soviet mini scooter that "could bear very favorable comparison with the most modern Japanese machines".[54] In particular, well-studied ergonomics and user comfort impressed Belgian journalists. Also, des Cressonnières admitted that she was pleasantly surprised by the exhibition because it did not reflect the traditional image of poor Soviet comfort.[55] Comfort, here, was understood as a scientific notion based on ergonomics. Convinced of this asset, the reviewer of the leading Catholic Flemish newspaper *De Standaard* viewed the exhibition as very instructive for the West.[56] It was argued that while Western designers have difficulty selling their ideas on "system-design" to the industry, in the Soviet Union design was simply imposed by the state, and this with excellent results. For the reviewer, the large amount of money spent by the Soviet government on scientific design research with regard to safety, hygiene and ergonomics fostered a real and admirable interest in the perspective of the user. This was in great contrast to Belgium where the consumer was, allegedly, at the mercy of the markets' whims.

The section dedicated to craft products was a different story. Artistic crafts, especially traditional, 'genuine' folk crafts, were accepted in the official Soviet discourse as a welcome complement to the industrially-produced environment. While des Cressonnières, in line with her Ulm opinion of rational and scientific design, disapproved of these Russian crafts, the Belgian press, on the contrary, found them to be the most surprising and satisfying part of the show.[57] The traditional copper Samovar, folkloristic embroidered cushion-cover, furs and other folk objects were seen as a "nostalgic testimony to a past described by Tolstoy".[58] In fact, they were considered to reveal to a greater degree the Russian soul that was understood as introvert, a little self-satisfied, but willing to learn from contact with foreign civilizations.[59] That interpretation perhaps hinted at hopeful prospects for détente between the West and the Soviet Union.

3.2 Belgian Design in a Socialist Realistic Jacket

In 1974 it was Belgium's turn to show its best face in Moscow. The Belgian counter-exhibition was in fact the icing on the cake of a large Belgian commercial mission to the Soviet Union. Anticipating the economic prospects prefigured by the Soviet-Benelux trade agreements and the ongoing Helsinki negotiations, Prince Albert and the minister of foreign trade traveled with industrialists from forty Belgian companies to explore commercial opportunities in the Soviet Union. The exhibition in Moscow was mounted in an inflatable pavilion of 600 square meters and displayed the familiar selection of "Belgian" design by the Design Centre, that is, rationalist industrial objects in line with the Ulm fashion and the doctrines discussed and preached in ICSID circles: a Barco television set, a Nova deep-frying pan, porcelain sets for hotels, a Meurop plastic child's chair, and pictures of larger machinery (figs. 4.16–17). The Russian press accordingly called the Belgian exhibition technical, rational, economic and novel.

What was quite unusual, however, was the remarkable design of the exhibition poster. While the posters of the Belgian Design Centre were normally characterized by a modernist look with abstract forms and Helvetica typeface, the poster for the Belgian exhibition in Moscow was a fine example of "realistic" graphic design (fig. 4.18). The bright green poster depicted a white bush with pink flowers representing the Belgian design concept: a sort of tree diagram with each branch representing a stage in the design process, such as 'production', 'distribution', 'consumption', etc. It was designed by the Ulm alumnus Jean-Jacques Stiefenhofer (1943-2013), who deviated from his habitual

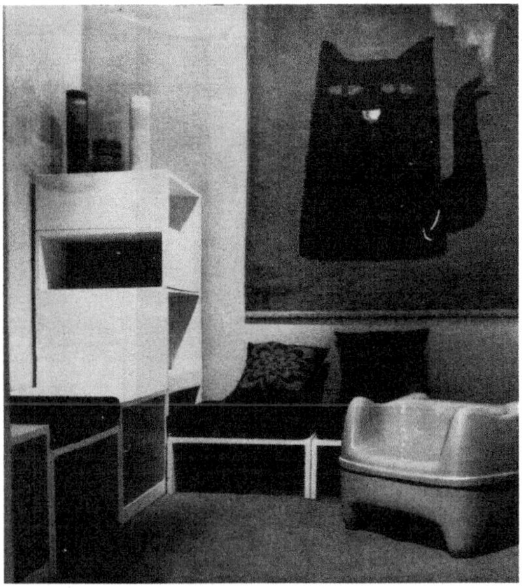

Fig. 4.16. View of the exhibition of Belgian industrial design in Moscow, 1974. Source: Brussels, State Archives, BE-A0545 / Fic 2000, Archive Design Centre, 392.

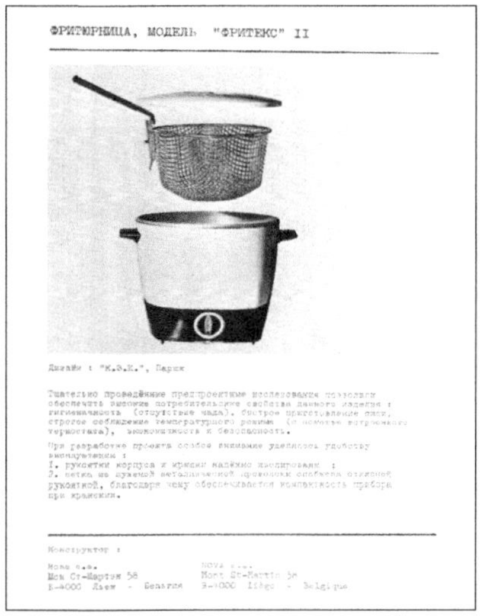

ФРИТЮРНИЦА, МОДЕЛЬ "ФРИТЕКС" II

Дизайн : "К.Э.К.", Париж

Тщательно проведённые предпроектные исследования позволили обеспечить высокие потребительские свойства данного изделия : гигиеничность (отсутствие чада), быстрое приготовление пищи, строгое соблюдение температурного режима (с помощью встроенного термостата), экономичность и безопасность.

При разработке проекта особое внимание уделялось удобству эксплуатации :
1. рукоятки корпуса и крышки надёжно изолированы ;
2. сетка из луженой металлической проволоки снабжена откидной рукояткой, благодаря чему обеспечивается компактность прибора при хранении.

Конструктор :

Нова с.а.
Мон Ст-Мартин 58
Б-4000 Льеж - Бельгия

NOVA s.a.
Mont St-Martin 58
B-4000 Liège - Belgique

Fig. 4.17. Russian index card of a Nova deep-frying pan for the Belgian exhibition in Moscow, 1974. Source: Brussels, State Archives, BE-A0545 / Fic 2000, Archive Design Centre, 381.

sleek formal language in a seemingly conscious attempt to appeal to the Russian public, via a more realistic, Soviet-style esthetic. Indeed, Socialist Realism — as an antipode to the capitalist International Style-modernism — had been the stylistic hallmark of Soviet design before Stalin's death.[60] By the 1970s, however, this Western gaze was actually quite outdated, since modernity had made its mark on the design culture of the Soviet Union by the early 1960s. With this anachronistic poster, Belgian industrial design was actually presented in the old-fashioned jacket of Socialist Realism, probably to avoid possible associations with capitalist, decadent or bourgeois notions. In that sense, the poster is "diplomacy by design" *par excellence*, since it developed a separate hybrid style to assimilate Belgian design to the assumed Russian esthetic conventions in order to bridge the Belgian and Soviet design worlds. Design in these Belgian-Soviet exchanges was then not only a simple Cold War propaganda tool, but also a means to find common ground to build commercial, professional and personal relations.

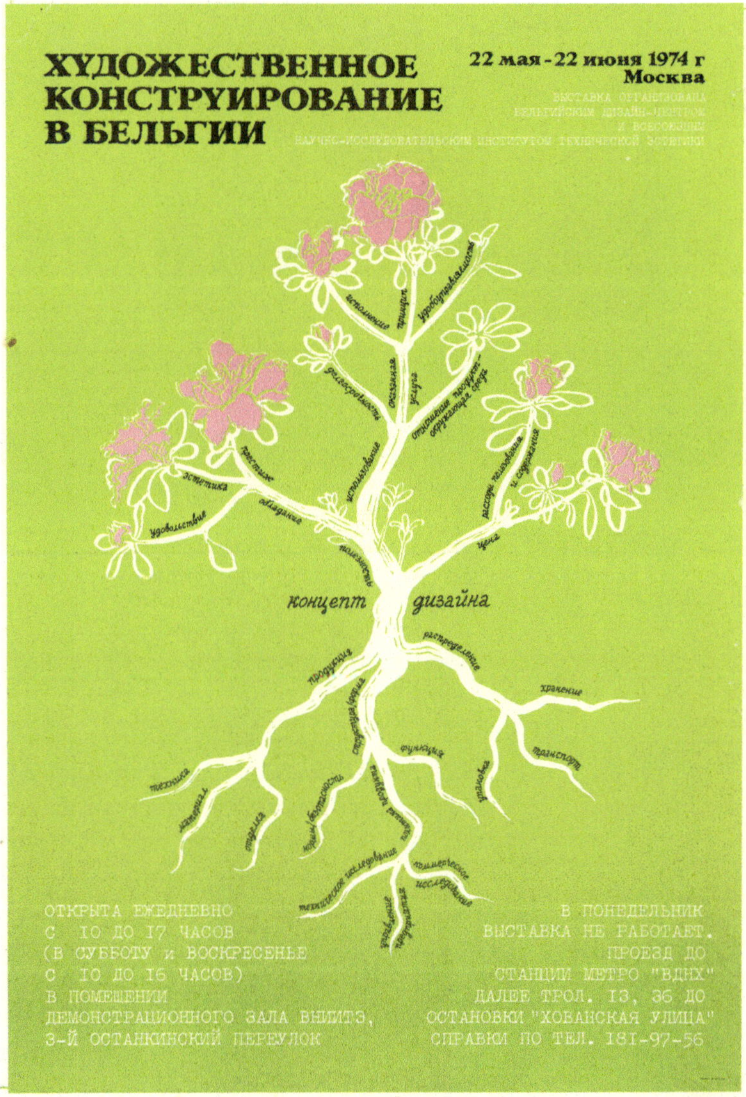

Fig. 4.18. Poster for the Belgian design exhibition in Moscow designed by Ulm alumnus Jean-Jacques Stiefenhofer (1943-2013). Source: © Jean-Jacques Stiefenhofer. Antwerp, Collection Flanders Architecture Institute – Collection Flemish Community, Archive Jean-Jacques Stiefenhofer.

4. CONCLUSION: "DES RELATIONS BIEN DESIGNÉES..."

The 1973 and 1974 design exchanges between Brussels and Moscow have to be understood in the symbiotic contexts of the transnational community of design bodies and of Belgian-Soviet détente diplomacy. On one hand, this design exchange was framed within the international institutional design landscape, where Cold War dynamics were much more ambiguous, blurred and complex. The ICSID included both organizations from the capitalist West and the communist East. It offered a stage where both worlds could meet on the topic of industrial design. This resulted in an intense transnational network, where design personalities such as the Russian Soloviev and the Belgian des Cressonnières could develop professional collaborations and close personal friendships.

On the other hand, the détente context was an equally important framework. Both VNIITE and the Brussels Design Centre, together with their transnational network, were mobilized by national governments in Cold War diplomacy. The bilateral exchange of these well-prepared design exhibitions was also intended to open up commercial possibilities. This complex symbiosis between industrial design and Cold War diplomacy was aptly described by Josine des Cressonnières as "well-designed relations".[61]

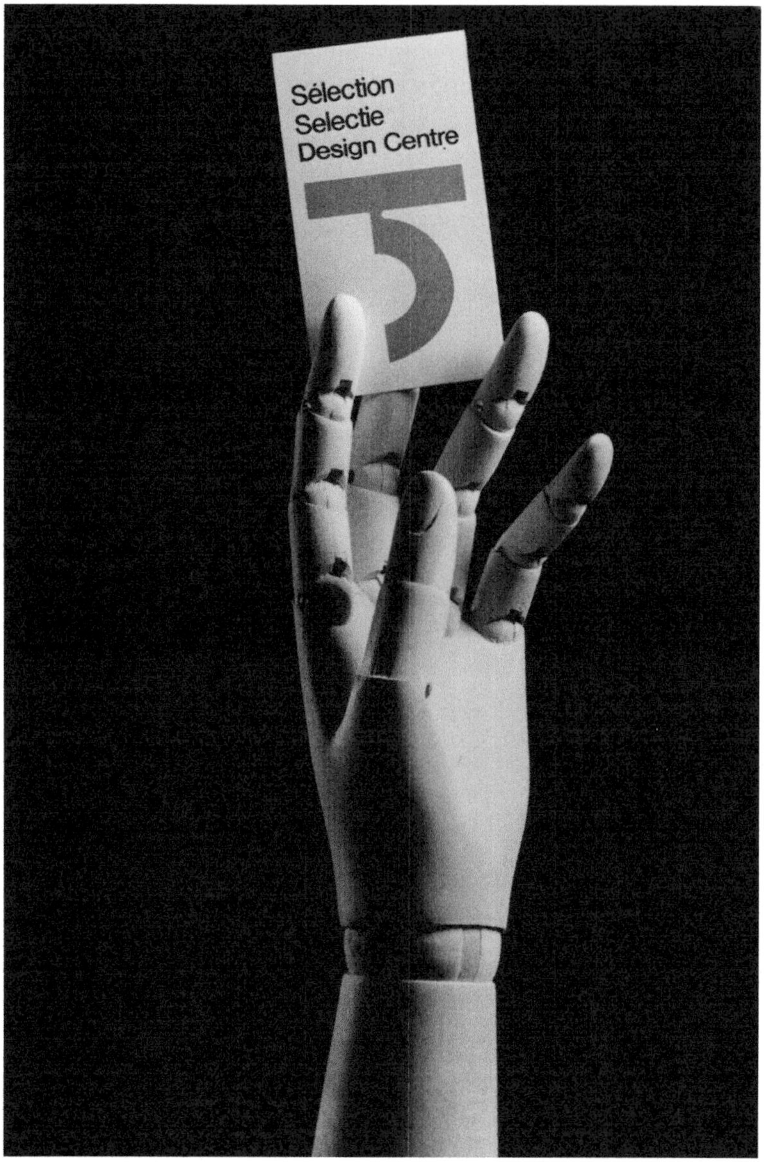

Fig. III.1. The prestigious Design Centre label designed by Michel Olyff held by wooden hands crafted by the Ateliers J. Bonhomme in Verviers. Source: Coutisse, Archives des Cressonnières.

PART III
Crafting “Belgian” Design

Industrial design became a marker for national identities on post-war international markets. One just has to think about Italian, Scandinavian, French or German design. But what about Belgian design? How did the Design Centre present the Belgian nation at home and abroad? How did it reconcile the international character of the modern design discipline with a national scope? This part explores the trope of “Belgian” design installed by the Design Centre to present Belgium as a fully-fledged modern competitive nation on the global markets. The neatly groomed modernist image advanced by the Design Centre was, however, not self-evident: few Belgian products were eligible for the Design Centre’s label, not much room was left for craft production, and the national doctrine produced in the capital could not count on much enthusiasm in the provinces, especially when the regional tensions between the Dutch- and French-speaking communities intensified in the late 1960s.

CHAPTER V

In Search of a "National Style" (1964–1970)

1. INTRODUCTION: "BELGIAN" DESIGN

> La Belgique est un pays formidable, merveilleux ... Tout devrait y être *redesigned*![1]

This exclamation was — so the story goes — the reaction of a famous Finnish architect who paid a visit to Belgium to see the director of the Brussels Design Centre, Josine des Cressonnières. Although this utterance is clearly a *boutade*, it nonetheless sketches the dominant image of Belgian design, namely a very poor one. Indeed, in striking contrast to the leading role of Belgian design institutions in international design circles, the image of Belgian design was seen by many as weak or even non-existent. But how is it possible, then, that Belgium, with one of the longest track records — 22 years! — among Design Centers, was not able to express itself as a design country?

Most studies of "Belgian" design get stumped by the fact that there is no Belgian identity, or they try to create a new one using very generic language, or repeat some outworn stereotypes without problematizing the agendas behind them. However, little attention has been paid to actual attempts to forge such an association — whether successfully or not — between the Belgian nation and industrial design. Many actors were involved in the field that mediated a link between design and Belgium: private companies, design personalities, world fairs, journalists, international trade fairs and many more. Take, for example, the famous biennial Interieur Kortrijk, which held its first edition in 1968 and became an internationally praised hotspot for modern domestic design.[2] And what about government-backed design bodies? Which "official" connections between Belgium and design were highlighted in its promotional activities?

This chapter looks at the formative processes induced by the Brussels Design Centre to advance industrial design as an effective calling card for Belgium at home and abroad. In the wake of the Second World War, the lack of a Belgian design image was a thorn in the side of Belgian trade aspirations, this in view of the growing market liberalization. Everyone talked about French taste, German precision and Italian fashion, but "Belgian" appeared to be associated with beer only. The Brussels Design Centre was established in 1962 to tackle this shortfall and provide Belgian industry with a positive image to boost exports. The Brussels Design Centre consciously aimed to advance a "national style" in the 1960s, one inspired, paradoxically, by international debates on modernist industrial design.

The following account will not look at any "Belgian" essence in a primordialist way, nor will it try to write counter-narratives of forgotten and neglected "Belgian" design, but it will rather analyze how storylines on Belgian design — whether successful or not — were constructed through public design promotion. The chapter more specifically examines the different building blocks used by the Brussels Design Centre to evoke a Belgian design image such as interior design, exhibitions, displayed objects and discourses. The chapter focuses mainly on how the Design Centre crafted "Belgian" design in its early years.

2. THE BRUSSELS DESIGN CENTRE AS AN INSTRUMENT OF NATIONAL PRESTIGE

In 1962 the Institute of Industrial Design and the Signe d'Or joined forces to create the Brussels Design Centre.[3] The former president and secretary general of the initiating organizations — Pol Provost (1907-1990) and Josine des Cressonnières (1926-1985) — respectively took on the roles of the Design Centre's president and director. Accordingly, the heyday of the Design Centre was a successful continuation of the reformed design policies of the Institute and the Signe d'Or.[4] Maintaining the same course, the founding of the Brussels Design Centre endorsed the close liaison between industrial design and national export interests that already prevailed during Jean Rey's political term in the 1950s. Its primary aim was to set up a permanent exhibition "to promote Belgian industrial prestige, especially on foreign markets" and to improve Belgian production with the introduction of "Industrial Design" in industry.[5] From its very beginning, it enjoyed the protection of the Belgian Foreign Trade Office, which also became its main sponsor and of which it became an ally in promoting Belgian design abroad.[6]

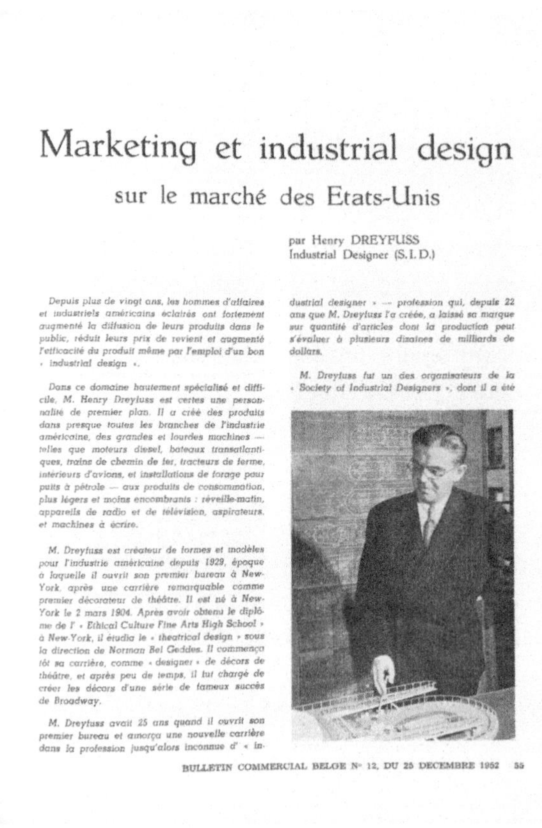

Marketing et industrial design

sur le marché des Etats-Unis

par Henry DREYFUSS
Industrial Designer (S.I.D.)

Depuis plus de vingt ans, les hommes d'affaires et industriels américains éclairés ont fortement augmenté la diffusion de leurs produits dans le public, réduit leurs prix de revient et augmenté l'efficacité du produit même par l'emploi d'un bon « industrial design ».

Dans ce domaine hautement spécialisé et difficile, M. Henry Dreyfuss est certes une personnalité de premier plan. Il a créé des produits dans presque toutes les branches de l'industrie américaine, des grandes et lourdes machines — telles que moteurs diesel, bateaux transatlantiques, trains de chemin de fer, tracteurs de ferme, intérieurs d'avions, et installations de forage pour puits à pétrole — aux produits de consommation, plus légers et moins encombrants : réveille-matin, appareils de radio et de télévision, aspirateurs, et machines à écrire.

M. Dreyfuss est créateur de formes et modèles pour l'industrie américaine depuis 1929, époque à laquelle il ouvrit son premier bureau à New-York, après une carrière remarquable comme premier décorateur de théâtre. Il est né à New-York le 2 mars 1904. Après avoir obtenu le diplôme de l' « Ethical Culture Fine Arts High School » à New-York, il étudia le « theatrical design » sous la direction de Norman Bel Geddes. Il commença tôt sa carrière, comme « designer » de décors de théâtre, et après peu de temps, il fut chargé de créer les décors d'une série de fameux succès de Broadway.

M. Dreyfuss avait 25 ans quand il ouvrit son premier bureau et amorça une nouvelle carrière dans la profession jusqu'alors inconnue d' « industrial designer » — profession qui, depuis 22 ans que M. Dreyfuss l'a créée, a laissé sa marque sur quantité d'articles dont la production peut s'évaluer à plusieurs dizaines de milliards de dollars.

M. Dreyfuss fut un des organisateurs de la « Society of Industrial Designers », dont il a été

BULLETIN COMMERCIAL BELGE N° 12, DU 25 DECEMBRE 1952 55

Fig. 5.1. The American designer Dreyfuss in the *Bulletin Commercial Belge* formulated advice to Belgian businessmen aiming to enter the American market. Source: Dreyfuss, H. (1952). Marketing et industrial design sur le marché des Etats-Unis. *Bulletin Commercial Belge, 66* (12), 55-62. © Koninklijke Bibliotheek van België/Bibliothèque Royale de Belgique

The underlying motive of this state support was thus export interests that were indeed, at that time, one of the Belgian government's top priorities. Belgian officials observed after all that a favorable national image was a trump card on the increasingly liberalized markets.[7] In contrast to its neighboring countries, Belgium seemingly lacked such a distinctive image. Minister of Foreign Trade Maurice Brasseur himself in this vein considered the Brussels Design Centre as "un outil de prestige" to create this urgently needed flattering Belgian image abroad.[8] Framed in this context, industrial design was seen as the expedient to enhance national prestige and to boost economic gain.

In 1952, the famous American designer Henry Dreyfuss had already pointed out this Belgian shortcoming on foreign markets.[9] At the request of the Belgian Foreign Trade Office, Dreyfuss formulated some advice to businessmen interested in entering the American market (fig. 5.1).[10] One of his main remarks was that "Belgium", unlike its neighboring countries, did not evoke many associations in the mind of the American consumer:

> Any mention of "Holland", for example, evokes in the mind of the average American an image of a country where one finds blossoming tulips, windmills, people wearing clogs or skates, cheeses and canals. Comparable familiar images come to the American mind when one mentions "France," "England," "Switzerland" and "Spain." But I doubt if this holds for "Belgium." The more so because many of my fellow Americans do not know as well as they should the fine textiles nor the machinery of excellent quality made in Belgium.[11]

In 1963 the lack of a clear Belgian profile on the increasingly liberalized markets was confirmed by a survey of Reader's Digest.[12] It showed that consumers from the Common Market and from Great Britain knew of hardly any Belgian products. Only consumers in France and the Netherlands associated Belgium with beer. In Great Britain, Italy and Germany, Belgium evoked nothing.

The Design Centre had to make this little known "industrial face" of Belgium more attractive. It had to serve as a signboard for Belgian industrial production at home and abroad. This diplomatic and representational role was highlighted in the importance paid by its organizers to every detail, from its location and interior design to the displayed objects, press releases and house style. It was a very carefully orchestrated whole, which on closer examination betrayed the Design Centre's initial agenda aligned with an export-minded public and internationally respected modernist standards.

3. BELGIAN DESIGN MYTHS

The Design Centre's discourses on "Belgian" design partly tapped into already existent narratives. To account for the lack of a distinctive Belgian design identity, various actors in the field of design lined up simplistic narratives with essentialist undertones that reduced "Belgian" design to the mere effect of its national culture. Most of these myths drew from old common Belgian stereotypes such as the merging of German and Roman cultures, a history of foreign tyrannies, the cultural crossroads of Europe, the individualist and surrealist character, and the main characteristics of moderation and modesty.

Also, the director of the Design Centre, Josine des Cressonnières, considered the poor Belgian image a reflection of Belgian culture. For her, the fact that Belgium was a country with no design tradition — des Cressonnières in this regard even called Belgium a "developing country" — was conditioned by the national spirit that contrasted strongly, in her vision at least, with the fundamentals of industrial design.[13] She explained that Belgians have "rich" taste, and as such love gaudy colors, luxurious materials, good food and everything sumptuous.[14] All this translated itself into a certain "joie de vivre" which, Des Cressonnières believed, conflicted with the primary principles of "design" such as purity and austerity (she clearly understood design in a modernist fashion). Different actors pronounced variations on this same theme, often referring to the twofold character of Belgian culture in the clichéd manner — the "Latin joy" of the South and the "lyrical character" of the North.[15] In his writings, poet and architect Albert de Bontridder, for example, considered this twofold culture to be the explanation for the development of a strong individual architecture.[16]

In 1994 Geert Bekaert would even go so far as to claim the absence of a Belgian identity as the biggest hallmark of "Belgian" design: "If we can call something Belgian, it is the quasi-innate awareness of identity as a limit."[17] The ability to escape any national or regional identification was thus — quite paradoxically — claimed by Bekaert as typically "Belgian": "Design made in Belgium could then stand for a design which leaves the national, regional or local categories empty, [...]"[18] Looking closely at Belgian design history, Bekaert distinguishes a design method that was rather "personal" — similar to Bontridder's characterization — and "universal" at the same time. This personal and pragmatic Belgian attitude was, in his view, also the reason why Belgians never could step into the limelight of the modern movement: "This instinctive opposition to any kind of dogmatism, the filtering of every theoretical proposition to assay the degree to which it meets reality directly, this rather drives Belgian design into a corner of the field of international design."[19]

venue was located in a monumental shopping arcade in the heart of Brussels that had been designed ten years earlier by architects Alexis and Philippe Dumont in the context of the radical modernization of Brussels (fig. 5.2).[32] This venue was strategically situated. After all, the gallery connected uptown, with the official buildings, and downtown, where the touristy city center and commercial buildings were located. Additionally, it was within easy reach of the central train station and the airline Sabena.

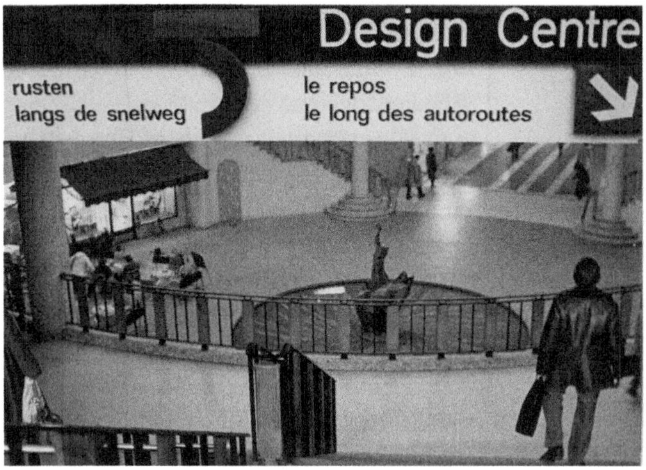

Fig. 5.2. The Design Centre was located in a monumental shopping arcade that had been designed ten years earlier by the architects Alexis and Philippe Dumont. Source: Coutisse, Archives des Cressonnières.

4.1.1 A FIVE-HEADED TEAM OF BELGIAN MODERNISTS

The Design Centre's exhibition space and offices were housed within this complex in a large trapezoidal space initially destined for a cinema but left unfinished at an early stage. Well aware of the importance of exhibition design and display techniques in the crafting of a Belgian design image, the organizers had entrusted the interior design of the space to a team of five respected Belgian modernist designers and architects. These were architect Constantin Brodzki (b. 1924), designer Pierre Génicot, graphic designer Michel Olyff (b. 1927), architect Walter Bresseleers (1927–1980) and designer André De Poerck (1920–1983).[33] The team approached the interior as the first industrial design object on display — a signboard of "Belgian" industrial design: "We wanted to turn the Design Centre into a work of Industrial Design in itself."[34]

The composition of this team and the interior design were just up the Design Centre's alley, when it was about to venture into the promotion of rational and scientific design methods in the manner of Ulm. The main architect, Constantin Brodzki, was an ardent rational modernist and a seasoned exhibition designer.[35] After graduating in 1948 from the Brussels La Cambre school, he went to the United States where he would start as a trainee at the construction site of the United Nations Secretariat Building in New York.[36] A member of a gigantic construction crew, Brodzki was confronted with the modernist functionalism that he later introduced into Belgium.[37] He was much impressed by the efficient and large organization of the American building practice — a "machine of war", to use Brodzki's words. The efficient and rational methods he encountered in New York would influence his later work and the rational and industrial composition of the Design Centre's interior.

The Design Centre commission perfectly fit Brodzki's expertise. In the first ten years of his career in Belgium, he became specialized in the design of interiors, exhibition and display units, mostly in close cooperation with graphic designer Corneille Hannoset (fig. 5.3).[38] Their designs were characterized by a highly modernistic esthetic and the use of American "International Style" furniture. In 1954 this duo, for example, mounted the first Belgian show on industrial design in Liège.[39] That same year, Brodzki assisted Marcel Breuer with the decoration of the Knoll shop in Brussels.[40] Also worth mentioning is his design for the famous pavilion of Congolese Fauna and Flora at the 1958 world exhibition 1958 (figs. 5.4–5). The oyster-shaped organic pavilion contained an impressive dramatic scenography with stuffed elephants and other exotic animals. One of the high points of his design was the total design of the Gallo-Roman Lapidaire Museum in Buzenol-Montauban, which he achieved in 1959 and which was selected by MoMA for the *Architecture of Museums* exhibition in 1969. What made him even more right for the job was his involvement in Belgian design education. From 1957 he had assisted in the development of the industrial design program at La Cambre.[41]

The other team members also fit the bigger picture of the Design Centre's objectives. The project coordinator — André de Poerck, nowadays a forgotten figure — was at the time a prominent architect and designer. He was, for example, honored for his exhibition design of the Belgian Section at the Milan Triennial in 1954. He was also closely involved in the first Belgian design institutes and the creation of the first design courses.[42] Although De Poerck's work has hardly been studied, we find an echo of his work methods in some of his correspondence and published articles. He had a profound aversion to an instinctive and emotional attitude and advocated a rational approach

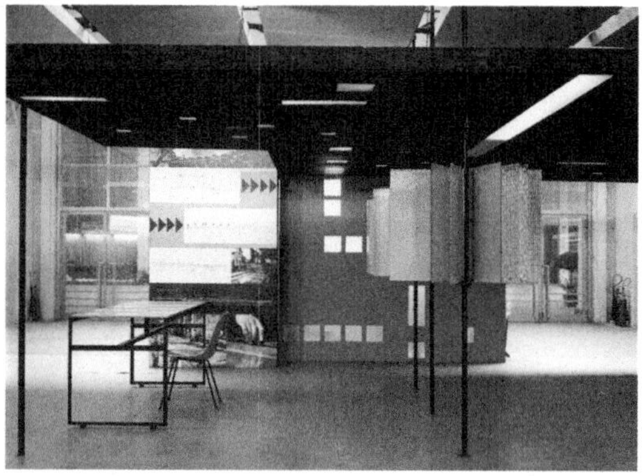

Fig. 5.3. Display designed by Constantin Brodzki and Corneille Hannoset with the help of Michel Olyff, Serge Vandercam and Louis Van Lint for the Belgian Union of Mechanical Glass Trade at the International Fair in Charleroi, 1954. Source: Fonds Hannoset © AML (Archives et Musée de la Littérature).

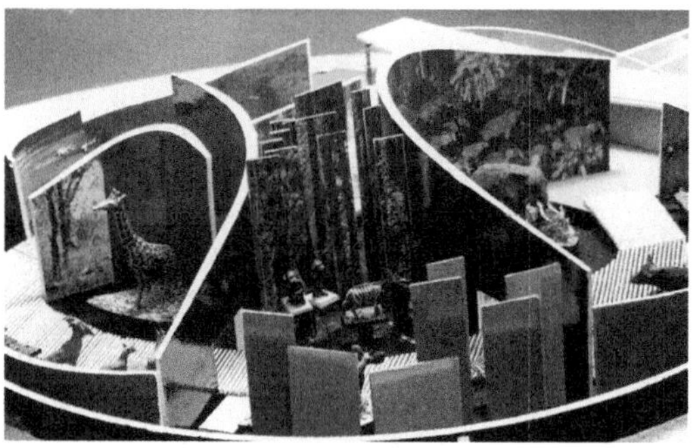

Fig. 5.4. Scale model of the Congolese Fauna and Flora pavilion by Brodzki at the Brussels World's Fair, 1958. Source: Fonds Hannoset © AML (Archives et Musée de la Littérature).

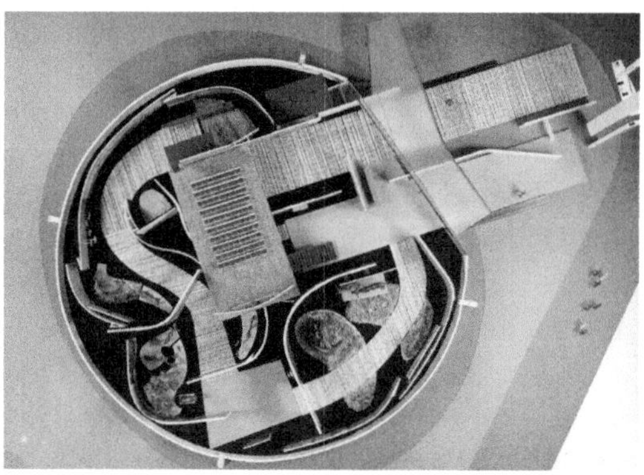

Fig. 5.5. Scale model of the Congolese Fauna and Flora pavilion by Brodzki at the Brussels World's Fair, 1958. Source: Fonds Hannoset © AML (Archives et Musée de la Littérature).

to industrial design.[43] The theoretical and practical issues of exhibition display particularly interested him: in 1954 he edited the tenth issue of the Belgian magazine *Architecture* on contemporary exhibition design and wrote a reflection on the different ways of presenting objects, making references to the Sao Paulo Museum of Lina Bo Bardi, MoMA in New York and Olivetti shops.[44]

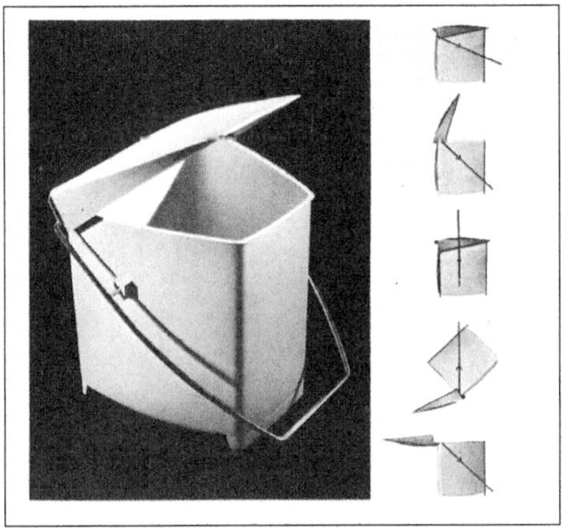

Fig. 5.6. Plastic pedal bin designed by Bureau d'Etude Génicot & Duqué and produced by Cie des Métaux d'Overpelt-Lommel et de Corphalie. Source: "Belgische Industrial Design-Selektie." *Informatieblad van het Instituut voor Industriële Vormgeving* 6, no. 2 (1963): 10-11.

The equally forgotten designer Pierre Génicot was also part of this prestigious crew. Together with his colleague Daniël Duqué, he ran one of the first industrial design bureaus in Brussels in the 1950s.[45] Their most successful design is probably the clean and functional plastic pedal bin for *Cie des Métaux d'Overpelt-Lommel et de Corphalie* that was also included in the opening exhibition of the Design Centre (fig. 5.6). For the interior of the Design Centre, Génicot worked closely with Constantin Brodzki on the design of the display units, lighting and other furnishings.

Graphic designer Michel Olyff conceived the Design Centre's graphic house style. He was an internationally respected graphic designer who studied under Joris Minne at La Cambre.[46] In the 1950s he was a member of the artistic group *Les Ateliers du Marais* with Olivier Strebelle, Pierre Alechinsky, Freddy Michiels, Reinhout D'Haese, André Jacqmain and Luc de Heusch. In his early career, he designed many posters for the *cinémathèque* in Brussels. Olyff is known for his emblematic logos for the national lottery, the Royal Museums of Fine Arts of Belgium and the BRT-RTB (Belgian Radio and Television) and the Crédit Communal. For the Design Centre, Olyff was concerned with every graphic detail from signposting and index cards to letter paper and logos — his logo for the Institute was also adopted for the Design Centre — as well as posters.[47] The bright red color he chose as a recurrent element in most of the Design Centre's printed matter became one of the Design Centre's hallmarks.

The last member of the team was architect Walter Bresseleers. He played only a minor role as "adviser" in the design process. According to Brodzki his profile was very much in vogue at that time.[48] He was the protégé of Léon Stynen and specialized in industrial buildings.[49] He designed, for example, the IBM Technical Centre in Diegem (1963-1966). He was also very active in the burgeoning industrial design scene in Belgium and participated in the first ICSID conferences and seminars.[50]

4.1.2 "A CLEVER ASSEMBLAGE OF PREFAB ELEMENTS"

The team's international and industrial profile struck the right note in the envisioned composition of the Design Centre's organizers. The result of this teamwork — with Brodzki, Génicot and Olyff as the main players — was a totally modular and flexible design arrangement of 800 square meters (fig. 5.7).[51] Every part of the design was based on a 1.22 meter module that allowed the different elements — from lighting, display units and partition walls to stairs and railings — to be rearranged at will. Pierre Puttemans later called it "a clever assemblage of prefab elements" and compared it to the serial music of Anton Webern.[52]

The whole was a transparent rational construction with many playful references to the unfinished cinema construction. A clear distinction was made between the given building site and the new elements added by the team. The walls once destined for the cinema complex were painted black, and as such disappeared into the background — a trick Brodzki also pulled off in the 1958 Congolese Fauna and Flora pavilion (fig. 5.8). The new brick constructions, on the contrary, were painted white, which brought them to the fore. A light wooden skeleton divided the sloping cinema floor into three levels. The skeleton supported fiberboards veneered with dark Congolese *wenge* wood. The floor on the lowest level was partly made of concrete in order to bear the weight of heavy industrial products, which were one of the

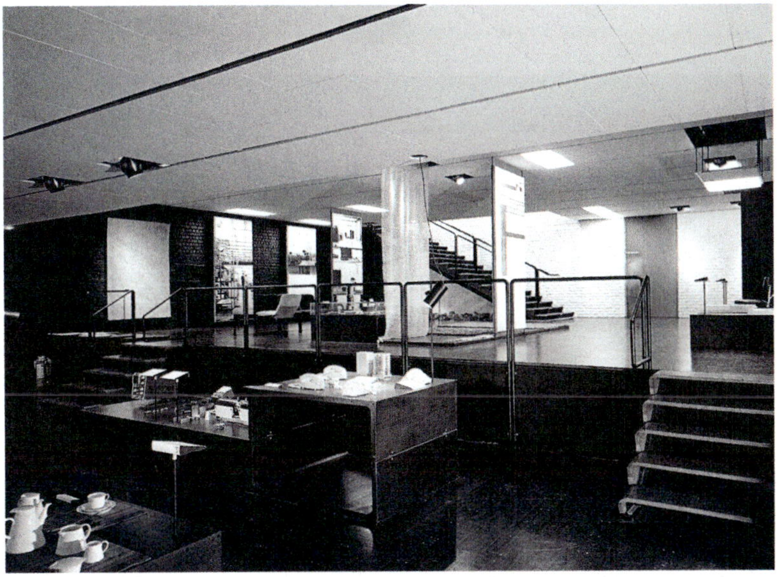

Fig. 5.7. Interior of the Design Centre, Brussels, 1964. Source: Linkebeek, Private archive Constantin Brodzki.

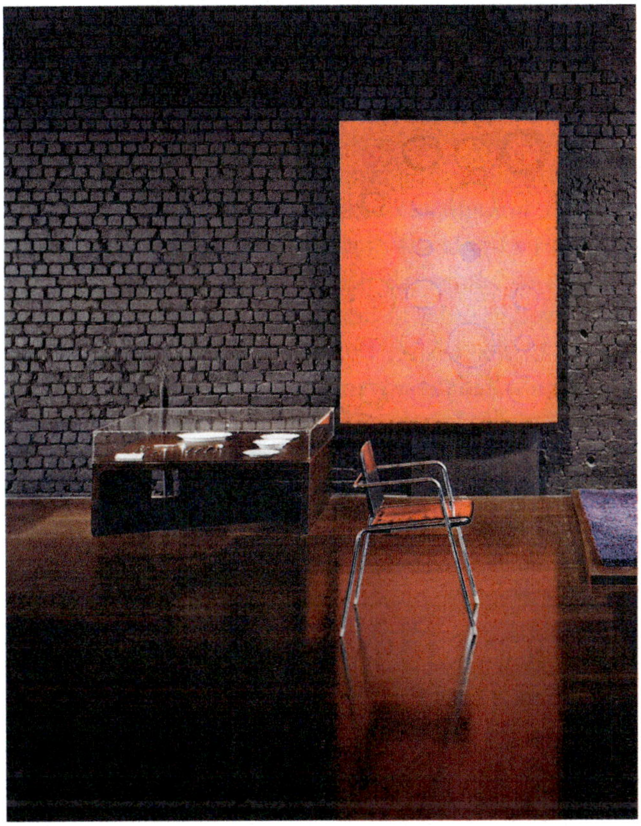

Fig. 5.8. Interior of the Design Centre, Brussels, 1964. In front is the TBA chair of Christophe Gevers (1928-2007). Source: Linkebeek, Private archive Constantin Brodzki.

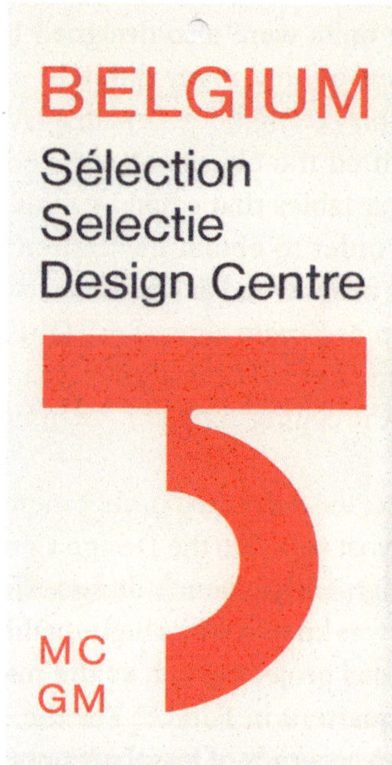

Fig. 5.10. The label of the Brussels Design Centre. Source: © Rijksarchief Antwerpen-Beveren, archief Philippe Neerman.

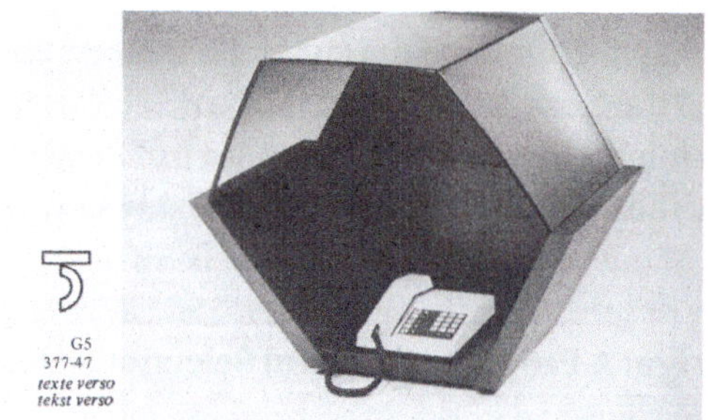

Fig. 5.11. The Design Centre's index card of a telephone cell by Jacques Deprince for Ets. Billen et Proburel (Brussels), ca. 1977. Source: Coutisse, Archives des Cressonnières.

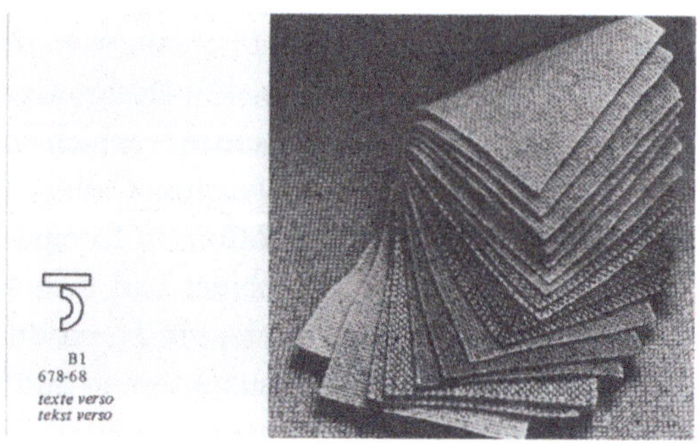

Fig. 5.12. The Design Centre's index card of a woolen furniture fabric collection designed by Marjatta Metsovaara (1927-2014) for Albert Van Havere (Sint-Niklaas) awarded the Signe d'Or label in 1968, ca. 1968. Source: Coutisse, Archives des Cressonnières.

equipment goods (such as machines and transport) and packaging. Chemical products, foodstuff, fashion and prototypes, on the other hand, were excluded. Artistic crafts were, in general, not accepted. In agreement with the Ministry of the Middle Class, however, a selected number of crafts were included in the permanent exhibition, but only those meeting the esthetic and functional criteria of industrial design, i.e. produced in small series — so no unique objects — and taking into account the technique, economy and the market. Third, the product had to be positively evaluated by a jury following strict criteria — based on the international modernist legacy — such as: technical quality, functional value, esthetics, reasonable price and creativity.[58]

The Design Centre thus strained the objects, as it were, through its modernistic filter. Accordingly, the Design Centre's version of "Belgian" design only accounted for a selected number of artifacts, namely those conforming to international modernist doctrines. This is hardly surprising since — as shown in the previous chapters — the Design Centre's policy was largely informed by international design tendencies. The selection policies were clearly marked by the Ulm fashion, which the Design Centre embraced from the start. From a critique on the artistic Bauhaus principles and American commercial "stylism", the Ulm School of Design focused especially on industrial design that had no strong ties to an artistic or craft tradition.[59] Domestic furniture and personal goods were passed over in favor of products that were of more of a technical and professional nature. Former student and teacher Gui Bonsiepe indeed testified that "the production typology of capital goods and work instruments (medical installations, for example)" was an important focus in the industrial design education at the Ulm School of Design.[60] Even when a student designed crockery, such as Nick Roericht in 1959 with his TC 100 set, it was conceived for public use such as in hotels or canteens. Stamped by this Ulm thought, the Belgian Design Centre prided itself on being the first Design Centre that surpassed the concept of design limited to consumer goods for domestic use, and also showed, from its very beginning, capital goods, engineering and professional instruments such as industrial milking — and milling — machines.[61]

Put differently, the Design Centre established a new trope of "Belgian" design, one conforming to modernist criteria in vogue within international circles. Thus, rather than being local, the tone was set by international design standards. This international attitude had much to do with the position of its director. As mentioned above, des

Cressonnières was for sixteen years the passionate secretary general of the ICSID, working together with heavyweights such as Tomas Maldonado, Eliot Noyes, Kenji Ekuan, Yuri Soloviev, André Ricard and more. In 1977 she even made an attempt at the ICSID presidency, rivaling her ICSID friend Yuri Soloviev.[62] The Belgian design policies were thus thoroughly grounded in the ongoing discussions at the heart of this transnational design community. Of course, this outward orientation was also closely tied to its government-endorsed agenda toward exports. The Design Centre was, after all, supported by the Belgian Foreign Trade Office, which aspired to position Belgium in international markets as a competitive force toward its neighboring countries. The aura of modernist industrial products had to contribute to this positioning game of Belgium as a modern nation on global markets.

As design historian Gimeno-Martínez states, in contrast to their "natural" appearance, these "national" labels are far from innocent, neutral or inclusive.[63] Indeed, the Design Centre's selection did not offer a "real" but rather a biased and distorted representation of the Belgian nation. Using particular conditions and criteria, only a small part of Belgian output was included to be presented at home and abroad under the label of "Belgian" design. Hence, the process of selection resulted in a very limited vision of the nation.

This practise was actually quite common when displaying national production in the postwar era. The image of a particular national material culture displayed at exhibitions contrasts in most cases with the everyday material fabric of that nation. Especially at international fairs or world exhibitions, objects were carefully selected by official organizations to represent the ideal notion of the national culture. Some design historians describe this diplomatic function entrusted to objects by the term "soft power", especially in the context of Cold War design.[64]

Displaying a part to represent the whole was not an exclusive practise of national government-endorsed institutions. Edgar Kaufmann did the same when organizing the *Design for Use* exhibitions at MoMA. The works on show displayed a monoculture of modernist rational design, concealing nostalgic designs that dominated the US market.[65] A similar example is provided by plastic manufacturer Tupperware, which in the 1940s exported a modernist esthetic through its exhibits in elitist museums like MoMA, from which its hand-painted floral patterns and whimsical motifs were excluded.[66] When official bodies implement these strategies, the result is a total national image based on a selection that underpins particular idealized notions of the national culture. In the Design Centre's case, this selection resulted in a "Belgian" canon composed of industrial objects suitable for export and meeting international design criteria as preached and discussed in international circles such as the ICSID.

4.3 The Royal Touch

To enhance the "Belgian aura" of the Design Centre's selection, the royal family was called on for assistance.[67] The royal family is often described as one of the only reference points of Belgian national identity. This Belgian symbol was mobilized to adhere Belgian identity to the Design Centre's activities. Prince Albert, as honorary president of the Belgian Foreign Trade Office, opened the Design Centre and became a stalwart friend at the Design Centre's activities. Princess Paola equally paid regular visits, but also King Baudouin and Queen Fabiola were regular guests at inaugurations, special occasions and award ceremonies. This was partly due to the good connections of Josine des Cressonnières with the royal family.[68] She was on various occasions one of the ladies-in-waiting of Queen Fabiola and was very well acquainted with the Belgian nobility.[69]

At Belgian exhibitions abroad, not only was Belgian royalty drummed up, but often also typical Belgian foodstuff, music and the official flag. For example, the Belgian design expo in the London Design Centre at Haymarket in 1969 made use of all kinds of Belgian symbols. Michel Olyff's poster was dominated by the Belgian colors black, yellow and red, fitting in with the Belgian flags waving to the visitors at the entrance.[70] Belgium's Prince Albert and Princess Paola opened the exhibition in presence of Britain's Princess Margaret and the Earl of Snowdon. In addition, Belgian food was distributed at the vernissage such as De Beukelaer cookies, Godiva chocolates and Rombouts coffee while Belgian music played in the background.

5. THE DESIGN CENTRE'S "BELGIAN" STYLE

After two years of careful preparation, the Brussels Design Centre finally opened its doors on 19 February 1964. It was a huge event. Headed by Prince Albert and Minister of Foreign Trade Maurice

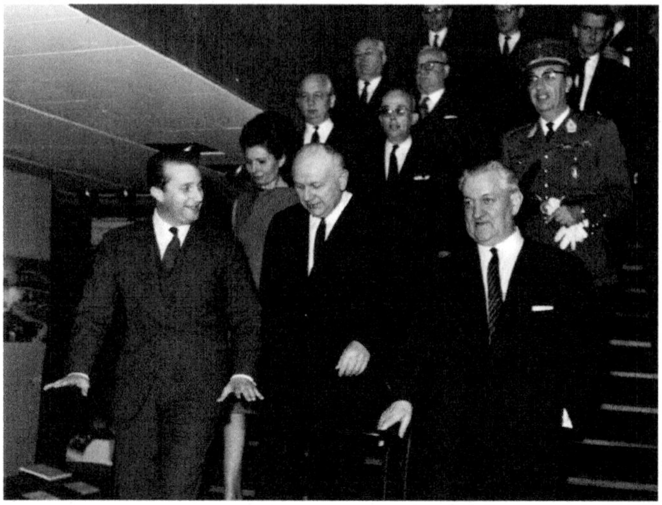

Fig. 5.13. The opening of the Design Centre on February 19, 1964 in the presence of Prince Albert of Belgium, Minister of Foreign Affairs Maurice Brasseur, the Design Centre's president Pol Provost and the Design Centre's director Josine des Cressonnières. Source: Brussels, State Archives, BE-A0545 / Fic 2000, Archive Design Centre, 427.

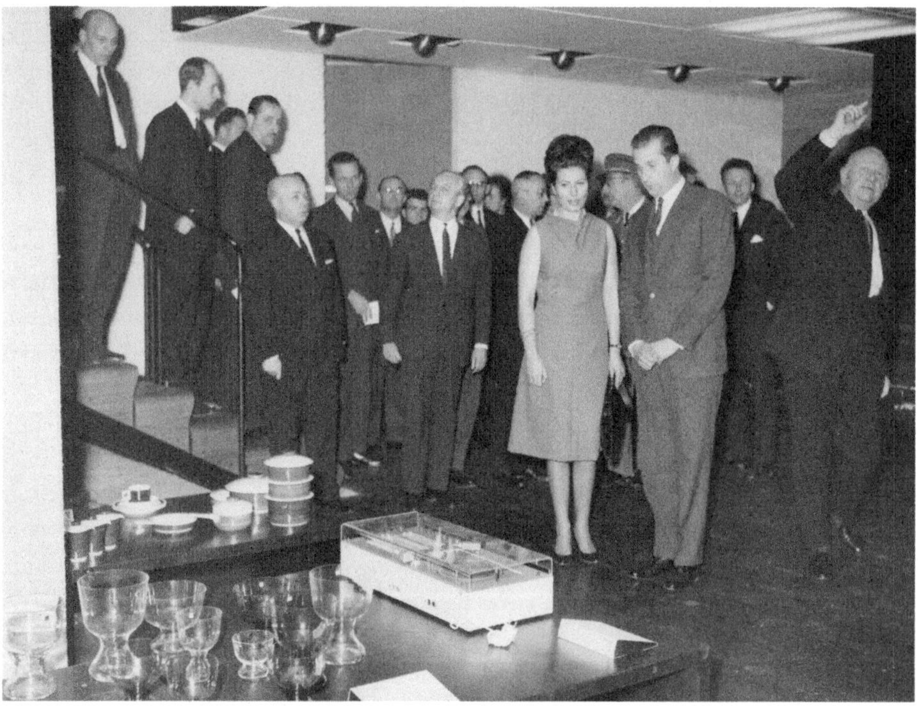

Fig. 5.14. Josine des Cressonnières and Prince Albert at the opening of the Design Centre, 1964. Source: Coutisse, Archives des Cressonnières.

Brasseur, a parade of Belgian political personalities, industrialists, press and international guests descended the stairs of the new Design Centre (fig. 5.13). While the host of black suits thronged the stairs, Prince Albert was guided by the new director of the Design Centre, Josine des Cressonnières, to set the first foot in this mecca of modernist Belgian design. Des Cressonnières thereupon took the time to introduce the Prince to 110 exemplary "Belgian" design objects at the first Design Centre show (fig. 5.14).

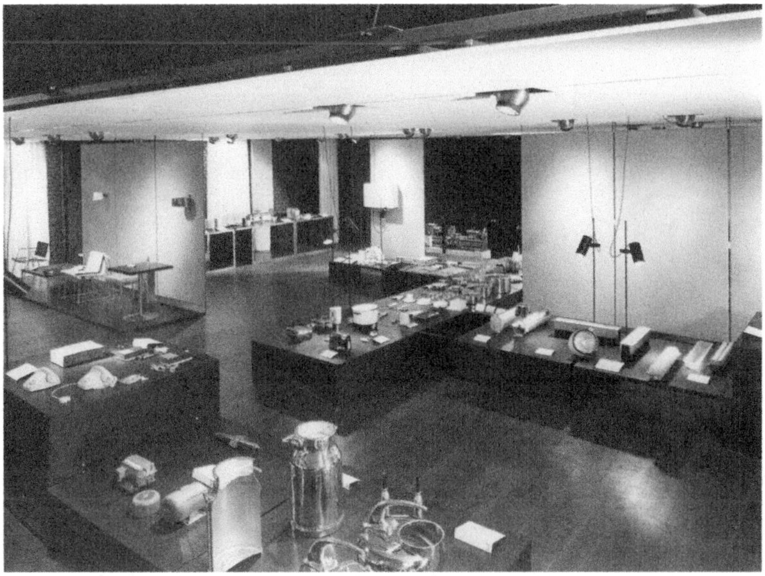

Fig. 5.15. View of the opening exhibition of the Design Centre, 1964. Source: Coutisse, Archives des Cressonnières.

The first stop was at the drinking glasses of Manufacture de Boussu glimmering on the dark *wenge* display furniture. On the same level, a *dactylo* chair by Barro and the modular furniture gamut *Decoplan* produced and designed by De Coene were neatly displayed next to many other objects in the industrial modernist interior. Behind the arrangement of the *Decoplan* furniture, index cards of each product on display were stacked in two turning racks and were at the disposal of the visitors. Descending to the second level, followed by a swarming crowd, Des Cressonnières showed the prince samples of professional equipment such as the *Resiblocs* of Usines Balteau and the electricity meters of Vynckier. Walking a bit further on the second level, the noble committee was introduced to a wide range of Belgian industrial design: Bell telephones, Nova kitchen appliances, FN milking tools, furniture by Jules Wabbes, a plastic beer crate by Vandenheubel, the textiles of Jeannine Coppens and a leather chair by Christophe Gevers (figs. 5.15–16). Des Cressonnières finally walked her royal company to the third and last level to present the more hardcore industrial design products (fig. 5.17). On the concrete floor, heavy machinery and bigger pieces were on show, such as the *Sanispray* water fountain produced by Mewaf and designed by the La Cambre trio Landuyt, Adins and Bossuyt, or the impressive *Europlex* seeding machine by Georges

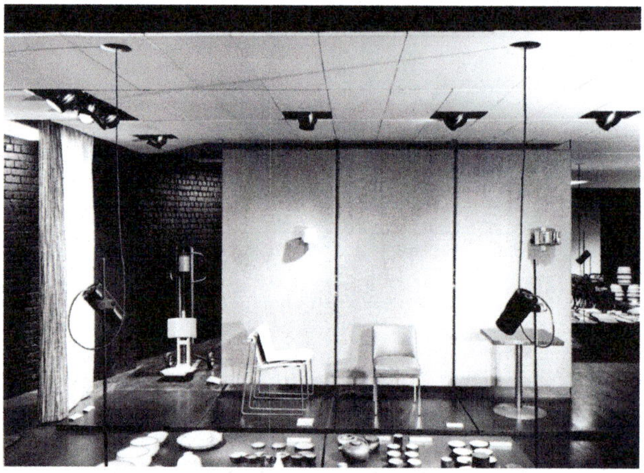

Fig. 5.16. View of the opening exhibition of the Design Centre, 1964. In the forefront is furniture by Christophe Gevers and Jules Wabbes. Source: Linkebeek, Private archive Constantin Brodzki. Photo by Kessels.

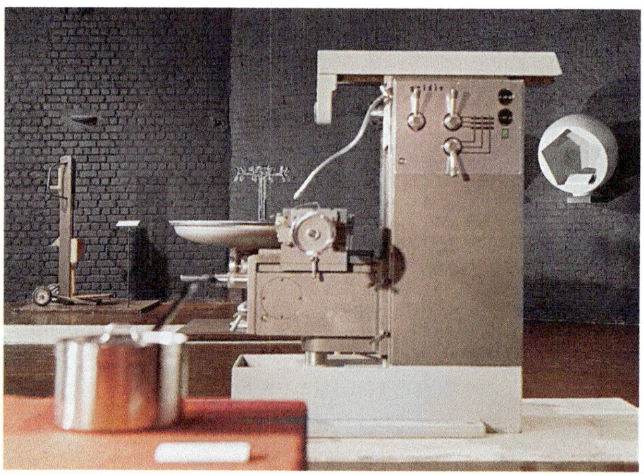

Fig. 5.17. View of the lower level of the Design Centre. In the forefront are a saucepan by sartel, the *Viking* milling machine by Roger Tallon (Technes) for Modiale and the *Sanispray* collective water tap by O. Landuyt, R. Adins, A. Bossuyt for Mewaf. Source: Linkebeek, Private archive Constantin Brodzki.

Frennet for Fabrieken Saint Hubert. One of the star pieces of this section was the highly praised *Gallic 16* produced by La Mondiale.

5.1 "A Little More Gaiety Would Not Go Amiss"

The rigid selection framed in the industrial interior design at the inaugural exhibition lent the Brussels Design Centre the image it hoped for: severe and technical. The engineering focus, the aversion to artistic industries and the focus on an export public are obvious. At the sight of the subdued colors of the small section of domestic furniture, the British Design Centre delegate even bluntly remarked that "a little more gaiety would not go amiss."

Among its foreign counterparts, the Belgian Design Centre stood out in particular because of its focus on engineering. It was known as the first to focus on equipment and capital goods instead of on consumer goods — which was in line with the Ulm philosophy adapted from its beginnings.[71] Others followed its lead, such as the "ancestor" in London which, shortly after the opening of the Belgian Design Centre, also shifted its emphasis from consumer goods towards engineering.[72] At the Design Centre's inaugural exhibition, for example, the domestic sphere was even rather hidden by foreign items selected by foreign counterparts, as the journalist of *Design,* a publication of the British Council of Industrial Design, observed:

> At the moment, the engineering emphasis is obvious: in its first exhibition, products for factories, laboratories and offices predominate, and although there is some furniture for the home, it has been carefully chosen so that it avoids any association with crafts industries. Even the textiles are subdued in colour, which is a pity: a little more gaiety would not be amiss. The impression is given that the domestic field is covered mostly by foreign items — cutlery and glassware from Germany, a wine set from Sweden, a dinner service from Finland, vases, textiles and an armchair from Holland — although there are Belgian condiment sets and tableware. In this setting, the Solitaire microscope from Britain and the Sony television set from Japan seemed more in keeping with the center's character.[73]

The meager domestic section not only reflected a belief in Ulm design, but also pointed out what kind of public the Design Centre had

this number had doubled. In the following years, the numbers kept rising: 64,000 visitors in 1968 and 92,000 visitors by 1970. The press coverage equally experienced an upward trend. The opening year was covered in 79 press articles and two television broadcasts. In 1965, 273 press articles and five television broadcasts had reported on the Design Centre's activities. This number only kept growing. In 1970 around 400 press reports and fourteen television shows had dealt with the Design Centre.

During its first five years of staging exhibitions, the Design Centre's values, esthetics and politics were further consolidated and refined. Besides the permanent selection it staged around sixty thematic exhibitions. Every object on show had passed the Design Centre's jury and thus fulfilled the modernist criteria. All these different kinds of objects, from kitchen utensils to agricultural tools, were displayed in the same clean modernist manner: neatly arranged on the dark *wenge* display furniture, artificially illuminated in the dim interior design

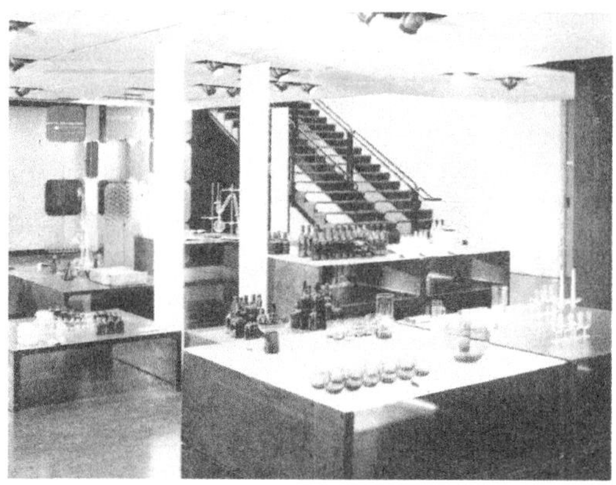

Fig. 5.22. View of the *Belgian Glass Today* exhibition (1965). Source: Josine des Cressonnières, «Le Design Centre a Cinq Ans,» *Industrie*, 1969, 356-80.

by well-directed lighting fixtures, and mostly backed by informative panels with additional information and announced by a poster which playfully linked the Design Centre's logo to the specific theme (figs. 5.18–21). In this setting, those selected Belgian products were imbued with an exclusive aura of "good" design.

The temporary exhibitions covered a more diverse program than the permanent exhibition in order to target different publics ranging from officials, salesmen and industrialists to consumers and designers.[79] An important part of the Design Centre exhibitions was organized in collaboration with national federations of industrial sectors. The federation of textiles was, for example, an enthusiastic partner. Textile exhibitions were mounted annually. The most successful was the 1967 *Textiles in the Home* exhibition that showed 163 textiles from 53 firms for domestic use, including textiles for the floor, walls, chairs, curtains and napery. This exhibition was said to have been a commercial success since many of the participating firms indicated having done "good business" and clamored for a sequel. The Belgian Society of the Glass Industry had its turn in 1965 with the exhibition *Belgian Glass Today*, which showed a wide range of glass: drinking

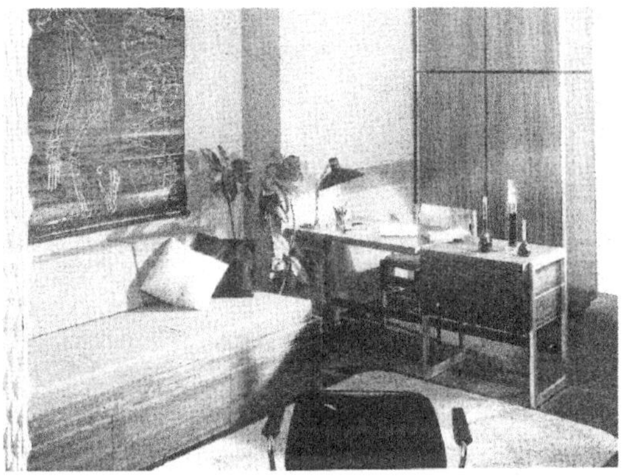

Fig. 5.23. View of the *Contemporary Belgian Furniture* exhibition (1967). Source: Josine des Cressonnières, «Le Design Centre a Cinq Ans,» *Industrie*, 1969, 356-80.

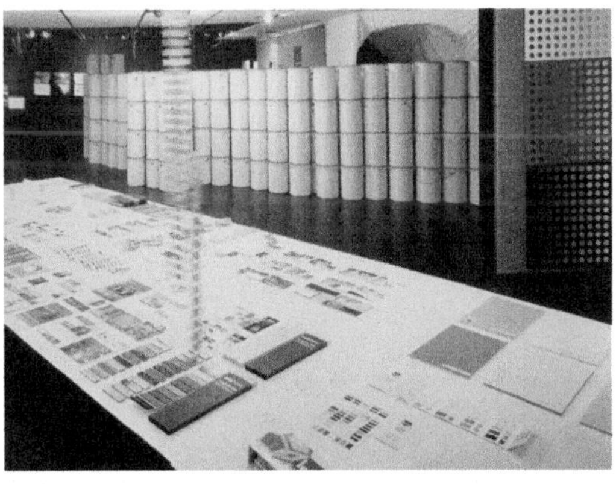

Fig. 5.24. View of the *Colour* exhibition (1969). Source: Coutisse, Archives des Cressonnières.

glasses, bottles, church windows, mirrors, laboratory tools, lighting and more (fig. 5.22). Glass in all sorts of colors, shapes and dimensions was staged on the dark display furniture and glittered in the spotlights.

Some transparent glass panels hung from the ceiling. The exhibition managed to attract around 10,000 visitors.

The wood federation had an important exhibition in 1967: *Contemporary Belgian Furniture* (fig. 5.23). The show staged 70 pieces of furniture for the living room and the bedroom that were produced by 26 Belgian firms. One ensemble on show was a well-designed furnished studio for an adolescent. After their visit, the visitors could vote for their favorite piece of furniture. In 1969 the Design Centre mounted the exhibition *Color* that put the Belgian producers of paint and ink in the limelight (fig. 5.24). Five large companies were selected to show their range, among them *Sikkens* and *Levis*. The Design Centre was, on that occasion, filled with samples of different colors, exemplary advertisements, brochures with beautiful layouts, and many paint pots.

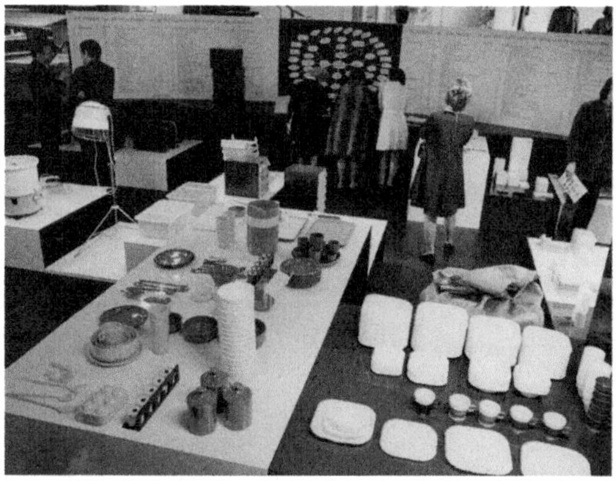

Fig. 5.25. View of the *In Real Plastic* exhibition (1970). Source: Coutisse, Archives des Cressonnières.

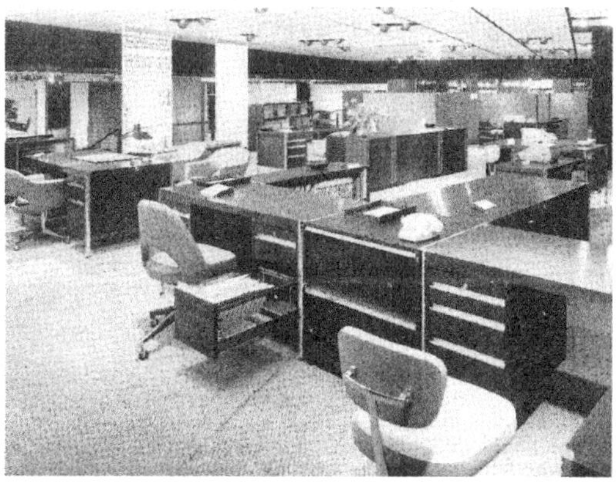

Fig. 5.26. The launch of the new De Coene collection of *EFAC* office furniture in the Design Centre, 1968. Source: Josine des Cressonnières, «Le Design Centre a Cinq Ans,» *Industrie*, 1969, 356-80.

In 1970 the booming Belgian chemicals industry was the center of attention with the *In Real Plastic* exhibition (fig. 5.25). The entrance was decorated with a curtain of plastic algae. Once visitors had passed through this plastic wall, a wonderful world of plastic revealed itself: dinnerware, mini washing machines, disposable plates and cups, *Meurop* furniture, telephones, *Samsonite* travelling bags, hair driers, floor coverings, and so on — in total more than 150 products, all in plastic. The show promoted the "authentic" and functional — read,

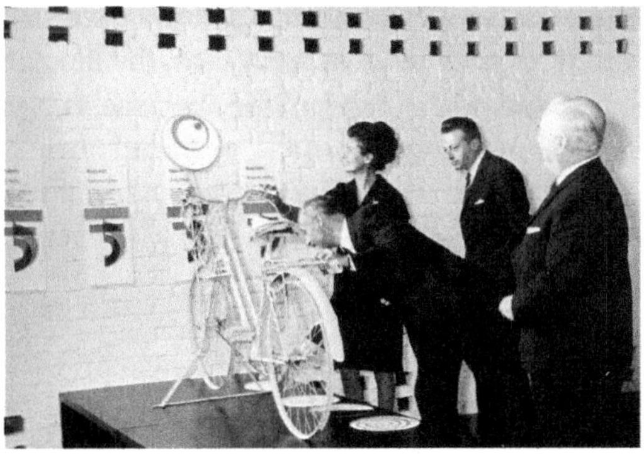

Fig. 5.27. View of the *Unissued Projects* exhibition (1966). Picture by Kessels. Source: Coutisse, Archives des Cressonnières.

modernist — application of plastic in order to combat the "wrong" use of plastic, namely the imitation of other more expensive materials. Therefore, a large didactic panel was mounted against the wall to explain the qualities and properties of different types of synthetic plastics and their possible uses in design.

The Design Centre also held "solo exhibitions", as it were, devoted to a single company. In most cases, the company in question covered all the costs of staging the exhibition. In 1968, for example, De Coene launched its new collection of *EFAC* office furniture in the Design Centre (fig. 5.26). The collection included desks, office chairs, accessories and separation walls designed by Pierre Génicot, the same designer who was involved in the interior design of the Design Centre.

Next to the focus on industry and commerce, the Design Centre equally paid attention to exhibitions appealing to the design community. In 1966, a whole exhibition, *Project Unissued,* was dedicated to projects that were not yet in production (fig. 5.27). Large-scale models, pictures and plans of designers and design students presented artistic handicrafts as opposed to the modern and innovative capital.[12] For example, the 1949 promotional brochure *Belgium To-Day* stated: "Handicrafts in Belgium seem to have flourished most in the provinces, and have been unusually brilliant in the past".[13] In the 1960s, this link between the provinces and handicrafts was still there and surfaced in provincial design policies. In the case of the Museum of Decorative Arts and Industrial Design in Ghent, modern handicrafts played an important role in its policy, which was focused on the promotion of a modern home culture.[14] The museum promoted modern handicrafts as necessary decor for furniture produced by machines. Through their unique and traditional character yet having a modern formal language, modern handicrafts brought a personal character to the domestic interior, something that was, in the museum's opinion, missing in most of the industrially produced products.

As early as in 1959 the museum's director described the status of artistic crafts versus industrial design within the framework of the newly reformed national commission for crafts under the Ministry of the Middle Class.[15] According to Van de Walle's statement, crafts stood closer to the "people" than art or industrial design.[16] Crafts could be considered as the desire of the "people" to be creative, and thus reflected at its best its national spirit "in spite of all international movements." Alphonse Botte, the president of the provincial committee for crafts and industrial design in West Flanders, attributed similar qualities to artistic crafts.[17] He was of the opinion that crafts would never disappear in favor of industrial design because people would always search for an escape route from the uniformization of mass production. Instead, crafts offered the possibility to affirm individuality and bring "more human warmth."

At the opening of the Design Centre, Josine des Cressonnières had anticipated this assumption that modernism with its abstract and geometrical forms would somehow erase national specificity or individuality because of its international and outward orientation. In the promotional brochure, she considers folk expressions only as a transitory phase in the modern project toward globalization and that these eventually would be replaced by design:

> To those who fear that the multiplication of these exchanges [exchanges with foreign initiatives] will give birth to uniformity, we can say that the old regional values will inevitably dissolve into a civilization that is becoming global, but that the folklore can be

replaced by man, not by man alone, but man in a team, in a firm, in a design studio.[18]

Accordingly, the Design Centre had, from its early years, a clear position toward handicrafts as being subordinate to industrial design. Only five to ten craft products were on display in the permanent exhibition that changed every three months.[19] These handicrafts were included only if they met the esthetic and functional criteria of industrial design. Crafts were not applauded for their unique character as in the provinces; on the contrary, craft items had to be produced in series and be available on the market at affordable prices. The Design Centre justified their occasional affiliation with crafts by emphasizing that they were an inspiration for industrial design and, moreover, that all the other design centers around the world were associated in one way or another with crafts.[20]

2.3 Counterattack: "Operation Decentralization"

To secure its national monopoly regarding industrial design promotion, the Design Centre launched a counterattack entitled "Operation Decentralization." With this plan, the Design Centre wanted to meet the demands of the regions and prevent the creation of independent, and in their eyes incompetent, initiatives.[21] The operation was expected to last four years and comprised eight exhibitions. The plan was to mount two twin exhibitions each year, one in Flanders and one in Wallonia. The aim was to be present on a local level and to collaborate with the provincial authorities.

To call this action "Operation Decentralization" was rather contradictory, given that the Design Centre still pulled all the strings and provincial initiatives were subordinate to those in the capital. One of the criteria explicitly mentioned in a report was that the local exhibitions had to be organized at the initiative of the Design Centre, and not by invitation of the province. And, more importantly, that the Design Centre had to hold authority over the selection of the products (i.e. the national canon) and their presentation, including the exhibition space, layout, poster, texts, and so on (i.e. discourse). Accordingly, "Operation Decentralization" can be understood as an attempt to centralize Belgian design promotion in the capital and to control related activities in the provinces and protect its version of "Belgian design."

3. "C'EST BELGE, ÇA!"

3.1 Political Tensions in the Unitary Belgian State

Indeed, the growing chasm in the 1960s between the Dutch- and French-speaking communities put pressure on the centralized unitary Belgian state.[32] Although calls for the federalization of Belgium were growing louder, the Design Centre protected — as "Operation Decentralization" shows — its "national" canon from fragmentation and tried to avoid any "regionalism" in its selection. However, these political sensitivities of the moment colored the debates and critiques on the promotion of Belgian design. In particular, the great dissatisfaction of Flemish intellectuals about the growing "Frenchification" of Brussels was projected in these critiques. The unilingual (French) speech at the opening by Minister of Foreign Trade Maurice Brasseur, a Walloon militant, fell on stony ground with the Flemish nationalists. Flemish journalists especially contested the French character of the Brussels Design Centre — "the French façade of the nevertheless nervous bilingual Design Centre."[33] They complained about the poor quality of the Flemish texts provided by the design institutes and the French atmosphere.[34] The sharpest pen was that of design critic K.N. Elno, who from 1964 to 1967 had a weekly or bi-weekly column on design in the pro-Flemish magazine *De Nieuwe*.[35]

However, it is difficult to accuse the Design Centre of being clear-cut French. Many of its key personalities were in fact Flemish militants. Pol Provost (1907–1990), cofounder and president of the Design Centre, was a French-speaking Flemish nationalist based in Kortrijk.[36] His *flamingantisme*, however, seemed to coexist peacefully with his plea for a national design style for export purposes. Frans Wildiers (1905–1986), cofounder and director of the Vlaams Economisch Verbond (VEV, the Flemish Economic Union) was also a Flemish nationalist. Both knew each other well from powerful pro-Flemish organizations such as the Dutch Flemish society Orde van den Prince and the VEV, of which they were both members.[37] Also, the new director of the Institute for Industrial Design, Hubert De Sy (1921–2010), was in the same Flemish circles and was well acquainted with Pol Provost and Frans Wildiers.[38] Josine des Cressonnières, the face of the Design Centre, was the only one of the three founding members who was not Flemish. Born in Brussels, her mother tongue was French but she mastered Dutch quite well and gave successful interviews on

Flemish television. Some Flemish journalists even made specific mention of her good knowledge of Dutch.

The "French" character can then perhaps be attributed to the fact that the Design Centre was created in the spheres of the Ministry of Foreign Trade, which was considered "French" at the time. From the 1950s, this ministry had been haunted by a negative image because of an imbalance between French- and Dutch-speaking officials. In particular, the higher ranks of the ministry were populated by French-speaking Belgians.[39] A second possible reason why the face of the Design Centre was perceived as "French" was the fact that industrial design seemed to be better infiltrated in the French-speaking milieus. In 1965, the president of the Institute for Industrial Design observed an unequal membership: only one third were Flemish, and two thirds were Walloons. K.N Elno explained this lack of Flemish members by criticizing the "national" character of the design promotional institutes, in the sense that they were Francophone and structured and based on the outdated "Belgian idea."[40] Elno subsequently argued that design was actually equated "with (real) nationalist character", cultural self-esteem and autonomy. He concluded his plea with the sharp question as to why they were waiting to federalize the Belgian design institutions.

3.2 Reinventing Belgian Souvenirs on the Eve of Belgian Federalization

On Christmas Eve 1970, the unitary Belgian state would indeed be reformed. The new constitutional amendment under Prime Minister Gaston Eyskens established cultural parliaments for both the Dutch- and French-speaking communities.[41] It thus laid the foundation for a decentralized federal state where the regions gained increased political autonomy. December 1970 is then considered the end of the "Belgique de papa" and a starting point of the political devolution of the Belgian State, which in 1993 would eventually become a fully-fledged federal state.

A couple of months earlier, the Design Centre pulled off a curious "nationalist" action: in collaboration with the Economic and Social Institute for the Middle Classes (ESIM), it initiated the "Belgian Image Products" or "BIP" campaign. The project aimed to boost the "shabby" image of Belgian souvenirs beyond the traditional Atomium,

Manneken Pis and examples of old lace. Josine des Cressonnières explained that souvenirs were too often "a piece of junk without any dignity" which did not have anything to do with "the national soul."[42] A call was made to designers and companies for new and improved souvenirs or objects related to Belgian folklore, culture, history and industry which could be sold in Belgium and abroad. The concept of "souvenir" was here broadly understood as both crafted and industrial objects that could represent Belgium. The Design Centre and the ESIM were no strangers to each other.[43] From 1968, ESIM annually occupied the exhibition gallery of the Design Centre to mount a show of artistic crafts. However, it was remarkably the first time that the Design Centre had cooperated so closely with the ESIM to set up a campaign in which both artistic crafts and industrial design were mobilized together to explicitly portray Belgium.[44]

The BIP campaign resulted in two Design Centre exhibitions: the 1970 exhibition *B.70: 300 Products as Image for Belgium*[45] and its sequel in 1971 *That's Belgian!* (figs. 6.1–2)[46] On the eve of the federalization, the Design Centre thus gave its all to renew Belgium's image, drawing on both artistic crafts and industrial design, a combination that in the past had clashed violently within the walls of the Design Centre. The status allotted to crafts in the Design Centre's parlance seems as such to have undergone a reversal for this occasion. While exhibiting artistic crafts together with industrial design used to be seen as a virtual act of heresy, and while folklore was previously diagnosed by des Cressonnières as only a transitory stage in the modern project, in 1970, however, in the midst of a political crisis, this seemed to have changed. In part, this was also inspired by the revitalization of crafts within the new flower power movement and DIY trends.

The interweaving of craft and design in one "Belgian" exhibition, in fact, proved to be particularly convenient. When the first *B.70* exhibition was inaugurated on the eve of ederalization, it was enthusiastically received by the press as evidence of "très grande classe" which was "mounted with imagination and humor".[47] Even the pro-Flemish magazine *De Nieuwe* wrote that the exhibition was well worth crossing the normally carefully avoided "Belgian threshold" of the Design Centre.[48] The French newspaper *Le Soir* called this exhibition "agréablement nationale" and a "success and revelation".[49] Only one press report made a subtle reference to the current political situation when eulogizing about the nostalgic image evoked in the Design Centre as one of "an unknown Belgium" without "any linguistic borders."[50]

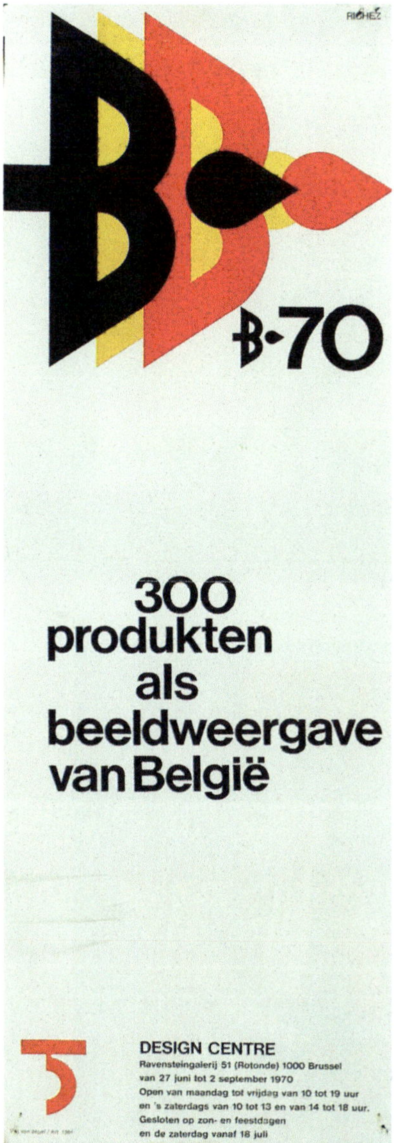

Fig. 6.1. Poster for the *B.70: 300 Products as Image for Belgium* exhibition (1970) designed by Michel Olyff. Source: © Michel Olyff. Letterenhuis, Antwerpen.

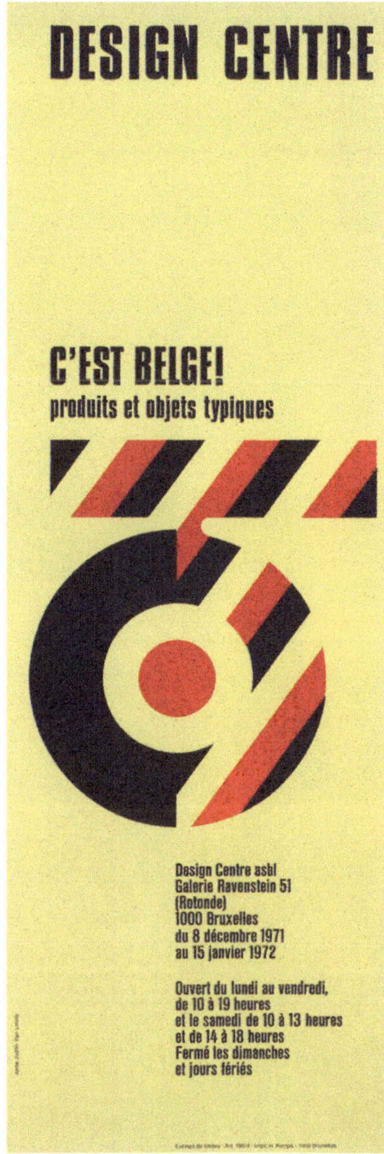

Fig. 6.2. Poster for the *That's Belgian!* exhibition (1971) designed by Anne Judith Van Loock. Source: Brighton, University of Brighton Design Archives, Icograda Archive (3-3-26).

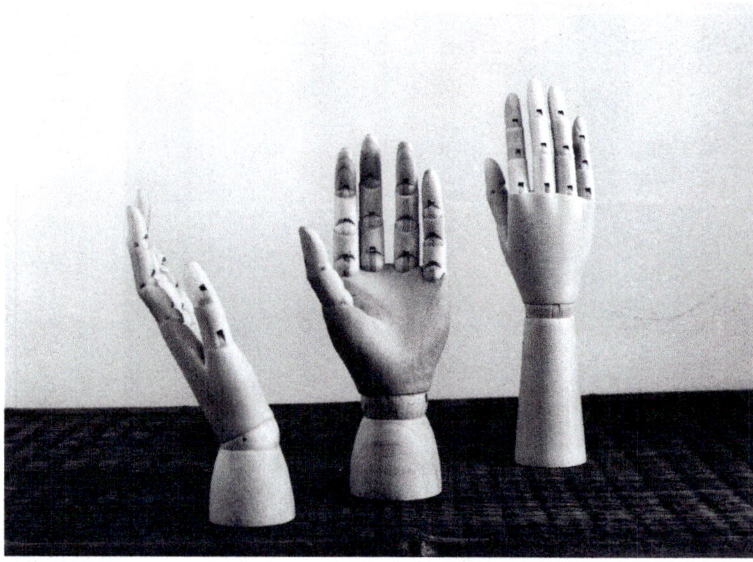

Fig. 6.3. Wooden model hands by Ateliers J. Bonhomme in Verviers. Source: Coutisse, Archives des Cressonnières.

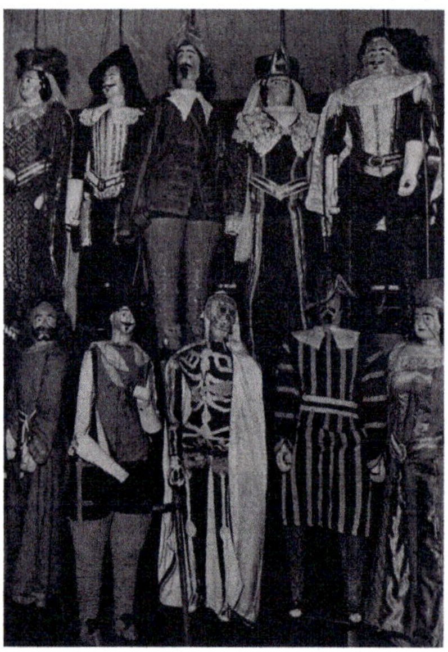

Fig. 6.4. Puppets from the Brussels Toone Theater in the Brussels Design Centre, 1970. Source: Brussels, State Archives, BE-A0545 / Fic 2000, Archive Design Centre, 282.

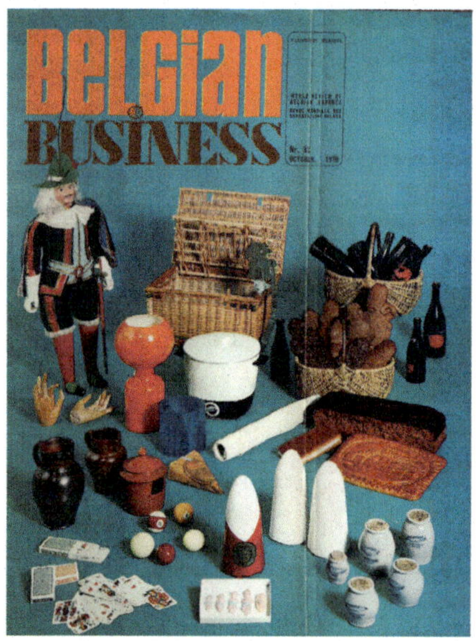

Fig. 6.5. Some products of the *B.70* show on the cover of *Belgian Business*. Source: Brussels, State Archives, BE-A0545 / Fic 2000, Archive Design Centre, 282.

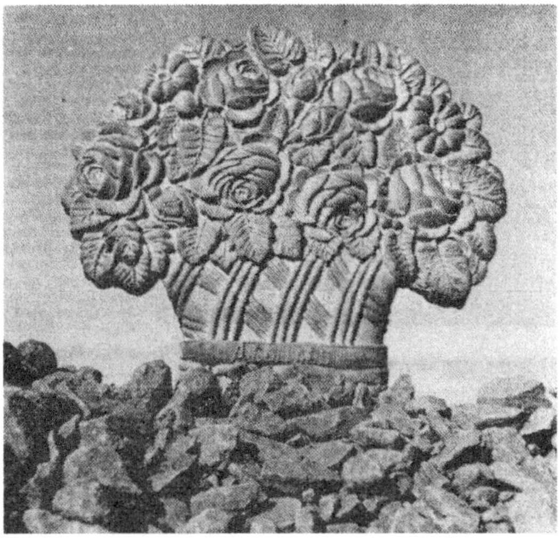

Fig. 6.6. The "Couques de Dinant" featured in the *B.70* show. Source: Brussels, State Archives, BE-A0545 / Fic 2000, Archive Design Centre, 282.

The show featured 300 products chosen to capture Belgium: playing cards from Turnhout, puppets from the Brussels Toone Theater, matches from the world's biggest match factory Union Allumettiere, billiard balls by Les Usines de Callenelle, wooden artificial hands by Ateliers J. Bonhomme in Verviers, comic books of the celebrated Tintin, tobacco cultivated in the Belgian Semois, famous Belgian linen and lace, and all kinds of things in the red-yellow-black colors of the Belgian flag (figs. 6.3–5).[51] The Belgian souvenirs also included children's toys and dolls based on legends such as the Bayard horse, Manneken Pis and Lion's Mound. What was remarkable was the large share of gastronomical products to represent the Belgian "soul."[52] Almost one third of the selected objects were foodstuffs: beer, chocolates, Ardennes ham and many sweets, biscuits and cookies, such as "katrienespekken", "kletskoppen", "Antwerpse handjes", "couques de Dinant," and sugary models of little Jesus (fig. 6.6). The reporter for the Flemish newspaper *Het Laatste Nieuws* was convinced that these foodstuffs were actually "our" best souvenirs.[53]

The specific political agenda of this BIP action coinciding with tensions in "la Belgique de papa" remains a matter for conjecture. One might assume that the action was set up to show that Belgium still served as a locus of identification — which was especially the case among the Francophones of Brussels. One other possible explication was that the then constitutional reform only occurred in the cultural realm, while crafts and design were rather part of economic policies. Altogether, it was apparently not controversial, since this "Belgian" intention of *B.70* was sponsored by both the Flemish and French ministries of culture.

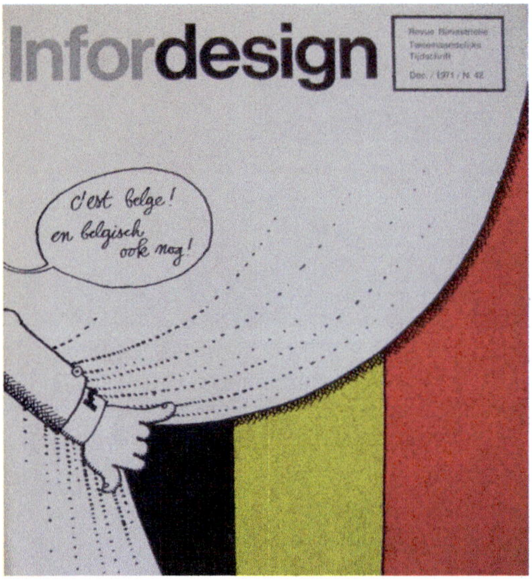

Fig. 6.7. One of the issues of *Infordesign* served as the catalog for the 1971 *That's Belgian!* exhibition. Source: Brussels, State Archives, BE-A0545 / Fic 2000, Archive Design Centre, 759, *Infordesign*, no. 42 (1971).

The *B.70* exhibition was, however, a solitary success. Its BIP sequel *That's Belgian!* in 1971 could not count on the same enthusiasm (fig. 6.7). The Belgian press received it with mixed feelings.[54] Some "Belgian" objects scored good reactions such as, for example, Boch Frères' service, Val Saint-Lambert's glass, the Elexir d'Anvers liqueur, Dujardin children's clothing and an Ardennes headscarf. The French newspaper *Le Soir* reported that the choice of objects was, however, less fortunate than the year before:

> Let us be clear: There ARE beautiful things (often traditional designs or inspired by old models). Yes, beautiful, unexpected, charming things which make for lovely "holiday souvenirs." However, there are also a bunch of far-fetched ideas (when they are not simply an offense against good taste) that seem even more out of place exposed under the name and with the seeming blessing of the Design Centre, which we have always considered as a kind of conservatory of good taste for the past, present and future.[55]

Accordingly, the varying quality of the objects sketched, according to *Le Soir*, multiple images of Belgium. Depending on which objects one chose "one can say that we are very artistic, original, spiritual ... or very dull, tormented and complex."[56] It was especially a bar of soap in the shape of Manneken Pis that was most contested in press reports.[57] Hence, the Design Centre felt the need to explain that this piece of "commercial pop-art" had to be understood in the context of the themed exhibition and would otherwise never have slipped through the nets of the Design Centre's jury.[58]

The two "Belgian Image Products" exhibitions — one more successful than the other -illustrate the changing role assigned by the Design Centre to design and crafts in the forging of a national identity at a turning point in Belgium's political history. One may consider the reclaiming of crafts for the molding of a Belgian image an illustration of what is called the Janus-like quality of nationalism: the combination of both references to a heroic past, while at the same time showing a future-oriented attitude.[59] In the case of the *B.70* and *That's Belgian!* shows, crafts reimagined the Belgian past, legacy and tradition while industrial design symbolized modernity and a prosperous future. Or, as the *That's Belgian* catalog put it, "Both [old and new things] give a timeless image of Belgium [...] The continuous thread that defines our singularity through the ages is important."[60]

4. DESIGN FOR THE STATE

In spite of the centrifugal forces that divided political Belgium, the Design Centre's scope kept on being national, promoting the prestige of Belgian products and creativity. However, there was a slight change of strategy around the mid-1970s. While before "Belgian" design was quite obviously constructed through a process of selection adhering to modernist criteria prevailing in international circles, the Design Centre now focused on introducing industrial design to public authorities and eventually creating a design policy at state level. Design was pitched as a way to make the "state" — whether federal, regional, provincial, municipal — tangible for the citizenry. The state was, after all, considered one of the largest potential clients for design activities. A large part of the gross national product was indeed spent on public works: transportation, city furniture, school equipment, and so on. The Design Centre argued that design could make a difference in “humanizing” the environment of the citizens, who after all paid for it through taxes, and could provide a favorable image of the state, and eventually this would feed back into the national economy.

This new strategy was no less “nationalist” than promoting exclusive selections of objects. On the contrary, it can perhaps be best described as a form of “banal” nationalism after the theory of social psychologist Michael Billig.[61] Around the mid-1990s Billig introduced the term “banal nationalism” to describe the ubiquitous forms of nationalism in the banal realm of everyday life which are quickly overlooked because they seem very natural. For example, simple actions such as going to the post office, paying with banknotes, filling in your tax form are all small reminders that one belongs to a specific nation. As Billig explains, “The metonymic image of banal nationalism is not a flag which is being consciously waved with fervent passion: it is the flag hanging unnoticed on the public building.” Actions of the Design Centre to enhance the public realm for the citizenry through the introduction of “design” are thus maybe less explicit than before but are certainly not deprived of a certain “nationalist” agenda. This kind of design promotion then — although it might not coincide directly with the agenda of the designers, industry or even promoters — is eventually still linked with the creation of national identities, whether Belgian, Flemish, Walloon, provincial, local or other.

Ideas on “design for the state” were already expressed in des Cressonnières' early ICSID activities.[62] She most explicitly developed these design strategies in 1977 when she wrote her statement of intent as a nominee for ICSID presidency.[63] She considered that the only way “to contribute to change in our Society and to the transformation of this economic and political structure” was to introduce industrial designers into these powerful milieus — understood by her as national governments but also large industries, management levels and non-governmental organizations — located at national and international levels.[64] In 1978 she even submitted a proposal for a fellowship to the Woodrow Wilson International Center for Scholars in Washington on the subject of a global policy of industrial design for public authorities.[65] The idea was to collect worldwide examples of national design policies and to formulate ideas for the future.[66]

In 1980 she held a very short public lecture amusingly entitled “5 minutes to convince” in which she gave a playful example of how to implement design in the “spheres of decisions” by capitalizing on the patriotic feelings of ministers to set up a stronger national economy, export and image.[67] The trick is that one first has to win them over on the topic of design. Des Cressonnières suggests starting as follows:

> Dear Minister.
>
> You have said — or written — that it is essential for our country to invest in innovation and creativity — (whether he actually said it or not is of no importance, he will agree to be credited with it) — creativity and innovation being some of the most precious assets which can compensate, on the internal markets, for the cost of our workforce, our strikes, our lifestyle, our growing social charges, etc. ...[68]

Des Cressonnières instructed to introduce "design" as a way to get the best results in the short term. There were several ways to introduce design. She explained that one can start by inviting the minister to declare in public the importance of design. One could go further and propose more fundamental contributions, namely to include design as one of the criteria for public contracts. But one can go even further ... and suggest imposing the use of designers for public commissions. As such, the state itself would become the motor of innovation. She instructed one to finish with a nationalistic note:

> This policy is expected to have an additional effect: a new image for our country, a possible quality for our export capacity. Think, for example, how much the image of Japan has changed on our markets in the past twenty years! One last point, dear minister, there is no other country in the world where there exists today a complete and coherent Industrial Design policy. Why shouldn't we be the first?[69]

4.1 Changing Design Scene

The shift from plainly selecting objects as examples of "good" Belgian design to the introduction of "design" on a state level had much to do with the changing design landscape in the 1970s. The modernist paradigm was exchanged for an environmental and activist paradigm. Most telling was the closure of the famous Ulm School of Design in 1968 at the height of student protests all over Europe. Instead, a smaller Institute of Environmental Studies was established as its successor in 1971.[70] Also, Maldonado's hardcore scientific modernist stance

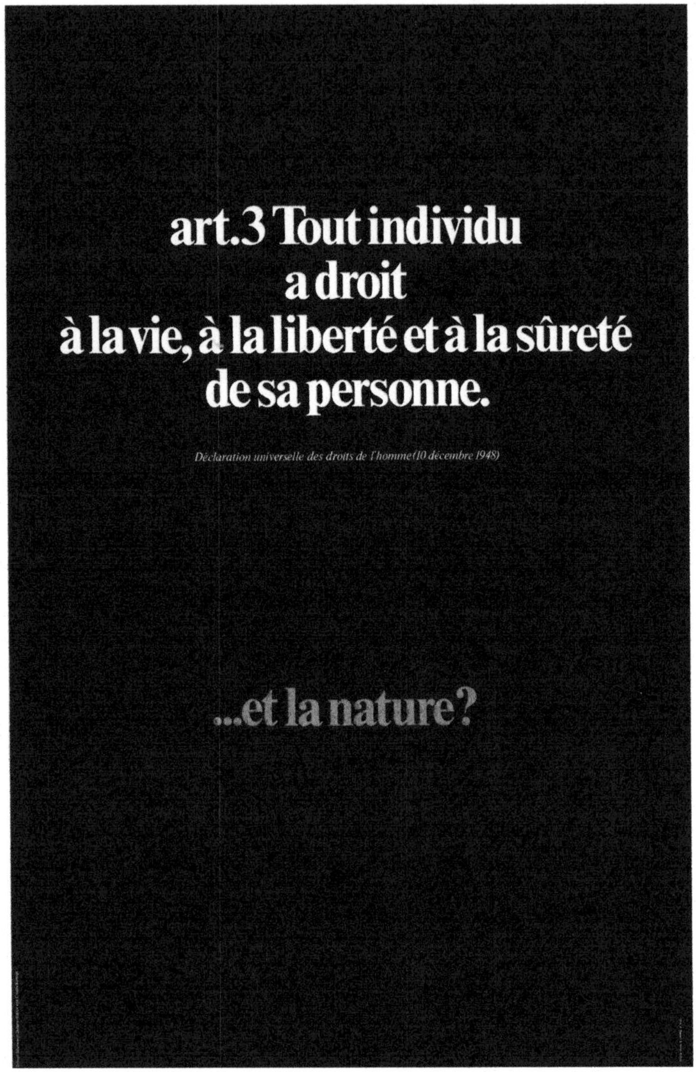

Fig. 6.8. Boudewijn Delaere, Poster "art.3 Everyone has the right to life, liberty and security of person ... and nature?", 1972. Source: © Boudewijn Delaere.

changed to a focus on the "human environment." In 1970 he published his book *Nature and Revolution: Toward a Critical Ecology*.[71] The debates in ICSID circles were equally colored by this new ecological sensitivity and, of course, Papanek's book *Design for the Real Word* (1971) kindled a strong debate on the future of design and its moral and social duties. In Belgium too this trend was visible when looking at the architectural magazine *La Maison,* which changed its name in 1969 to *L'Environnement*. Also in the early editions of the Biennale

Interieur in Kortrijk the validity of modern design was questioned in a true '68 revolutionary spirit.[72] The critic Geert Bekaert made an appeal in the very first catalog in 1968 to no longer 'design' man or woman, but rather to encourage their creativity. While the old design stars Raymond Loewy and Gio Ponti — representing 'good' functional design — were invited as honorary guests in 1970 and 1972, the younger generation started to criticize design as part and parcel of capitalist paternalism and manipulation. In line with early ecological thinking, the consumer society and its influence on the natural world were duly questioned. The Interieur 72 theme was nature: 250 trees were placed inside the exhibition halls decorated with posters by graphic designer Boudewijn Delaere, featuring the inscription: "Art.3 Everyone has the right to life, liberty and security of person ... and nature?" (fig. 6.8). For that same edition Bellinni's bright-green Kar-a-Sutra concept car was flown over from New York where it had created waves at MoMA's famous exhibition *Italy: The New Domestic Landscape*. In 1974 the guest of honor Verner Panton proposed his new slogan "forward to the natural". Two editions later he suits the action to the word and he creates an 'anti-interior' with bales of straw and scratching chickens.

With these new paradigms sneaking in, Design Centers seemed to become an easy target for criticism on the politics of "good design" modernism. Design Centers were associated with a kind of top-down modernism: people with "taste" chose the right things for consumption by the common man. Around the mid-1970s, design critic K.N. Elno, for example, began challenging this modernistic practise. At the tenth anniversary of the Design Centre in 1974, he looked back on a decade of Belgian design promotion and accused the center of having realized old ideals on the eve of a new area.[73] For him, the center failed to anticipate new developments in the decade 1960–1970 when the new generation of "Beatles" initiated the revolution of May 1968 and ornamentation was reintroduced in architecture and design. In that same year, Ulmer Gui Bonsiepe, who was active as a designer in Latin America, expressed in personal correspondence his great dissatisfaction concerning Design Centers and their modernist premise.[74] According to him, the name "recalled connotations with the tendency of 'good' design and its links with the well-educated middle class and the hypertrophy of the formal and aesthetic characteristics of products."[75] He believed that although functionality was in high esteem to evaluate a product, "form" was the most important criterion that was singled out.[76]

Also, within the transnational Design Center community itself, the role of Design Centers in relation to societal changes was questioned. In 1974, twenty-five directors of design promotional organizations from twenty different countries gathered in Berlin to evaluate the role of Design Centers and look for new criteria for a policy of design promotion.[77] One of the important conclusions of the gathering was that "objective" product criteria used for selection were considered paternalistic. Thereupon it was decided that "good" or "bad" products did not exist.[78] A general agreement was also reached on the future of Design Centers: they had to serve as instruments of societal change. The high ambitions for Design Centers to improve our industrial society were, however, in sharp contrast with reality. The three rapporteurs of the event — the German François Burkhardt, the Belgian Jean-Jacques Stiefenhofer and the Dutch Loek van de Sande — were very disappointed with the results of the questionnaire issued on this occasion.[79] One of the greatest disillusions was that most Design Centers still understood design from a functionalist point of view and that its main activities were limited to commercial aims: the export promotion of national design.

While the Belgian Design Centre had some difficulty balancing the commercial agenda of its patrons and the utopian ideals of the discipline, it fully subscribed to these new developments.[80] During its first decades of existence, it slowly moved away from its original commercial objectives and incorporated more and more social and cultural concerns. While the Design Centre's permanent exhibition remained faithful to its original *raison d'être*, namely its commercial aim, the temporary exhibitions were open to experiments and reflected international design trends and the evolving philosophy of the Design Centre. The attention given to the industrial object alone was soon considered a constraint, and the Design Centre's actions included architectural, urban, ecological and social issues.

In 1970, a sixth criterion was added, drawn from a new environmental awareness: the influence on the environment. Around its tenth birthday, the Design Centre was proud to state that it enlarged its scope from the single product to the "product environment." Design was now seen as an attitude and the buzzword "creativity" entered the Design Centre's discourse. This heightened attention for the public space, human environment and social issues was reflected in the Design Centre's exhibitions on public transport (*Level -1* in 1972, *Traffic in Historical Cities* in 1976), public or semi-public spaces (*Hotel Design* in 1973, *Airport Design* in 1975, *Design in the Hospital* in 1977), and sports and children (*Design and Sport* in 1978, *A Design Environment for the Child* in 1979) (figs.6.9–11).

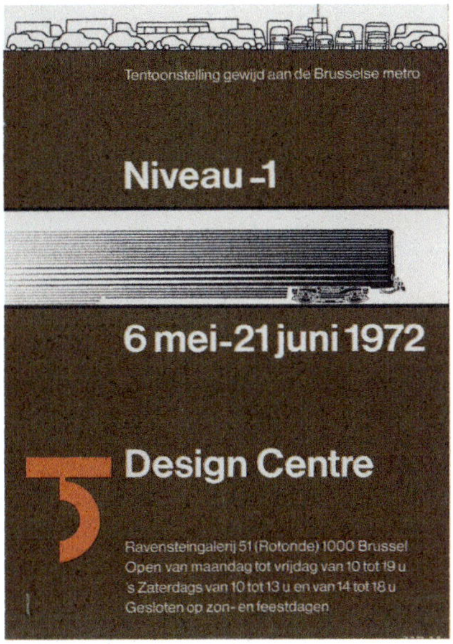

Fig. 6.9. Poster for the *Level-1* exhibition (1972) designed by Michel Olyff. Source: © Michel Olyff. Rijksarchief Antwerpen - Beveren.

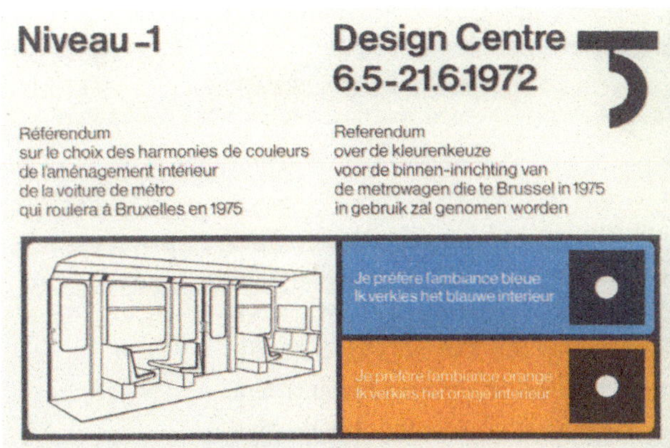

Fig. 6.10. Form for the visitor to the Design Centre exhibition *Level-1* to indicated their preferred color for the interior of the metro carriage, 1972. Source: Rijksarchief Antwerpen - Beveren.

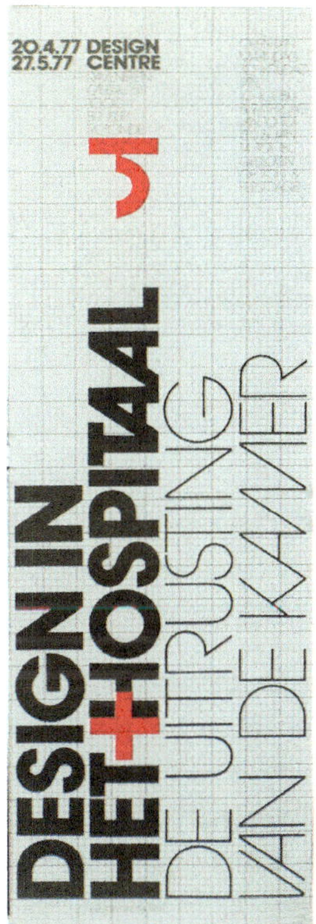

Fig. 6.11. Poster for the *Design in the Hospital* exhibition (1977). Source: Letterenhuis, Antwerpen.

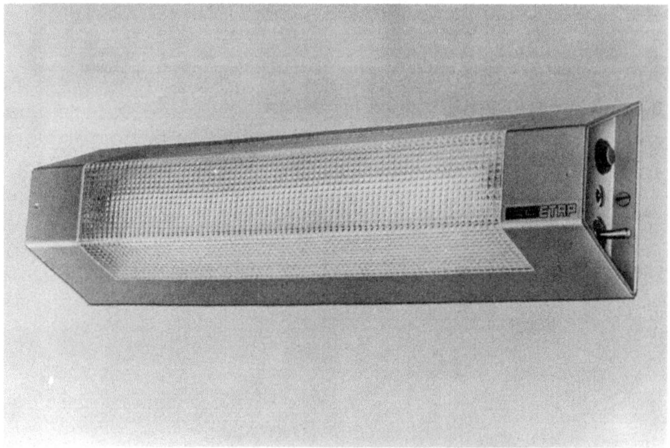

Fig. 6.12. ETAP emergency lighting shown at the *Habitat and the Third Age* exhibition (1980). Source: Brussels, Documentatiecentrum Design Vlaanderen, box 23/C. 235.

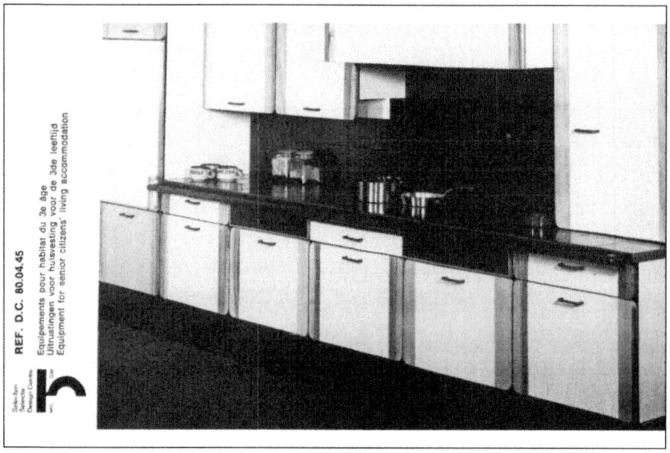

Fig. 6.13. Index card of the kitchen unit designed by Frank Smout for the *Habitat and the Third Age* exhibition (1980). Source: Brussels, State Archives, BE-A0545 / Fic 2000, Archive Design Centre, 235.

In the 1980s, the focus subsequently shifted to "the social implications of our material surroundings." New allies for projects were introduced such as the King Baudouin Foundation and the National Institute for Housing. For example, in 1980 the Design Centre, with the King Baudouin Foundation, staged an exhibition on *Habitat and the Third Age.*[81] The show displayed all kinds of objects that would give comfort to the everyday life of the elderly, such as a wooden bench for a sitting shower, a fire alarm and video intercom by Siemens, emergency lighting by ETAP, a humidifier by Philips and a kitchen unit designed by Frank Smout (figs. 6.12–13). The exhibition was completed by a special series of index cards and the monograph *Housing Accommodation and Senior Citizens* published by the King Baudouin Foundation.

4.2 Design Commission

The most concrete result of this new focus on introducing design policies on a state level in Belgium was the establishment in 1977 of the "Inter-ministerial Design Commission" — in short, the Design Commission. [82] It was set up under the protection of the Belgian Ministry of Public Works and was headed by its secretary general, Robert De Paepe. The Design Centre in Brussels served as the secretariat. The Commission's aim was to harmonize all public equipment. More concretely, it described its responsibility as "design-coordination" between governmental agencies and ministries. If one ministry, for example, needed street signs and the other streetlights, the Commission would figure out the possible overlaps or omissions or cooperation. Therefore, the Commission not only included officials of the Ministry of Public Works, but also representatives of the Ministries of Transport, Postal Services, Education and Public Health. Along with its role as coordinator, the Commission also proposed pilot projects for public services induced by current debates or needs. One of these proposed projects was a totally new concept for urban equipment by designer Bruno Stoppa, who designed a modular system based on an aluminum pole. The whole included traffic lights, road signs, bus stops, kiosks and telephone booths. The Commission's ultimate aspiration was to include design criteria and methods in purchasing orders from the Belgian State and, as such, to introduce a consistent design policy at state level. After all, as the Design Commission reasoned, the mass

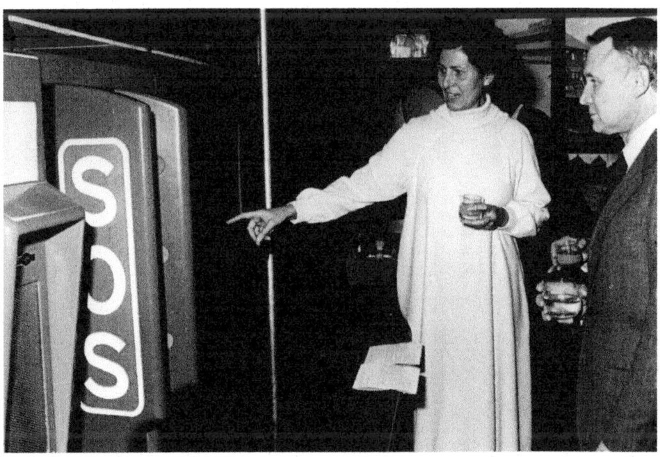

Fig. 6.14. Josine des Cressonnières and Robert De Paepe, secretary general of the Ministry of Public Works and president of the Design Commission, at the opening of the *Rest Areas along Motorways* exhibition (1978). Source: Brussels, State Archives, BE-A0545 / Fic 2000, Archive Design Centre, 220.

of equipment which the state buys represented more than a quarter of the national production — and as such there was great potential for design to humanize the everyday public environment.

The Design Commission's initiatives resulted in some Design Centre exhibitions or publications: *Rest Areas along Motorways* (1978), *Design and Equipment of Public Spaces in the City Center* (1980) and *Furnishing the City* (1982) (figs. 6.14–18). But the absolute zenith was the 1983 *Design and the State* exhibition. The show presented an overview of the Design Commission's achievements and was paired with an international symposium on the same subject (fig. 6.19). The prestigious event included lectures by Josine des Cressonnières, Jens Nielsen from Denmark, Jean François Grunfeld from France and two of the leading ICSID figures, Loek van der Sande from the Netherlands and Kenji Ekuan from Japan.[83]

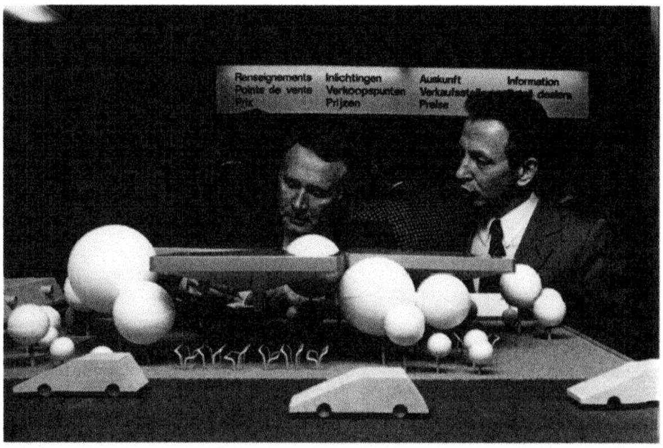

Fig. 6.15. Robert De Paepe (left) at the opening of the *Rest Areas along Motorways* exhibition (1978). Source: Brussels, State Archives, BE-A0545 / Fic 2000, Archive Design Centre, 220.

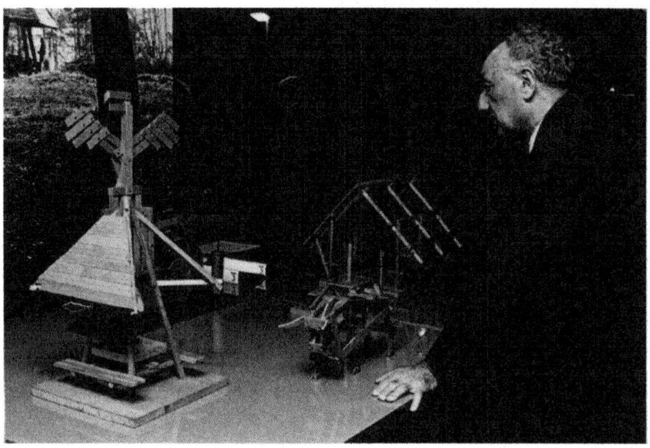

Fig. 6.16. View of the *Rest Areas along Motorways* exhibition (1978). Source: Brussels, State Archives, BE-A0545 / Fic 2000, Archive Design Centre, 220.

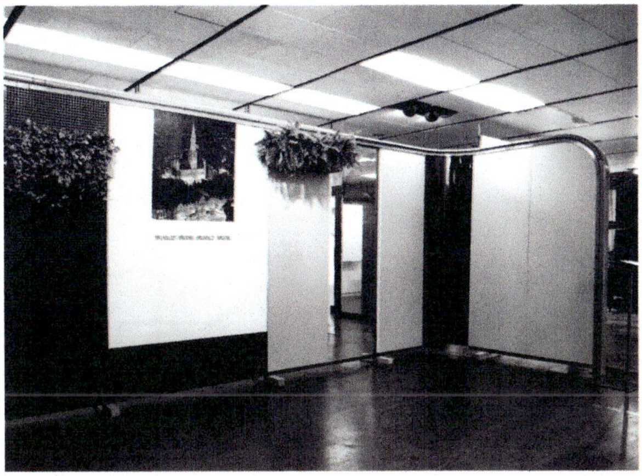

Fig. 6.17. View of the *Furnishing the City* exhibition (1982). On show were acoustic panels and other urban furniture designed by Bruno Stoppa. Source: Brussels, State Archives, BE-A0545 / Fic 2000, Archive Design Centre, 251.

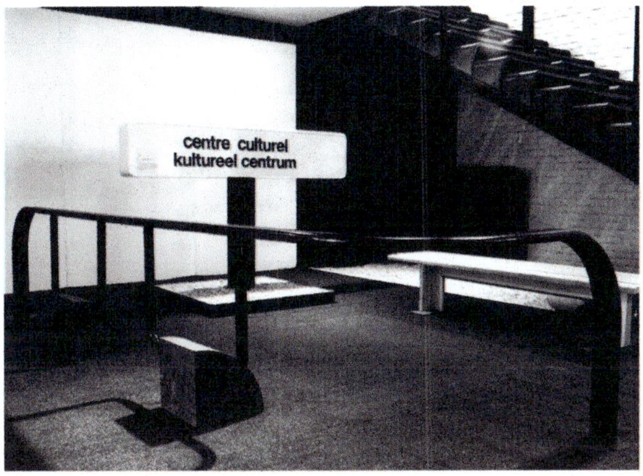

Fig. 6.18. View of the *Furnishing the City* exhibition (1982). Source: Brussels, State Archives, BE-A0545 / Fic 2000, Archive Design Centre, 251.

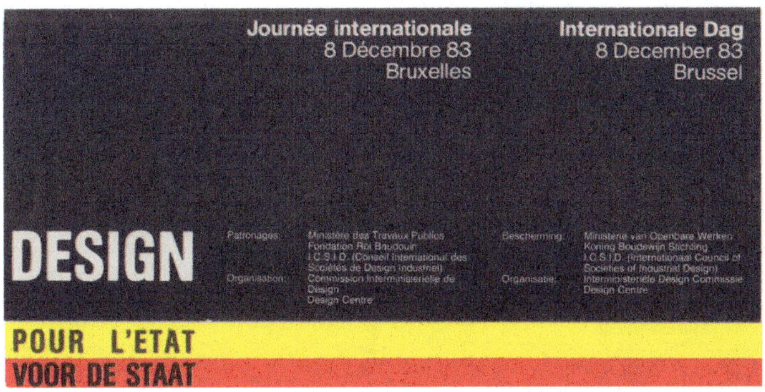

Fig. 6.19. Brochure of the international symposium organized by the Design Centre on the theme 'Design for the State', 1983. Source: Rijksarchief Antwerpen – Beveren.

Each speaker presented a national case of "design for the state" for comparison. The exhibition itself presented pilot projects of urban furniture, school furniture, graphic design for public transport, and other objects developed by designers for public services on all levels in the federalizing Belgian state (figs. 6.20–23).[84]

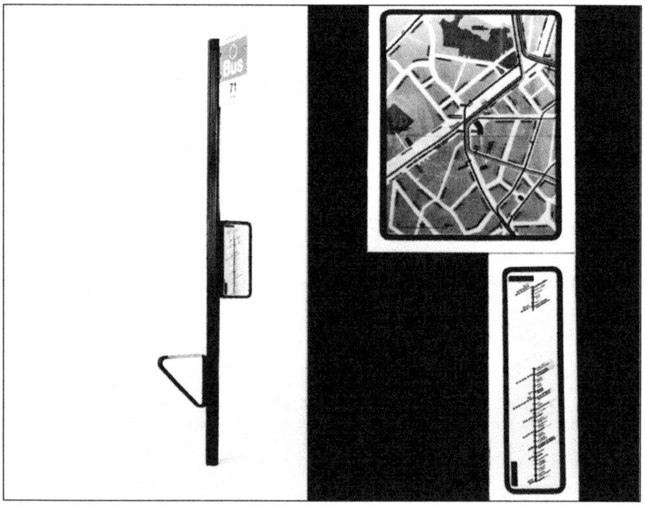

Fig. 6.20. Michel Olyff, Herman Lampaert and Jacques Tilman, pilot project of a "Passenger information system at public transport stops", commissioned by the Ministry of Transport and Public Works through the Belgian Design Centre and the Inter-ministerial Design Commission, 1983. Source: Brussels, State Archives, BE-A0545 / Fic 2000, Archive Design Centre, 263.

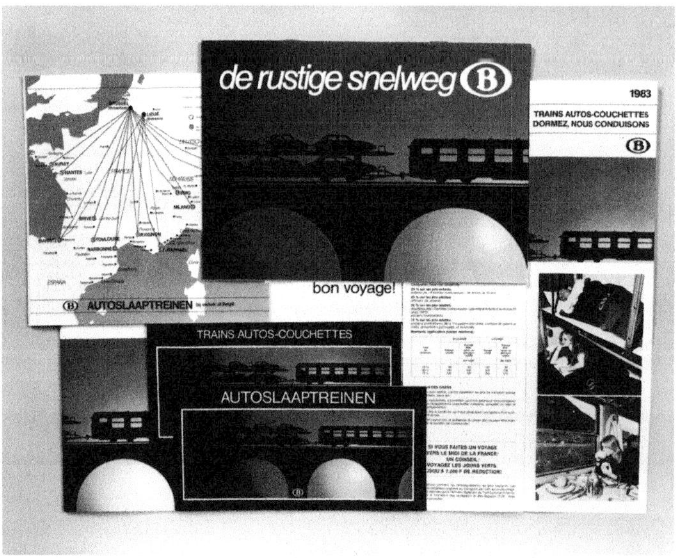

Fig. 6.21. Julian Key, poster design for the Ministry of Traffic and the NMVB, 1983. Source: Brussels, State Archives, BE-A0545 / Fic 2000, Archive Design Centre, 263.

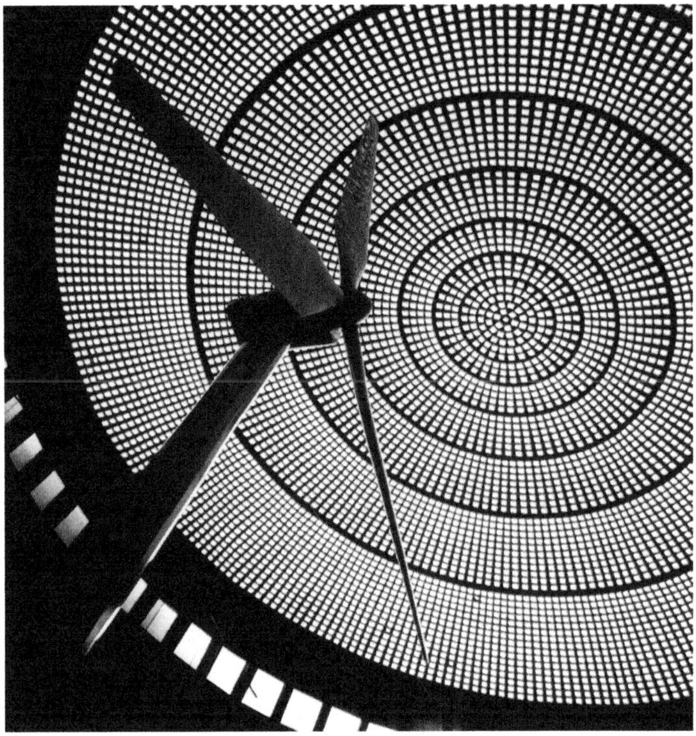

Fig. 6.22. *Windmaster* wind turbine designed and manufactured by HMZ (Sint-Truiden) and ordered by the Department of Public Traffic, on display in the Ravenstein Gallery, 1983. Source: Brussels, State Archives, BE-A0545 / Fic 2000, Archive Design Centre, 427.

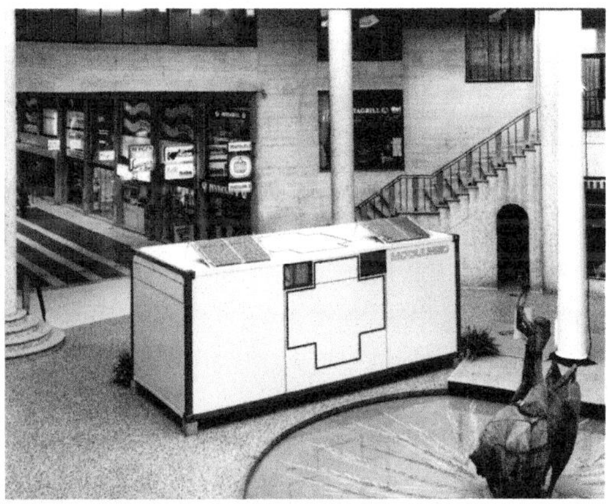

Fig. 6.23. The *Modul-Med* on display in the Ravenstein Gallery, 1983. The *Modul-Med* was designed by Bruno Stoppa and houses a dispensary in a movable container. Source: Brussels, State Archives, BE-A0545 / Fic 2000, Archive Design Centre, 263.

5. CONCLUSION: THE PERSISTENCE OF THE BELGIAN DESIGN LABEL

With the death of Josine des Cressonnières in October 1985, the Design Centre lost one of its main driving forces. It had to close its doors in 1986 after twenty-two intense years of promoting "Belgian" design. The vacuum left behind was filled only in the 1990s by regional institutions such as Design Vlaanderen (now Flanders DC), Wallonie-Bruxelles, and recently also Mode and Design Centre MAD Brussels. Notwithstanding federalization, which turned design promotion into a regional affair, the "Belgian design" label was not thrown away. On the contrary, it is still currently in use and quite popular, even with regional institutions. Conclusive proof is the "Belgium is design" label under which the three current regional design bodies and their respective export promoting organizations work together on crucial design events such as Milan Design Week. The website describes it as follows:

> "Belgium is design" is a declaration [...] highlighting the unique characteristics of Belgium's creative production. A joint declaration between the three different regions in the country, it recognizes that Brussels, Flanders and Wallonia are host to various talented designers that transcend cultural, linguistic and sociological differences to form a unique style.[85]

Thus, the label of "Belgian" design has not met with the same tragic fate as the Design Centre. On the contrary, it seems to be alive and kicking. However, the confined and limited "Belgian" label crafted by the Design Centre as rational, scientific and industrial was not a lasting hit, neither was the strategy to establish a distinct state policy. Rather, the flexible vagueness of the "Belgian" label and the old recurrent mantras of the puzzling inability of "Belgian" design to adhere to a strong national identity, have proved to be the persistent tropes in design discourses to this day.[86] In 2013, the famous British design critic Justin McGuirk wrote a piece in *Domus* on "The Belgian Affair" which tellingly articulates the ingrained paradoxes of this Belgian cliché: "Can such a thing as 'Belgian design' exist, in a country where the lack of a sharply defined common identity is arguably an identity in itself?"[87]

Conclusion

The great postwar narrative of design in Belgium ends in the turbulent year of 1986. The Chernobyl disaster caused great shockwaves all over the world. Sandra Kim was singing the high notes of her winning song "J'aime la vie" in the Eurovision Song Contest, and the Belgian economy was tumbling into recession. Moreover, that year the Design Centre closed its doors after the death of its driving force, Josine des Cressonnières. It would take about a decade before the regional institutions would pick up the thread of design promotion again. But what was so significant about design in this period from 1950 until 1986?

Design was a maverick in the cultural and political field of postwar Belgium. This, in the sense that it did not follow the lines of the strong ideological "pillars" (catholic, socialist and liberal) and the language conflicts between Dutch-speaking and French-speaking groups that dominated postwar Belgian society. While the discourses on social housing and home culture were expressed in that period along the confines of the different ideological compartments, industrial design was hovering above this stiff grid and seemed to be above linguistic, political and gender categories. The new, rather exotic phenomenon seemed attractive enough for different stockholders to zero in on its discourse and promotion. The Institute for Industrial Design, the Signe d'Or award and the Design Centre could entice royalists, communists, Catholics, Liberals, Flemish nationalists, industrialists, artists and Walloon militants. Not that it was without political or ideological friction. In the case of the contested figure Henry van de Velde, the tensions between the capital and the regions, and the sporadic attacks on the "French" character of the Design Centre show that it was absolutely a politically charged matter. However, it was not inscribed in a clear ideological family or sociopolitical line but anchored in different backgrounds.

An equally curious feature is the outward orientation of design promotion in Belgium. The Belgian design bodies were not only national agents but also acted as inspired representatives of the international discipline of industrial design. The Brussels Design Centre, under its director Josine des Cressonnières, was one of the pivots of the International Council of Societies of Industrial Design (ICSID).

Unified by the universal message of modern design, this organization incorporated national design organizations from around the world. Even the Soviet counterpart joined this international community in the midst of Cold War tensions. This outward orientation is also notable in the Belgian efforts to frame and promote industrial design beyond the national. Indeed, the Belgian-funded bodies were characterized, curiously enough, by a supranational scope in the 1950s and early 1960s. The Institute for Industrial Design encompassed the territory of both Belgium and the Grand Duchy of Luxembourg. The Signe d'Or award in turn rewarded products not only from Belgium but also from the Netherlands and Luxembourg. In 1961 the Belgian design bodies would go even further by reuniting the design organizations of the European "six" into a "Liaison Committee." Although the plan was a damp squib, its aims indicate the geopolitical ambitions grafted onto design: to accommodate the design practice of the new European prospects and to promote a panorama of "European" design versus the American capitalist and Soviet communist designs.

Why, then, did industrial design achieve this singular position in postwar Belgium? First of all, industrial design acquired this collective momentum in Belgium in 1954 because of a shared interest in its commercial potential. That year the design critic Léon-Louis Sosset deemed that of all the outstanding questions of the day, industrial design was the most important. Indeed, it was in 1954 that La Cambre organized the first Belgian "industrial design" training, that the famous MoMA exhibition *American Design for Home and Decorative Use* came to Belgium, and that the De Coene timber merchant began its production of the renowned Knoll furniture. More importantly, however, 1954 was the year in which industrial design was endorsed by the Belgian government: Minister of Economic Affairs Jean Rey claimed "industrial design" as one of the spearheads of its policy. Later that same year he funded the first Belgian industrial design events in Liège and at the X Milan Triennial. This enthusiasm would eventually lead to the creation of two new design bodies in 1956: the Institute for Industrial Design and the Signe d'Or award. While prewar modernist efforts and their resurgence in the transition period of the early 1950s were largely infused with utopian ideals to improve everyday life, this official backing of design was driven by commercial motives. Especially in view of the European market, industrial design was considered an essential tool to face the new and feared competition. The philanthropic attitude was thus largely overshadowed after

1954 by an economic and political rationale. "Good" design became synonymous with good trade.

This commercial line was only strengthened when in 1964 a third important design institute, the Brussels Design Centre, opened its doors. Under its legendary director, Josine des Cressonnières, it concentrated on the promotion of industrial design inspired by the Ulm School of Design and its rector, Tomas Maldonado. The modernism of Ulm was characterized by a hardcore rational and scientific approach and allocated a central position to the industry. Capital goods, medical equipment and professional tools became the new favorites. The Design Centre even fit a part of its exhibition space with a concrete floor to be able to display heavy machinery.

Secondly, industrial design was considered an obvious hallmark of modernity. Design became immensely important, especially in the Cold War between the United States and the Soviet Union, as a marker for modern and advanced nations at world's fairs and international exhibitions. Think of the legendary 1958 Brussels World's Fair, where nations were measured by their kitchens, televisions, teapots and chairs. As such, the Belgian government felt the need to catch up with its neighboring countries and to prepare for the future vicious showdown on the increasingly liberalized international markets with the help of this new modern phenomenon. Everyone talked about French taste, German precision and Italian fashion, but Belgium appeared to be associated only with beer. Reaching out to industrial design had to tackle this shortfall and provide Belgian industry with a positive and more sophisticated image in order to boost its exports. Thus, to connect with its modern counterparts, "Belgian design" was declared by means of an international recognizable modernist language. Through a strict process of selection based mainly on the modernist principles of "good" design, certain groups of objects were endorsed in the postwar period by the Belgian State and were subsequently staged at exhibitions and trade fairs, and published in high-profile catalogs and brochures. These objects then gave shape to an idealized notion of Belgium as a modern nation.

Thirdly and finally, this unifying character of industrial design in the postwar period should also be attributed to the singular personality of Josine des Cressonnières. She was not only key to the institutionalization of Belgian design, but also occupied a central position in the ICSID as its renowned secretary general for sixteen years, from 1961 until 1977. Her exceptional career and impressive transnational network — way beyond national borders and the ideological stratification of Belgian society — made her a sublime mediator for industrial design on a Belgian and international level. Because of her outstanding political and diplomatic talents, she was able to unite many a priori opposite contexts. A Brussels-born Catholic who was closely affiliated to the Belgian royal family, she knew, as one of the rare females in these male-dominated circles, the importance of establishing connections with Flemish nationalists like Pol Provost and Frans Wildiers, Belgian nobles like Count Boël and the d'Ursel family, and international designers like the Argentinian Tomas Maldonado and the Japanese Kenji Ekuan. With her good contacts, she even managed to get the Soviet design exhibition to Brussels in 1973. In this way, she played a leading role in advancing industrial design in a wide range of contexts: governments, international organizations, industrial magnates and large companies.

In addition, des Cressonnières' flexible, open and conciliatory attitude towards ideas and people of different backgrounds enabled her easily to absorb new tendencies in politics, economics and design. The emerging environmental concerns in the 1970s were, for example, immediately implemented in the as yet rather strict modernist discourse of the Design Centre. She had no problem combining the philosophy of Ulm with ideas from the new counterculture and the focus on the urban environment. In the mid-1970s, the program of the Design Centre even featured the economy-driven "good" design next to critical and activist exhibitions on the human environment. Neither did Belgium's federalization process throw a spanner in her mission. The newly created governments were seen rather as excellent clients for design since the public commissions of graphic identities, public transport and urban furniture were considered important design opportunities.

The framing of design policies in Belgium was governed, in that regard, not so much by specific ideological convictions as by the general excitement about design's potential to turn Belgium into a modern and prosperous nation which was considered indispensable while awaiting a globalized postwar world. As Josine des Cressonnières phrased it in 1964: "the old regional values will inevitably dissolve in a civilization which is becoming planetary, and where folklore will be replaced by man, not by man alone, but man in a team, in a firm, in a design studio." [1]

Notes

Acknowledgements

1 I defended my PhD thesis at the University of Antwerp on 16 June 2016 obtaining the degree of Doctor of Interior Architecture. The assessment commission included Prof. dr. Rajesh Heynickx, Prof. dr. ir.arch. Els De Vos, Prof. dr. Javier Gimeno-Martínez, Prof. dr. ir.arch. Fredie Floré, Prof. dr. Harriet Atkinson and Prof. dr. Henk de Smaele. See: Katarina Serulus, "Design & Politics. The Public Promotion of Industrial Design in Postwar Belgium (1950–1986)" (Unpublished PhD Dissertation, University of Antwerp, 2016).

Introduction

1 Jan-Albert Goris, ed. *Belgian American Survey 1958* (New York: Belgian Chamber of Commerce, 1958), 122.

2 The novelist John Gardner in his first James Bond novel, *Licence Renewed* (1981), replaced Bond's former gun with a Belgian Browning pistol because it was more reliable. The FN pistol is shown in the original hardback cover illustration, http://jamesbond.wikia.com/wiki/List_of_firearms.

3 On the occasion of the 20th birthday of the Brussels Design Centre in 1984, the Royal Museums of Fine Arts of Belgium established a permanent section on design. After the Design Centre closed its doors in 1986, this section faded away. See Jo Haerens and Josine des Cressonnières, "Design in België 1940–1984," ed. Museum voor Moderne Kunst van Brussel (Brussel: Koninklijke Musea voor Schone Kunsten van België, 1984).

4 Javier Gimeno-Martínez, "The Signe d'Or Award Scheme from 1956 to 1960: The Economic Reasons for "Good Design"," *Konsthistorisk Tidskrift/Journal of Art History* 79, no. 3 (2010): 127; Jonathan Woodham, *Twentieth-Century Design* (Oxford/New York: Oxford University Press, 1997), 171–77; "Formulating National Design Policies in the United States: Recycling the "Emperor's New Clothes"?," *Design Issues* 26, no. 2 (2010); Penny Sparke, *An Introduction to Design and Culture. 1900 to the Present*, 2 ed. (London/New York: Routledge, 2004), 198–206.

5 Rik Coolsaet, *België en zijn buitenlandse politiek 1830–2000*, 3rd ed. (Leuven: Van Halewyck, 2001), 449–70; Els Witte, Jan Craeybeckx, and Alain Meynen, *Politieke geschiedenis van België. Van 1830 tot heden* (Antwerpen: Standaard, 2005).

Revue Bimestrielle/Infordesign. Tweemaandelijk Tijdschrift, (1979-1972), Design News. Zeswekelijks tijdschrift/Design News. Périodique paraissant toutes les six semaines (1975-1986).

16 Some accounts do mention the Belgian design bodies such as the Signe d'Or award scheme, the Institute for Industrial Design and the Design Centre, but only casually and without delving deeper into the topic. See for example: Monique Bucqoye, Lieven Daenens and Mil De Kooning, *Forms from Flanders. From Henry van de Velde to Maarten Van Severen 1900-2000* (Gent/Amsterdam: Ludion, 2001); Lise Coirier, *Design en Belgique/in Belgium/in België 1945-2000* (Brussels: Racine, 2004).

17 See: Marie Ferran-Wabbes, *Jules Wabbes. Architecte d'intérieur 1919-1974* (Tournai: La renaissance du livre/Dexia Banque, 2002); Iwan Strauven, Fredie Floré and Ferran-Wabbes Marie, *Jules Wabbes. Furniture Designer* (Brussels: A+ editions/Bozar Books, 2012); Pierre Loze et al., *Christophe Gevers Designer* (Brussels: Editions d'art Laconti, 2008); Mil De Kooning, *Willy Van der Meeren*, vol. 21-24 (Damme: Laat-XXe-eeuws genootschap, 1993); Gillo Dorfles, Christian Kieckens and Eva Storgaard, «Pieter De Bruyne 1931-1987: Pionier van het Postmoderne,» ed. Design Museum Gent (Brussels: ASP, 2012). However, many important Belgian designers are still neglected in the specialist literature, to name but a few: Philippe Neerman, Michelle Olyff, Claude Blondel, Charles Déthier, Bob Daenen and many others.

18 For example the Belgian firm De Coene: Frank Herman, Terenja Van Dijk, and Marc Dubois, *Kortrijkse Kunstwerkstede Gebroeders De Coene. 80 jaar ambacht en industrie: meubelen, interieurs, architectuur* (Kortrijk: Groeninghe, 2006); Fredie Floré, "Architect-Designed Interiors for a Culturally Progressive Upper-Middle Class: The Implicit Political Presence of Knoll International in Belgium," in *Atomic Dwelling. Anxiety, Domesticity, and Postwar Architecture*, ed. Robin Schuldenfrei (London and New York: Routledge, 2012); Fredie Floré and Rika Devos, "Modern Wood: De Coene at Expo 58," *Construction History* 24(2009).

19 Steven Jacobs, *Henry van de Velde. Wonen als een kunstwerk. Een woonplaats voor kunst*, 2nd ed. (Leuven: Van Halewyck, 2005); Fredie Floré, *Lessen in goed wonen. Woonvoorlichting in België 1945-1958* (Leuven: Universitaire Pers Leuven, 2010); Els De Vos, *Hoe zouden we graag wonen? Woonvertogen in Vlaanderen tijdens de jaren zestig en zeventig* (Leuven: Universitaire Pers Leuven, 2012); Katrien Theunis, "De zoektocht naar een Belgisch woonproject, 1965-1975: Toenaderingen tussen ontwerpers en overheid in de praktijk van het private wonen" (K.U.Leuven, 2008); Hilde Bouchez, "'I Don't Care What It Is for, I Want It!': The Flow of Meaning of 'High Design' in Flanders from 1980-2010, through the Dynamics between Producer, Media and Consumer" (KU Leuven, 2013); Bucqoye, Daenens, and De Kooning, *Forms from Flanders. From Henry van de Velde to Maarten Van Severen 1900-2000*; Frans Defour, *Belgische meubelkunst in de XXe*

Eeuw. Van Horta tot heden (Tielt: Lannoo, 1979); Leblanc, *Art Nouveau & Design. Sierkunst van 1830 tot Expo 58.*

20 Mil De Kooning, Fredie Floré and Iwan Strauven, *Hedendaags design. Alfred Hendrickx en het fifties-meubel in België* (Mechelen: Stedelijke Musea, 2000).

21 Floré, "Wabbes and the Milan Triennales: Representing the Belgian Nation;" Javier Gimeno-Martínez, "The Introduction and Dissemination of the English Word 'Design' in the Belgian Context," in *Words for Design. Comparative Etymology and Terminology of Design and Its Equivalents. The First Step Towards a Full Collection of Words for Design from All over the World*, ed. Haruhiko Fujita (Japan Society for the Promotion of Science, 2009); "The Signe d'Or Award Scheme from 1956 to 1960: The Economic Reasons for "Good Design"; "Industrial Design in the Museum. The Case of the FN Milking Machine (ca. 1947)," *The Burlington Magazine* 152, no. 1290 (2010); "Redefining Social Design in 1970s Belgium: Affordable Design vs. Elite Design," *Interiors* 2, no. 2 (2011).

22 "The Role of the Creative Industries in the Construction of Regional/European Identities (1975–2002)" (Katholieke Universiteit Leuven, 2006).

23 See for example the recent special issue of *The Radical Designist* on design policies, edited by Helena Souto with an interesting contribution by Anna Calvera: Souto, "Editorial. Design Policies: Between Dictatorship and Resistance. Special Issue of the Radical Designist; Anna Calvera, "From the East Side of the Iberian Peninsula: Meanings of Design When It Arrived During the 50s and Early 60s to Barcelona," ibid.

24 Sparke, *An Introduction to Design and Culture. 1900 to the Present*; Widar Halén and Kerstin Wickman, eds., *Scandinavian Design Beyond the Myth. Fifty Years of Design from the Nordic Countries* (Stockholm: Arvinius Förlag/Form Förlag, 2003); Tony Fry, *Design History. Australia. A Source Text in Methods and Resources* (Sydney: Hale & Iremonger, 1988); Penny Sparke, *Japanese Design* (New York: Museum of Modern Art, 2009); Jeremy Aynsley, *Nationalism and Internationalism. Design in the 20th Century* (London: Victoria & Albert Museum, 1993); Frederike Huygen, *Brits design, imago en identiteit* (London/Rotterdam: Thames and Hudson/Museum Boymans-Van Beuningen, 1989); Woodham, *Twentieth-Century Design*; Greg Castillo, "Marshall Plan Modernisme in Divided Germany," in *Cold War Modern: Design 1945–1970*, ed. David Crowley and Jane Pavitt (London: V&A Publishing, 2008); *Cold War on the Home Front: The Soft Power of Midcentury Design*; Gay McDonald, "The 'Advance' of American Postwar Design in Europe: MoMA and the Design for Use, USA Exhibition 1951–1953," *Design Issues* 24, no. 2 (2008); "The Modern American Home as Soft Power: Finland, MoMA and the 'American Home 1953' Exhibition," *Journal of Design History* 23, no. 4 (2010); Paul Betts, *The Authority of Everyday Objects: A Cultural History of West German Industrial Design* (Berkeley: University of California Press, 2004); Souto, Satar and Gomes, "Portuguese Design History (1960–1974). Shaping the Future:

The Role of the Art, Industrial Architecture and Industrial Design Nuclei of the National Institute of Industrial Research (I.N.I.I.); Souto, "Editorial. Design Policies: Between Dictatorship and Resistance. Special Issue of the Radical Designist; Anna Calvera, "From the East Side of the Iberian Peninsula: Meanings of Design When It Arrived During the 50s and Early 60s to Barcelona," ibid; James-Chakraborty, *Bauhaus Culture. From Weimar to the Cold War*; Harriet Atkinson, *The Festival of Britain. A Land and Its People* (London/New York: I.B. Tauris, 2012).

25 Patrick Maguire and Jonathan Woodham, eds., *Design and Cultural Politics. The Britain Can Make It Exhibition of 1946* (London: Leicester University Press, 1997); Catherine Moriarty, "A Backroom Service? The Photographic Library of the Council of Industrial Design, 1945–1965," *Journal of Design History* 13, no. 1 (2000); Michelle Jones, "Design and the Domestic Persuader. Television and the British Broadcasting," ibid. 16, no. 4 (2003). Artemis Yagou, "Unwanted Innovation: The Athens Design Centre (1961–1963)," ibid. 18, no. 3 (2005); Kjetil Fallan, "How an Excavator Got Aesthetic Pretentions - Negotiating Design in 1960s' Norway," ibid. 20, no. 1 (2007); Christopher Thompson, "Modernizing for Trade: Institutionalizing Design Promotion in New Zealand," ibid. 24, no. 3 (2011). Ozorio de Almeida Meroz, Joana, "Transnational Material Politics. Constructions of Dutch Design, 1970–2012." (Unpublished PhD Dissertation, Vrije Universiteit Amsterdam, 2018).

26 Grace Lees-Maffei, "The Production-Consumption-Mediation Paradigm," ibid. 22, no. 4 (2009); Kjetil Fallan, *Design History. Understanding Theory and Method* (Oxford/New York: Berg, 2010).

27 "Our Common Future. Joining Forces for Histories of Sustainavle Design," *Technoscienza. Italian Journal of Science & Technology Studies* 2, no. 5 (2014): 17–18.

28 Fry, *Design History. Australia. A Source Text in Methods and Resources.*

29 John Walker, *Design History and the History of Design* (London: Pluto Press, 1989), 65–67.

30 Anna Calvera, "Local, Regional, National, Global and Feedback: Several Issues to Be Faced with Constructing Regional Narratives," *Journal of Design History* 18, no. 3 (2005): 376.

31 Guy Julier, "Re-Drawing the Geography of European Design: The Case of Transitional Countries", paper given at the 2nd European Academy of Design Conference, 'Contextual Design – Design in Context', Stockholm, April 1997. Available at: http://www.designculture.info/reviews/ArticleStash/GJReDrawing1997.pdf.

32 Maguire and Woodham, *Design and Cultural Politics. The Britain Can Make It Exhibition of 1946*, 14.

33 John Heskett, *Industrial Design* (London: Thames and Hudson, 1980), 182.

34 Castillo, "Marshall Plan Modernisme in Divided Germany; *Cold War on the Home Front: The Soft Power of Midcentury Design*; Crowley and Pavitt,

Cold War Modern: Design 1945–1970; Betts, *The Authority of Everyday Objects: A Cultural History of West German Industrial Design*; McDonald, "The 'Advance' of American Postwar Design in Europe: MoMA and the Design for Use, USA Exhibition 1951–1953; "The Modern American Home as Soft Power: Finland, MoMA and the 'American Home 1953' Exhibition; Betts, "The Bauhaus as Cold War Legend: West German Modernism Revisited; James-Chakraborty, *Bauhaus Culture. From Weimar to the Cold War*; Wharton, *Building the Cold War: Hilton International Hotels and Modern Architecture*; Eva Cockcroft, "Abstract Expressionism," *Artforum* 15, no. 10 (1974); Jan Loeffler, *The Architecture of Diplomacy: Building America's Embassies* (New York: Princeton Architectural Press, 1998); Harriet Atkinson and Verity Clarkson, "Editors' Introduction," *Design and Culture* 9, no. 2 (2017).

35 Alison Clarke, "Design for Development, ICSID and UNIDO: The Anthropological Turn in 1970s Design," *Journal of Design History* (2015).

36 Gavin Grindon, "Cruel Designs at Banksy's Dismaland," available at: http://gavingrindon.net/?p=246.

37 Paola Antonelli and Jamer Hunt, eds., *Design and Violence* (New York: The Museum of Modern Art, New York, 2015); Catherine Flood and Gavin Grindon, *Disobedient Objects* (London: V&A Publishing, 2014).

38 Kjetil Fallan, "Nordic Noir: Deadly Design from the Peacemongering Periphery," *Design and Culture* 7, no. 3 (2015).

39 Carlos Bártolo, "The Importance of Being 'Fascist', a Design History of 'Bad' People," in *PhD Workshop/DHS Day Symposium "Design History in Practice: Theory, Method, and Materials", organized by the Department of Philosophy, Classics, History of Art and Ideas of the University of Oslo and the Design History Society* (Oslo2015).

40 Ellen Van Impe and Eva Van Regenmortel, eds., *Vormgevingserfgoed in Vlaanderen na 1945* (Antwerp: Art Paper Editions & CVAa, 2016).

41 Calvera, "Local, Regional, National, Global and Feedback: Several Issues to Be Faced with Constructing Regional Narratives," 374.

42 Tony Fry, Clive Dilnot and Susan Stewart, *Design and the Question of History* (London/New York: Bloomsbury, 2015).

43 Calvera, "Local, Regional, National, Global and Feedback: Several Issues to Be Faced with Constructing Regional Narratives," 375.

44 Fallan, *Design History. Understanding Theory and Method*, X.

45 Ibid.

46 D.J. Huppatz, "Globalizing Design History and Global Design History," *Journal of Design History* 28, no. 2 (2015); Denise Whitehouse, "The State of Design History as a Discipline," in *Design Studies. A Reader*, ed. Hazel Clark and David Brody (Oxford/New York: Berg, 2009).

47 Victor Margolin, "A World History of Design and the History of the World," *Journal of Design History* 18, no. 3 (2005); Anna Calvera, "Local,

Regional, National, Global and Feedback: Several Issues to Be Faced with Constructing Regional Narratives," ibid; Jonathan Woodham, "Local, National and Global: Redrawing the Design Historical Map," ibid; Christopher Bailey, "The Global Future of Design History [Introduction]," ibid.

48 Victor Margolin, "A World History of Design and the History of the World," ibid.: 235.

49 *World History of Design*, vol. 3 (London: Bloomsbury Academic, 2015). Huppatz will critique Margolin's approach for being modelled after the historic narrativies McNeil and Braudel. Huppatz, "Globalizing Design History and Global Design History," 189.

50 Pat Kirkham and Susan Weber, eds., *History of Design. Decorative Arts and Material Culture, 1400–2000* (New York/New Haven and London: Bard Graduate Center: Decorative Arts, Design History, Material Culture/Yale University Press, 2013).

51 Glenn Adamson, Giorgio Riello, and Sarah Teasley, eds., *Global Design History* (London/New York: Routledge, 2011).

52 Guy Julier, *The Culture of Design*, 2 ed. (London: Sage, 2010).

53 Rod Hague and Martin Harrop, *Comparative Government and Politics*, 7th ed. (Hampshire and New York: Palgrave Macmillan, 2007); Robert Goodin, ed. *The Oxford Handbook of Political Science* (New York: Oxford University Press, 2009); David Marsh and Gerry Stoker, eds., *Theory and Methods in Political Science*, 3th ed., (Hampshire and New York: Palgrave Macmillan, 2010).

54 Peter-Paul Verbeek, "Politics at Issue. On Art and the Democratization of Things," *Open: Politics of Things. What Art & Design do in Democracy*, no. 24 (2012).

55 Paul Betts, "Science, Semiotics and Society: The Ulm Hochschule für Gestaltung in Retrospect," *Design Issues* 14, no. 2 (1998).

56 Greg Castillo, "The Bauhaus in Cold War Germany," in *Bauhaus Culture. From Weimar to the Cold War*, ed. Kathleen James-Chakraborty (Minneapolis/London: University of Minnesota Press, 2006).

57 Tony Fry, *Design as Politics* (Oxford/New York: Berg, 2011).

58 Ibid., 6.

59 Loeffler, *The Architecture of Diplomacy: Building America's Embassies.*

60 Michel Foucault and Paul Rabinow, "Space, Knowledge, and Power," in *Power*, ed. James Faubion, *The Essential Works of Foucault 1954–1984* (New York: The New Press, 2001).

61 Ibid., 349.

62 Ibid., 350.

63 Langdon Winner, *The Whale and the Reactor. A Search for Limits in an Age of High Technology* (Chicago and London: The University of Chicago Press, 1986).

64 James C. Scott, *Seeing Like a State. How Certain Schemes to Improve the Human Condition Have Failed* (New Haven and London: Yale University Press, 1998).

65 Tony Bennett and Patrick Joyce, eds., *Material Powers. Cultural Studies, History and the Material Turn* (New York: Routledge, 2010).

66 Joseph Nye, *Soft Power. The Means to Success in World Politics* (New York: PublicAffairs, 2004), 1–32; Castillo, *Cold War on the Home Front: The Soft Power of Midcentury Design*; McDonald, "The Modern American Home as Soft Power: Finland, MoMA and the 'American Home 1953' Exhibition."

67 This part is partly based on two papers I wrote and presented together with other scholars. The first one is the paper entitled "Design and the State. A Theoretical Account of the Role of the State in the Creation of Design Cultures" which I wrote and presented together with dr. Javier Gimeno-Martínez at the international symposium *Raison D'Etre 2: Crafting a Global History of Interior Decoration and Design* on March 9 and 10 in New Orleans (Louisiana, US). The symposium was organized by the *Journal of Interior Design* and *Interiors: Design, Architecture, Culture* in the framework of the annual conference of the Interior Design Educators' Council (IDEC). The second is the paper entitled "A Theoretical Straddle: Design Culture between National Structures and Transnational Networks" which I wrote and presented together with Joana Ozorio de Almeida Meroz (VU Amsterdam) at the *Design Culture: object, discipline and practice* conference, held from 18–19 September 2014 at the University of Southern Denmark, Kolding, Denmark. Some elements of this part were also mentioned in the paper "National Design Policies through a Transnational Lens: The case of the Belgian Design Centre (1964–1986)", paper presented on June 9 at the PhD Workshop/DHS Day Symposium entitled *Design History in Practice: Theory, Method, and Materials*, organized by the Department of Philosophy, Classics, History of Art and Ideas of the University of Oslo and the Design History Society, Oslo, Norway.

68 Betts, *The Authority of Everyday Objects: A Cultural History of West German Industrial Design.*

69 Gimeno-Martínez, "The Role of the Creative Industries in the Construction of Regional/European Identities (1975–2002)," 38.

70 Huppatz, "Globalizing Design History and Global Design History."

71 Pierre-Yves Saunier, "Transnational," in *The Palgrace Dictionary of Transnational History*, ed. Akira Iriye and Pierre-Yves Saunier (Basingstoke: Palgrave Macmillan, 2009); Anna Amelina et al., "Methodological Predicaments of Cross-Border Studies," in *Beyond*

Terminology of Design and Its Equivalents. The First Step Towards a Full Collection of Words for Design from All over the World, ed. Haruhiko Fujita (Japan Society for the Promotion of Science, 2009).

3 "En Belgique, les industriels ont généralement adapté leurs produits fabriqués à l'évolution des goûts, étant poussés à agir ainsi par la nécessité de conserver leur place sur les marchés extérieurs. L'aspect extérieur des produits fabriqués a donc retenu l'attention de nos industriels. De ce fait, la nécessité de créer une association du genre «Industrial Design» n'est pas jusqu'à présent apparue." See: Brussels, State Archives, BE AGR/ARA Prijzendienst (510–2331), 609: P. Van Lint [Secretary General of the Federation of Belgian Industries], Letter to Buyse Inspecteur Général de direction general de l'Industrie et de Commerce, (18/08/1950).

4 The original Dutch and French names of this institute were: *Instituut voor Industriële Vormgeving voor België en het Groothertogdom Luxemburg/Institut d'Esthétique industrielle pour la Belgique et le Grand-Duché de Luxembourg.* In this book we will use the English translation of these original names: Institute for Industrial Design for Belgium and the Grand Duchy of Luxembourg.

5 This institute was originally established in October 1956 under the name: *Société pour promouvoir la qualité dans la création industrielle/Vereniging ter bevordering der kwaliteit in de Nijverheidscreaties* (Foundation for Promoting Quality in Industrial Creation). At the general assembly of February 6, 1958, it was decided to change the name of the organization to *Signe d'Or/Gouden Kenteken* (Golden Signet) after the award scheme it organized from 1957 to 1960. In this book we will use its popular French name, Signe d'Or. See Beveren, State Archives, BKM collection (Archives of the Department of Fine Arts and Museums, Ministry of the Flemish Community/Transfer 2005 [1945–2004]): BKM 2013: Assemblée Générale de la "Société pour Promouvoir la Qualité dans la Création Industrielle", (6/02/1958); Javier Gimeno-Martínez, "The Signe d'Or Award Scheme from 1956 to 1960: The Economic Reasons for "Good Design"," *Konsthistorisk Tidskrift/Journal of Art History* 79, no. 3 (2010).

6 Fredie Floré, *Lessen in goed wonen. Woonvoorlichting in België 1945–1958* (Leuven: Universitaire Pers Leuven, 2010), 178–88; Mil De Kooning, Fredie Floré and Iwan Strauven, *Hedendaags design. Alfred Hendrickx en het fifties-meubel in België* (Mechelen: Stedelijke Musea, 2000), 16–27. The French name was used next to the Dutch name *Nieuwe Vormen.* In this book, we will use the French designation: Formes Nouvelles.

7 "Korte Nieuwsberichten. 'Nieuwe Vormen'," *Kunstambachten en Kunstnijverheden. Maandelijks Inlichtingsblad der Provinciale Commissies voor Kunstambachten en Nijverheidsbedrijven*, no. 5 (1950).

8 *Jaarboek Nieuwe Vormen 1955. Annuaire Formes Nouvelles 1955*, (Brussels: Formes Nouvelles/Nieuwe Vormen, 1955); Eric Lemesre, "Nieuwe Vormen," *Kunstambachten en Kunstnijverheden. Maandelijks Inlichtingsblad*

der Provinciale Commissies voor Kunstambachten en Nijverheidsbedrijven, no. 7 (1950).

9 For example, the Federation of Belgian Industries categorized Formes Nouvelles as a more artistic organization concerned about crafts. See Brussels, State Archives, BE AGR/ARA Prijzendienst (510-2331), 609: P. Van Lint [Secretary General of the Federation of Belgian Industries], Letter to Buyse Inspecteur Général de direction general de l'Industrie et de Commerce, (18/08/1950).

10 Leon-Louis Sosset, "1951 Design Congress," *Kunstambachten en Kunstnijverheden. Maandelijks Inlichtingsblad der Provinciale Commissies voor Kunstambachten en Nijverheidsbedrijven*, no. 23 (1951); Arthur J. Pulos, *The American Design Adventure, 1940-1975* (Cambridge: MIT Press, 1988), 209-11.

11 Phil Mertens and P. Roberts-Jones, "Naar een zuiver beelden. De eerste Belgische abstracten, 1918-1930," ed. Koninklijke Musea voor Schone Kunsten van België (Brussels: Koninklijke Musea voor Schone Kunsten van België, 1972); De Kooning, Floré and Strauven, *Hedendaags design. Alfred Hendrickx en het fifties-meubel in België*, 13, 67-77; Nell Andrew, "Living Art: Arakova and the Belgian Avant-Garde," *Art Journal* 68, no. 2 (2009).

12 An Paenhuysen, *De nieuwe wereld. De wonderjaren van de Belgische avant-garde (1918-1939)* (Antwerpen: Meulenhoff Manteau, 2010), 143-64, 271-73.

13 «Objet, Principes, Tactique,» *7 Arts*, no. 1 (1922).

14 Maurice Casteels, «Les Arts industriels,» ibid., no. 3; «Les Arts industriels,» *7 Arts*, no. 4 (1922); «Les Arts industriels.»

15 Marcel Baugniet, "Industrial Design," *La Maison* 10, no. 12 (1954)."; ""Schoner Wonen"," *Kunstambachten en Kunstnijverheden. Maandelijks Inlichtingsblad der Provinciale Commissies voor Kunstambachten en Nijverheidsbedrijven*, no. 5 (1950).

16 Floré, *Lessen in goed wonen. Woonvoorlichting in België 1945-1958*, 178-79.

17 Marcel Baugniet, *Essai sur la psychologie des formes* (Dilbeek (Brussels): Editions de la Maison du Poète, 1963).

18 "Un 'inventeur de structures', un 'ensemblier' qui ait de l'imagination et du talent, dont les réalisations ne soient ni formule stricte, ni fantaisie pure, gratuite, n'est-ce pas plus rare et plus précieux qu'un bon peintre." Paul Fierens, *Marcel Baugniet*, ed. Paul Fierens and Luc Haesaerts, La Gerbe D'or (Brussels/Paris: Editions Marion, 1942), 12.

19 Pierre-Louis Flouquet, «Situation des métiers d'art à Bruxelles. Interview de Marcel Baugniet, Administrateur-délégué de la coopérative «Form»,» *La Maison* 11, no. 9 (1955).

20 Léon-Louis Sosset was a critic of design, art and literature. He became Formes Nouvelles' president from 1956 until 1965 when the

non-profit organization was shut down. Concerning his activities as a design critic, he was the editor of the magazine *Kunstambachten en Kunstnijverheden. Maandelijks Inlichtingsblad der Provinciale Commissies voor Kunstambachten en Nijverheidsbedrijven* and wrote regularly on design in the *Les Beaux-Arts. Hebdomadaire d'Information Artistique, Guide Intérieur Gids, Architectue* and *Wonen.* See: Floré, *Lessen in goed wonen. Woonvoorlichting in België 1945–1958*, 179–80. In the early 1950s, he devoted many articles to industrial design, for example: Sosset, "1951 Design Congress; "Eindelijk Belgische verwezelijkingen in het seriemeubel. "Tubax" voert modellen uit van Willy Van der Meeren," *Kunstambachten en Kunstnijverheden. Maandelijks Inlichtingsblad der Provinciale Commissies voor Kunstambachten en Nijverheidsbedrijven*, no. 51 (1954); "Gaat de "Industrial Design" in België wortel schieten?," *Kunstambachten en Kunstnijverheden. Maandelijks Inlichtingsblad der Provinciale Commissies voor Kunstambachten en Nijverheidsbedrijven*, no. 53 (1954); "Gaat de 'Industrial Design' in België wortel schieten? II," *Kunstambachten en Kunstnijverheden. Maandelijks Inlichtingsblad der Provinciale Commissies voor Kunstambachten en Nijverheidsbedrijven*, no. 54 (1954); "Wij hebben aan lectuur gedaan," *Kunstambachten en Kunstnijverheden. Maandelijks Inlichtingsblad der Provinciale Commissies voor Kunstambachten en Nijverheidsbedrijven*, no. 55 (1954).

21 Mil De Kooning, *Willy Van der Meeren*, vol. 21–24 (Damme: Laat-XXe-eeuws genootschap, 1993); De Kooning, Floré, and Strauven, *Hedendaags design. Alfred Hendrickx en het fifties-meubel in België*, 85–95; Iwan Strauven and Maurice Culot, *De gebroeders Bourgeois. Architecture et plastique pure* (Brussels: AAM, 2005), 38.

22 In 1955 he compiled a theme issue on "Esthétique Industrielle" for the periodical *Architecture 55*. The issue (number 14) included articles by architect Albert Bontridder and graphic designer Corneille Hannoset.

23 Sosset, "Eindelijk Belgische verwezelijkingen in het seriemeubel. 'Tubax' voert modellen uit van Willy Van der Meeren."

24 Fredie Floré, "Sociaal modernisme. De designkritiek van K.-N. Elno (1920–1993)," *De Witte Raaf*, no. 89 (2001); "Elno, K.-N., Pseudoniem van Karel Horemans, architectuur-, design- en kunstcriticus," in *Nationaal biografisch woordenboek* (Brussels: Koninklijke Academiën van België, 2011); "Design Criticism and Social Responsibility: The Flemish Design Critic K.-N. Elno (1920–1993)," in *Writing Design: Words and Objects*, ed. Grace Lees-Maffei (Oxford: Berg, 2011).

25 The *L'Art Monumental dans les établissements publics et industriels* exhibition was held at the Palais de Beaux-Arts arts center from 8 until 29 October 1952. See Pierre Harmel and Lucien Christophe, «'Art et travail'. L'Art monumental dans les établissements publics et industriels,» ed. Palais des Beaux-Arts Bruxelles (Brussels: Editions de la Connaissance, 1952); Virginie Devillez, *Kunst aan de orde. Kunst en politiek in België 1918–1945* (Brussels/Gent: Dexia Bank/Snoeck, 2003), 117–18.

26 Harmel and Christophe, « 'Art et travail'. L'Art monumental dans les établissements publics et industriels,» 3.

27 Leon-Louis Sosset, «du 8 au 29 octobre, au Palais des Beaux-Arts, Exposition L'Art monumental dans les etablissements publics et industriels,» *Les Beaux-Arts. Hebdomadaire d'Information Artistique* 17, no. 580 (1952).

28 Exhibiting artists on show who were affiliated with the interwar organization Art and Industry were Charles Counhaye, Julien Van Vlasselaer, Elisabeth de Saedeleer, Floris Jespers, Joris Minne, Marc Rau and Rodolphe Strebelle.

29 Sosset, «du 8 au 29 octobre, au Palais des Beaux-Arts, Exposition L'Art Monumental dans les etablissements publics et industriels.»

30 Exhibiting artists affiliated with the interwar organization Art and Industry were Charles Counhaye, Julien Van Vlasselaer, Elisabeth de Saedeleer, Floris Jespers, Joris Minne, Marc Rau and Rodolphe Strebelle.

31 *Art et Industrie. Kunst en Nijverheid* (Brussels: Art et Industrie. Kunst en Nijverheid, 1938); Böröcz, "A Remarkable Continuity between 1930s Ideas and Reconstruction after the Second World War. New Stained-Glass Windows in War-Damaged Churches in the Diocese of Namur (Belgium)," in *Living with History* ed. Nicholas Bullock and Luc Verpoest (Leuven: Leuven University Press, 2011), 180; Devillez, *Kunst aan de orde. Kunst en politiek in België 1918–1945*, 95–6, 362.

32 The full executive committee and the board of directors, according to the 1938 Art and Industry brochure, was as follows:
Exectutive committee: President Louis Piérard (literary man), Vice-presidents Marcel de Fraipont (industrialist and director of the Christalleries Val Saint-Lambert) and Marcel Schmitz (architect), Secretary Marc-Antoine Pierson (advocat), Treasurer Arthur Devos (industrialist) and as members: Charles Counhaye (artist), Jan-Albert Goris (principal private secretary of the ministry of Economic Affairs), F. Parentani, Charles Stockmans, Julien Van Vlasselaer (interior architect).
Board of directors: Sadi Asiel (director Société linotype belge), Roger Avermaete (director of the School voor Kunstambachten in Antwerp), Victor Bourgeois (architect), Edmond Colin (director of the Faïenceries de Saint Ghislain), Florent Colpaert (artist), Charles Counhaye (artist), Robert Decerf (interior architect), Jean Delcouffre (trader), Jean De Ligne (architect), Marcel de Fraipont (industrialist and director of the Christalleries Val Saint-Lambert), Elisabeth de Saedeleer (Professor at the School of Architecture and Applied Arts La Cambre/Terkameren in Brussels), Arthur Devos (industrialist), Gaspar De Wit, Ernest Dumolin (industrialist), Paul Feron, Paul Fierens (Professor at the Liège University), Jean-Albert Goris, R. Gruslin, Alex Guillon, B. Helman, F. Jacques, Floris Jespers, Henri Kerels (industrialist), J. Leonard, G. Marlier, Ernest Melot (journalist), Joris Minne (Professor at the School of Architecture and Applied Arts La Cambre/Terkameren in Brussels), Jozef Muls, F.

Parentani, Paulus Paulus (artist), Louis Piérard (literary men), Marc-Antoine Pierson, Marc Rau, John Salt, Schmitz, Stockmans, Rodolphe Strebelle (artist), Marcel Wolfers (sculptor), Van Vlasselaer.

33 Strauven and Culot, *De gebroeders Bourgeois. Architecture et plastique pure*, 40.

34 As a reaction to his own observations in the report, Lucien Christophe launched a project to set up a permanent bureau that had to coordinate art and industry, after the German model of the *Werkbund*. See Virginie Devillez, "Een Belgische Werkbund? Het streven naar een nationale kunstnijverheid," in *Art Nouveau & Design. Sierkunst van 1830 tot Expo 58*, ed. Claire Leblanc (Tielt/Brussels: Lannoo/Racine, 2005); *Kunst aan de orde. Kunst en politiek in België 1918-1945*, 94.

35 Subsidies of 100.000 Belgian Francs were granted to this new organization. See *Kunst aan de orde. Kunst en politiek in België 1918-1945*, 362.

36 For more information on the work of Victor Bourgeois see Iwan Strauven, "Victor Bourgeois, 1897-1962: Radicaliteit en pragmatisme, moderniteit en traditie" (Université libre de Bruxelles, 2015). For more information on the artistic career of Elisabeth de Saedeleer see: Caterina Verdickt, "The Case of Elisabeth De Saedeleer (1902-1972): The Influence of Welsh Hospitality in the Great War on Belgian Modernist Interior Design," *The British Art Journal* 15, no. 3 (2015).

37 *Art et Industrie. Kunst en Nijverheid.*

38 "De Belgen mogen niet langer de sjouwers van Europa blijven die half en onafgewerkte producten op de markt brengen." See ibid., 5. On May 14, 1938 Louis Piérard would say something similar at the Cercle Artistique et Littéraire in Brussels: "On ne cesse de répéter que notre pays doit tendre de plus en plus vers une production de qualité. Nous ne pouvons rester les manoeuvres de l'Europe." See Louis Piérard and Adrien Van der Bruch, *Manifestation en l'honneur de M. Louis Piérard. Fondateur et Président de l'Association Art et Industrie. Organisée le 14 mai 1938 au Cercle Artistique et Littéraire Bruxelles* (Liège: Georges Thone, 1938), 24.

39 Devillez, *Kunst aan de orde. Kunst en politiek in België 1918-1945*, 363.

40 Ibid.

41 Pierre Harmel, *Culture et Profession*, «Batir». Collection d'études pour la restauration de la société chrétienne (Tournai/Paris: Casterman, 1944).

42 "Il faudrait que le travail soit aussi source de beauté. Que l'art s'inspire de la vigueur et de la grandeur du travail et que le travailleur s'ouvre à la beauté de l'art. Que l'art lui révèle non seulement la noblesse de sa vie, mais aussi les aspects les plus gratuits de la culture. C'est de l'agrément des travailleurs qu'il s'agit surtout, de leur initiation à l'art et à la culture." See «Art - Travail,» *Bulletin hebdomadaire d'information et de liaison edité par la federation des entreprises de l'industrie des fabrications métalliques*, no. 335 (1952): 950.

43 Ibid., 951.

44 Lionel Giraud-Mangin, «Lettre ouverte à M. Pierre Harmel Ministre de l'instruction publique et des beaux-arts,» *Les Beaux-Arts. Hebdomadaire d'Information Artistique* 17, no. 580 (1952).

45 "[...] il serait souhaitable de créer un organisme qui veillerait à l'entretien de la flamme et à ce que des vents mouvais ne l'inclinassent pas dans d'absurdes directions. La bonne volonté d'une grosse majorité de nos industriels doit être non seulement entretenue, mais guidée, si l'on ne veut pas qu'elle s'égare vers de médiocres et que les lieux de travail s'enlaidissent des équivalents de ces monuments aux morts qui foisonnent un peut partout en compagnie de bustes du roi Albert qui font injure à cet éminent monarque." Ibid.

46 The not-for-profit organization Art et Travail was founded on July 8, 1953. See «L'esthétique industrielle,» *Bulletin de la Foire internationale de Liège. Mines, métallurgie, mécanique, électricité, industrielle*, no. 1 (1953).

47 "Il s'agit à la fois d'embellir les lieux du travail et de rechercher la beauté dans les objets fabriqués." See Brussels, Archives & Musée de la Litérature (AML), Fonds Henry van de Velde, 339: Jean del Marmol, Letter to Henry van de Velde, (7/10/1953).

48 Jean del Marmol mentions in a letter to Henry van de Velde that it was the minister who asked him to become president: "Je vous ai peut-être dit un jour que le Ministre m'avait demandé de prendre la présidence d'une association qui s'appelle 'Art et Travail'. Il s'agit à la fois d'embellir les lieux du travail et de rechercher la beauté dans les objets fabriqués. J'ai naturellement demandé Mariamé parmi mes collaborateurs et avec lui, Stynen, Moutschen, Albert Dasnoy et l'un ou l'autre, nous avons formé un petit comité de travail." See Brussels, Archives & Musée de la Litérature (AML), Fonds Henry van de Velde, 339: Jean del Marmol, Letter to Henry van de Velde, (7/10/1953).

49 According to Roger Avermaete, Jean del Marmol was on the occasion of the 1937 Paris exhibition an active and enthusiastic member of the technical committee under the directorship of Henry van de Velde. After the Paris exhibition, in October 1937, the liberal Minister of Public Education Julius Hoste created a special commission under the direction of Marius Renard to discuss the relation between craft, artistic and technical education. Jean del Marmol was one of the twelve members of this commission together with C. Bernard, L. Christophe, E. De Bruyn, P. Fierens, J. Goris, R. Gruslin, P. Lambotte, F. Met den Ancxt and M. Rau. The commission was supported by G.A. Bogaerts. On the role of Jean del Marmol in the organization of the Belgian sections on world exhibitions in 1935, 1937 and 1939 see Devillez, *Kunst aan de orde. Kunst en politiek in België 1918–1945*, 72, 112, 364. Tanja Boon, "Ter Kameren (1927–1936)" (Katholieke Universiteit Leuven, 2007), 205. Roger Avermaete, "Kanttekeningen bij het werk van A.M. Hammacher 'De wereld van Henry van de Velde'," *Mededelingen van de Koninklijke Vlaamse Academie voor Wetenschappen, Letteren en Schone Kunsten van België* 29, no. 2 (1967): 11–12.

50 The correspondence between Jean del Marmol and Henry van de Velde mentions that their friendship started in 1935. See the letter: Brussels, Archives & Musée de la Littérature (AML), Fonds Henry van de Velde, 339: Jean del Marmol and Mary del Marmol, Letter to Henry van de Velde, (5/01/1936).

51 For the correspondence between Jean del Marmol and Henry van de Velde between 1936 and 1956 see Brussels, Archives & Musée de la Littérature (AML), Fonds Henry van de Velde, 339: J. del Marmol.

52 "Comme rien n'existe en Belgique depuis notre déconvenue due au départ de van Isacker (avez-vous lu ses mémoires qui viennent de paraître?) notre pays se trouve une fois de plus en retard." See Brussels, Archives & Musée de la Littérature (AML), Fonds Henry van de Velde, 339: Jean del Marmol, Letter to Henry van de Velde, (7/10/1953).

53 In 1953, the administrative board comprised Gaston de Formanoir de la Cazerie (president), Baron Léon Lambert (vice-president), Jean del Marmol (delegate administrator), Philippe van der Plancke (general secretary) and administrators Jacques Bemberg, Louis Camu, Camille Gutt, Guy de Lhoneux, Alain de Rothdschild and Louis Zurstrassen. See «Compagnie d'Afrique pour l'Industrie et la Finance. Bilan et compte de profits et pertes exercise 1952/1953,» (Brussels Weissenburch: Compagnie d'Afrique pour l'Industrie et la Finance., 1953). In 1956 the *Compagnie d'Afrique pour l'Industrie et de la Finance* merged with *Compagnie Européenne pour l'Industrie et la Finance* and the *Société Immobilière de la Chancellerie en Immobilière Lambert.* The new company that comprised these three was called the *Compagnie d'Outremer pour l'Industrie et la Finance.* This organization would later become the *Compagnie Bruxelles Lambert pour la Finance et l'Industrie* and eventually the *Groupe Bruxelles Lambert.* See Frans Buelens, *Congo 1885–1960. Een financieel-economische geschiedenis* (Berchem: epo, 2007), 563.

54 *Congo 1885–1960. Een financieel-economische geschiedenis*, 562.

55 Jean del Marmol married the daughter of Count Maurice Lippens, Mary Lippens, with whom he had three children.

56 At that time Jean del Marmol was on the board of many colonial firms affiliated with the Lambert concern. He was, among others, president of *Belgika, Bouteillerie de Léopoldville, Brasseries, Limonaderies et Malteries Africaines (Bralima)*; vice-president of *Compagnie du Kasaï*; delegate administrator of *Compagnie d'Outremer pour l'Industrie et la Finance;* administator of *Belgikamines, Société Minière du Lualaba; Compagnie Financière Africaine (the old Crégéco).* See Buelens, *Congo 1885–1960. Een financieel-economische geschiedenis*, 480.

57 R., "La 6e Foire Internationale. Ouverture des journées d'étude,» *La Dernière Heure (Ed. Liège)*, 4 mai 1954.
In a letter to Henry van de Velde, Jean del Marmol mentions that in his capacity of president he had invited some more people, such as Albert

Dasnoy and Moutschen, next to Stynen and Mariamé, to join the committee. However, these names are not mentioned again in letters or press reports. It thus remains unclear if they actually participated in Art et Travail. See the letter Brussels, Archives & Musée de la Litérature (AML), Fonds Henry van de Velde, 339: Jean del Marmol, Letter to Henry van de Velde, (7/10/1953).

58 «La 6e Foire Internationale. Ouverture des journées d'étude,» *La Dernière Heure Bruxelles*, 4 mai 1954; «L'esthétique industrielle au Musée De Gand,» *Le Peuple Bruxelles*, 30 juin 1954.

59 Jean del Marmol was involved in the organization of the 1939 world exhibition in New York, where Léon Stynen together with Victor Bourgeois and Henry van de Velde conceived the Belgian pavilion.

60 Albert Bontridder, *Gevecht met de rede.* Léon Stynen: Leven en werk (Antwerpen: Comité Léon Stynen, 1979), 25.

61 Delevoy and De Poerck were appointed to prepare a course of "L'esthétique industrielle" drawing on the concept of the Bauhaus and ideas of Van de Velde. In 1954, the section *Institut supérieur d'esthétique industrielle* was founded and subsequently headed by Robert Delevoy. See Eugénie De Keyser, «Delevoy, Robert,» in *Nouvelle biographie nationale* (Académie Royale des sciences, de lettres et des beaux-arts de Belgique, 2007); Léon Stynen, «Nécessité de l'enseignement de l'esthétique industrielle en Belgique. Intervention de l'architecte Léon Stynen au récent colloquium d'industrial design,» *Habitat et Habitation. Documents d'architecture et d'urbanisme* 14, no. Juilliet-Aout (1954); «Chronique. Institut supérieur d'esthétique industrielle,» *L'Ossature Métallique* 19, no. 11 (1954).

62 «Rapport général du Congrès international d'esthétique industrielle,» *Esthetique Industrielle*, no. 10/11 (1953): 200.

63 R. «La 6e Foire Internationale. Ouverture des journées d'étude.»

64 Pierre-Louis Flouquet, «Plaidoyer pour les formes nouvelles,» *La Maison* 9, no. 9 (1953); Eric Lemesre, «'Schoner Wonen',» *Kunstambachten en Kunstnijverheden. Maandelijks Inlichtingsblad der Provinciale Commissies voor Kunstambachten en Nijverheidsbedrijven*, no. 5 (1950); Eva Van Regenmortel, "Paul Mariamé," (2013).

65 In the context of the 1953 international Paris conference on industrial design, the French term "L'esthétique industrielle" was mostly used.

66 The lecture was given by Paul Lejeune during the plenary session on September 16, 1953.

67 Lejeune referred to this exhibition as the "présentation de 300 objets sélectionnés les plus représentatifs des dernières tendances de la forme industrielle aux Etats-Unis." The MoMA's *American Design for Home and Decorative Use* exhibition was composed of 300 objects and had opened a couple of months before as a travelling exhibition in Europe. Thus most likely Lejeune was alluding to this exhibition. For more information on

Syracuse University Libraries, Collection IDSA, Box 54, Folder "Liege Fair – Exhibits 1955": Letter from P.M. Lejeune (Manager of Foire Internationale de Liege) to Peter Müller-Munk, (22/10/1954).

77 Marc Reynebeau calls this cabinet under Achille Van Acker most unusual in the political history of Belgium: Marc Reynebeau, *Een geschiedenis van België* (Tielt: Lannoo, 2009), 216.

78 "Les organisateurs de la section Belge se sont inspirés de la première de ces propositions et ils l'ont en quelque sorte dépassée en choisissant l'"industrial design" comme thème particulier. Peut-être ceci appelle-t-il un commentaire. On ne manquera pas d'objecter d'abord que ce sont des considérations d'ordre commercial qui font se développer le movement de l'"industrial design"; que les industriels recourent à cette technique pour accroître leurs ventes et non par amour de l'Art; que, des lors, placer sous le signe de l'esthétique une méthode dont on apprécie l'efficacité à la lumière des statitiques de vente revient à desservir et à dénaturer l'objet du design." See Luc Van Bellinghen, «"Industrial Design" Belge à la dixième Triennale de Milan,» *Bulletin Commercial Belge* 11, no. 68 (1954).

79 Brussels, Archives & Musée de la Litérature (AML), Fonds Henry van de Velde, 211/15: Maurice Paquay, *Association Belge des Industrial Designers. Historique des débuts de l'association.*

80 Jonathan Woodham, *Twentieth-Century Design* (Oxford/New York: Oxford University Press, 1997), 171.

81 Frederike Huygen, "Vechten tegen de bierkaai? De promotie van industriële vormgeving via instituten en overheid," in *Holland in vorm: Vormgeving in Nederland 1945–1987*, ed. Geert Staal and Hester Wolters ('s-Gravenhage: Stichting Holland in Vorm, 1987).

82 Paul Betts, *The Authority of Everyday Objects: A Cultural History of West German Industrial Design* (Berkeley: University of California Press, 2004), 178–211.

83 Jocelyne Le Boeuf, "Jacques Viénot and the "Esthétique Industrielle" in France (1920–1960)," *Design Issues* 22, no. 1 (2006).

84 Ḥelena Maria Souto, Anjoom Satar and Márcia Gomes, "Portuguese Design History (1960–1974). Shaping the Future: The Role of the Art, Industrial Architecture and Industrial Design Nuclei of the National Institute of Industrial Research (I.N.I.I.)" (paper presented at the 9th International Committee Design History and Design Studies. Tradition, transition, trajectories: major or minor influences?, Aveiro, 2014); Helena Maria Souto, "Editorial. Design Policies: Between Dictatorship and Resistance. Special Issue of the Radical Designist," *The Radical Designist*, no. December (2015).

85 Gimeno-Martínez, "The Signe d'Or Award Scheme from 1956 to 1960: The Economic Reasons for "Good Design"; Souto, "Editorial. Design Policies: Between Dictatorship and Resistance. Special Issue of the Radical Designist; Kjetil Fallan, "How an Excavator Got Aesthetic

Pretentions - Negotiating Design in 1960s' Norway," *Journal of Design History* 20, no. 1 (2007); Artemis Yagou, "Unwanted Innovation. The Athens Design Centre (1961-1963)," ibid.18, no. 3 (2005).

86 Rik Coolsaet, *België en zijn buitenlandse politiek 1830–2000*, 3rd ed. (Leuven: Van Halewyck, 2001), 344–46, 66–69, 408; Reynebeau, *Een geschiedenis van België*; Els Witte, Jan Craeybeckx and Alain Meynen, *Politieke geschiedenis van België. Van 1830 tot heden* (Antwerpen: Standaard, 2005), 455–70.

87 Coolsaet, *België en zijn buitenlandse politiek 1830–2000*, 371.

88 "Quatre ans d'action économique," 81–87.

89 Michael Ryckewaert, *Building the Economic Backbone of the Belgian Welfare State. Infrastructure, Planning and Architecture 1945–1973* (Rotterdam: 010, 2011), 176–97.

90 The 6th International Liège Fair was held from April 24 until May 9, 1954. For press coverage on this fair see Brussels, Koninklijke Musea voor Schone Kunsten van België, Archief voor Hedendaagse Kunst in België, Fonds Léon-Louis Sosset, Caisse l'Esthétique industrielle: L. Heinen, *L'industrial design*, [ca. 1954]; "La Foire International de Liège et le Congrès scientific," *Le Bulletin du Grand Liège*, no. 14 (1954); Sosset, "Gaat de "Industrial Design" in België wortel schieten?; *Foire 6e Internationale de Liège: Mines, métallurgie, mécanique, électricité, 24 Avril-9 Mai 1954. Catalogue Officiel* (Liège1954); «Chronique. 6e Foire Internationale de Liège,» *L'Ossature Métallique* 19, no. 5 (1954); «Chronique. Colloquium d'»Industrial Design»,» *L'Ossature Métallique* 19, no. 5 (1954); *Foire 6e Internationale de Liège: Mines, métallurgie, mécanique, électricité, 24 Avril-9 Mai 1954. Catalogue Officiel.*

91 This show was the last visible manifestation of the *Art et Travail* organization. After 1954 *Art et Travail* is not mentioned anymore in accounts or documents. Only Emiel Langui (then director of Fine Arts at the Ministry of Public Affairs) hinted in a letter to Josine des Cressonnières (Secretary General of the Signe d'Or Industriel) about its unfortunate financial breakdown. According to Langui, its funding granted by the Ministry of Economic Affairs was blocked at the Department of Finances and finally cancelled. See Beveren, State Archives, BKM collection (Archives of the Department of Fine Arts and Museums, Ministry of the Flemish Community/Transfer 2005 [1945–2004]): BKM 2013: Letter from Emiel Langui to Josine des Cressonnières, (12/11/1957).

92 Robert Bosmans, «L'industrial design et la dixième Triennale de Milan,» *Bulletin Commercial Belge* 68, no. 11 (1954).

93 Lejeune, «La Foire Internationale de Liège a fait le point de l'esthétique industrielle.»

94 The exhibition was probably rather documentary since the official catalog describes it as follows: "selon des documents gracieusement envoys par la Society of Industrial Designers de New-York." The catalog also mentions

that the exhibition was designed by a certain C. Burton. See Syracuse, Special Collections Research Center, Syracuse University Libraries, Collection IDSA, Box 54, Folder "Liege Fair - Exhibits 1955": *Newsletter Supplement*, (01/07/1955); *Foire 6e Internationale de Liège: Mines, métallurgie, mécanique, électricité, 24 Avril-9 Mai 1954. Catalogue Officiel*, 481; «Chronique. Colloquium d'»Industrial Design».»

95 This exhibition also toured to Darmstadt in 1952 (Schönheit in Stahl), to Düsseldorf in 1953 (Schönheit in Stahl), to Liège in 1954 (La Beauté de l'Acier) and to Sao Paulo in 1954 (Belleza de Aço). See «Le beauté de l'acier,» *L'Ossature Métallique* 18, no. 5 (1953).

96 E. Greiner, «La béauté de l›acier,» *Bulletin de la Foire internationale de Liège. Mines, métallurgie, mécanique, électricité, industrielle*, no. 4 (1953).

97 The *Beauty of Steel* exhibition is only one illustration of the interest taken by C.B.L.I.A. in industrial design. Its magazine *L'Ossature Métallique* devotes attention to the issue from 1952 and in 1954 a special section is even reserved for industrial design. Moreover, its deputy-director, Jean-Jacques Thiry, was affiliated with *Formes Nouvelles* and was a member of the first industrial design institutes. See «Industrial Design. Villa Hügel,» *L'Ossature Métallique* 19, no. 12 (1954); «Le beauté de l'acier;» Robert L. Delevoy, «Actualité de l'industrial design,» ibid.19, no. 4 (1954); Herwin Schaefer, «Maison et mobilier de Charles Eames,» ibid.1è, no. 7–8 (1952); J.-J. T, «Industrial Design,» ibid.19, no. 6 (1954).

98 The frame was made by Paquay, designer at the Usines à Tubes de la Meuse, after the idea of Paul Mahlberg.

99 Delevoy, «Actualité de l'industrial design,» 174.

100 S. Sasson, «Exposition d'industrial design Foire Internationale de Liège,» *Architecture 54. Revue bimestrielle d'architecture et d'urbanisme*, no. 10 (1954); Lejeune, «La Foire Internationale de Liège a fait le point de l'esthétique industrielle.»

101 The conference was organized on May 3, 1954 under the protection of the Federation of Belgian Industries and by the Grand Liège organization, the federation for metal industries, Fabrimetal, the Belgo-Luxembourg Information Center of Steel and the Art et Travail organization. See Sosset, «Gaat de «Industrial Design» in België wortel schieten?»

102 An account of Stynen's lecture was published in *habitat et habitation*: Stynen, «Nécessité de l'enseignement de l'esthétique industrielle en Belgique. Intervention de l'architecte Léon Stynen au récent colloquium d'industrial design.» Stynen's lecture is also discussed in Javier Gimeno-Martinez's article: Gimeno-Martínez, "The Introduction and Dissemination of the English Word 'Design' in the Belgian Context."

103 Brussels, Archives & Musée de la Litérature (AML), Fonds Henry van de Velde, 211/15: Maurice Paquay, *Association Belge des Industrial Designers. Historique des débuts de l'association.*

104 Betts, *The Authority of Everyday Objects: A Cultural History of West German Industrial Design*, 186.

105 Fredie Floré, "Wabbes and the Milan Triennales: Representing the Belgian Nation," in *Jules Wabbes. Furniture Designer*, ed. Iwan Strauven and Marie Ferran-Wabbes (Brussels: A+ editions/Bozar Books, 2012); "The Belgian Participation in the Milan Triennials," in *Design Frontiers - Territories, Concepts, Technologies, 8th Conference of the International Committee of Design History and Design Studies*, ed. Farias Priscila et al. (Sao Paulo: Blucher, 2012); Fallan, «Annus Mirabilis: 1954, Alberto Rosselli and the Institutionalisation of Design Mediation.»

106 The section was organized by the General Commissioner J. Hamels (Inspecteur Général, Chef de l'Administration du Commerce au Ministère des Affaires Economiques), assisted by J. Geubel (Services des Expositions et Foires du Ministère des Affaires Economiques) and in collaboration with J. Van Lerberghe (délégué du Ministère de l'Instruction publique).

107 "Agissant en liaison avec l'Administration des Beaux-Arts, le Ministère des Affaires Economique a pris l'heureuse initiative d'orienter la participation de la Belgique à la prochaine Exposition Triennale de Milan (août-novembre) vers la forme utile de nature industrielle. Rompant ainsi avec une tradition qui l'avait amené à se cantonner dans le domaine de l'artisanat (et l'on sait combien furent et demeurent médiocres les expériences tentées à ce niveau) la Belgique fait, pour la première fois, un effort en vue de mettre en évidence la valeur, non seulement technique, mais aussi esthétique de nos produits industriels." See: "Industrial Design."

108 Fallan, "Annus Mirabilis: 1954, Alberto Rosselli and the Institutionalisation of Design Mediation."

109 A more precise description of the products on display: theatre projector (A. De Backer), camera for amateurs (Usines Gevaert), glasses (Usines du Val-Saint Lambert), table (Baucher-Feron), electricity transformers (Usines Balteau), optic crystals (Société d'Optique et d'Instruments de Précision), typography (Marc Séverin), guns (Fabrique Nationale d'Armes de Guerre), binocular microscope (Société d'Optique et d'Instruments de Précision), coffee and tea set by Oliver Strebelle (Cerabel), a portable post for industrial radiography Baltographe 200 (Usines Balteau).

110 The diplomas of honor were awarded to architect André De Poerck for the exhibition display, to De Coene for the laminated timber beams, and the bookbinder Beenkens. The two gold medals were won by Société d'Optique d'Instruments de Précision for its microscopes and Cristalleries du Val Saint-Lambert for its glasses. The silver medal went to Cerabel for Olivier Strebelle's tableware.

111 K.N. Elno, "Belgische pointering 1956," *Scheppend Ambacht* 7, no. 6 (1956); Leon-Louis Sosset, "De lessen der Xe Driejaarlijkse van Milaan," *Kunstambachten en Kunstnijverheden. Maandelijks Inlichtingsblad der*

123 Brussels, Archives & Musée de la Littérature (AML), Fonds Henry van de Velde, 211/14: Secretary Desamory and Vice-President Paquay, Letter to Henry van de Velde, (7/01/1956); Brussels, Archives & Musée de la Littérature (AML), Fonds Henry van de Velde, 211/14: Secretary Desamory and Vice-President Paquay, Letter to Henry van de «Velde, (28/01/1956).

124 Brussels, Archives & Musée de la Littérature (AML), Fonds Henry van de Velde, 211/15: Maurice Paquay, *Association Belge des Industrial Designers. Historique des débuts de l'association.*

125 The seventh edition of the Liège International Fair was held from April 23 until May 8, 1955.

126 Syracuse, Special Collections Research Center, Syracuse University Libraries, Collection IDSA, Box 54, Folder "Liege Fair - Exhibits 1955": *Note for Mr. Roy Williams re the organization of an exhibition of American industrial design as a section of the 7th. Liège International Fair in 1955*; Syracuse, Special Collections Research Center, Syracuse University Libraries, Collection IDSA, Box 54, Folder "Liege Fair - Exhibits 1955": *Newsletter Supplement*, (01/07/1955).

127 It was estimated that 60,000 people visited the American exhibition. See Syracuse, Special Collections Research Center, Syracuse University Libraries, Collection IDSA, Box 54, Folder "Liege Fair - Exhibits 1955": Letter from Peter Müller-Munk (President of the Society of Industrial Designers) to André De Poerck, (6/01/1955, dictated 4/01/1955); Syracuse, Special Collections Research Center, Syracuse University Libraries, Collection IDSA, Box 54, Folder "Liege Fair - Exhibits 1955": *Newsletter Supplement*, (01/07/1955).

128 Brussels, Archives & Musée de la Littérature (AML), Fonds Henry van de Velde, 211/14: Secretary Desamory and Vice-President Paquay, Letter to Henry van de Velde, (18/02/1956); Brussels, Archives & Musée de la Littérature (AML), Fonds Henry van de Velde, 211/15: Maurice Paquay, *Association Belge des Industrial Designers. Historique des débuts de l'association.*

129 N. 3356. Association belge des Industrial Designers, à Liége. Modification des statuts, Annexe au Moniteur Belge. Associations sans but lucratif et établissements d'utilité publique/Bijlage tot het Belgisch Staatsblad. Verenigingen zonder winstgevend doel en instellingen van openbaar nut [Attachment to the Belgian Official Journal. Non-profit Organizations and Public Utility Foundations], September 8, 1956, pp. 1365–1366. "Il faut aller vite, il faut voire grand. L'heure n'est plus aux initiatives restreintes et isolées." See Brussels, Archives & Musée de la Littérature (AML), Fonds Henry van de Velde, 211/14: *A.B.I.D. Programme d'activité pour l'année 1956, Premier exercise social de l'association. Raport présenté par le Secrétaire Général, M. L. Desamory à l'Assemblée Général extraordinaire du 27.2.1956*, (27/2/1956).

130 Note that while the two previous organizations chose the English designations "Industrial Design" and "Industrial Designers", the new institute opted for the French designation of "Esthétique Industrielle" or the Dutch designation "Industriële Vormgeving." The headquarters of this national institute were based in Liège. In 1957, its documentation center would be moved to the Montoyerstraat 47 in Brussels. In 1958, the head offices would follow. See "Het Instituut voor Industriële Vormgeving," *Berichten van het Instituut voor Industriële Vormgeving*, no. 1 (1957); "Secretariaat-Adresverandering," *Berichten van het Instituut voor Industriële Vormgeving*, no. 2 (1958).

131 This was the situation on May 8, 1956 as described by Maurice Paquay to Henry van de Velde. See Brussels, Archives & Musée de la Litérature (AML), Fonds Henry van de Velde, 211/15: Maurice Paquay, Letter to Henry van de Velde (8/5/1956). The constitution of the board changed quickly. At the beginning of the year 1957 the engineer Philippe Questienne (head of department of the Fabrique Nationale d'Armes de Guerre, Herstal) became the president. The rest of the board stayed the same. Only Fleerackers was no longer appointed as secretary and the publicity technician Victor Salme (Commercial director at the Publisher George Thone, Liége) became the treasurer. See "Het Instituut voor Industriële Vormgeving."

132 This part is large based on the findings of Javier Gimeno-Martínez: Gimeno-Martínez, "The Signe d'Or Award Scheme from 1956 to 1960: The Economic Reasons for "Good Design"."

133 N. 3919. Associations sans but lucratif et etablissements d'utilité publique. Annex au Moniteur Belge/Verenigingen zonder winstgevend doel en instellingen van openbaar nut. Bijlage tot het Belgisch Staatsblad [Attachment to the Belgian Official Journal. Not-for-profit Organizations and Public Utility Foundations], 2-3-4 November 1956, pp.1601–1602.

134 "L'association a pour but, en dehors de tout esprit de lucre, de promouvoir la fabrication et l'expansion commerciale d'objects usuels, produits en grande ou petite série, et répondant au triple critère de qualité, de prix adéquat et d'esthétique." See N. 3919. Associations sans but lucratif et etablissements d'utilité publique. Annex au Moniteur Belge/Verenigingen zonder winstgevend doel en instellingen van openbaar nut. Bijlage tot het Belgisch Staatsblad [Attachment to the Belgian Official Journal. Not-for-profit Organizations and Public Utility Foundations], November 2, 3 and 4,1956, pp.1601–1602.

135 *Art et Industrie. Kunst en Nijverheid.*

136 Paul Hassaert and Henri d'Ursel knew each other well. Together they founded the precursor the *Musée du cinéma de Bruxelles.*

137 Brussel, Koninklijke Musea voor Schone Kunsten van België, Het Archief voor Hedendaagse Kunst in België (AHKB), Fonds Haesaerts, 3.8. Architect, 3.8.1. Realisaties als Architect, Nr. 260.

138 He consulted on this issue Leon Stynen, then artistic advisor of the Belgian Railway Company.

139 Josine des Cressonnières - her maiden name was Naus - was married on November 13, 1951 with Denis des Cressonnières (1919–1952). Together with her husband she moved to the Belgian Congo. After his death on July 18, 1952, Josine stayed behind with a little daughter. In October 1953, she was appointed by the Belgian department store À *l'Innovation* to become Belgian's first "stylist." In the department store's magazine, she was called "La pionnière de la pêche aux nouveautés", which means as much as the pioneer in the search of novelties. Her contributions as a stylist were very successful and soon a second stylist was hired. In 1956, with the erection of the Signe d'Or, she quitted this job. Interview by the author with Geraldine des Cressonnières on November 8, 2013, Coutisse; «La pionnière de la pêche aux nouveautés,» *Entre Nous. Le Magazin de l'Inno* 32(1954).

140 "Toen ik daar werkte - waarbij ik overigens veel heb geleerd - kwam ik eigenlijk tot de ontdekking van mijn tijd. Ik werd naar heel wat landen gezonden, op zoek naar nieuwe strekkingen in alles wat de huishouding betreft. Met verontrusting ben ik tot de vaststelling gekomen van een gemis aan smaak, durf en creativiteit bij onze eigen industrie. [...] Deze ervaring heeft mijn interesse opgewekt en ik was zó ontdaan over deze toestand, dat ik me gedwongen voelde er iets aan te gaan doen." See «Mevrouw des Cressonnières, wie bent U?,» *Distributie*, no. juni-juli (1977): 15.

141 Sosset, «Aux Beaux-Arts, Exposition des oeuvres sélectionnées pour «Le Signe d'Or" label de qualité pour le style industriel.»

Chapter II: Reconfiguring Henry van de Velde (1956–1964)

1 The third edition of the *Industrial Design Show* (Salon d'Industrial Design) was held from April 28 until May 13, 1956. On this occasion, the institute organized an international conference on industrial design on May 1956. Prominent international guests gave the event an international luster: Egidio Bonfante (director of publicity at Olivetti) lectured on *Le facteur social de l'Industrial Design*, J. Penraat (secretary of the Kring voor Industriële Ontwerpers in Nederland) lectured on *Het standpunt van de Nederlandse ontwerper*, Peter Muller-Münk lectured on *De industriële vormgeving in de Verenigde Staten* and René Smeets (director of the design academy at Eindhoven) lectured on *l'enseignement de l'esthétique industrielle à la Dagschool voor Industriële Vormgeving* in *Eindhoven*. For more information on the *Industrial Design Show* see Brussels, Archives & Musée de la Litérature (AML), Fonds Henry van de Velde, 488: Maurice Paquay, Letter to Henry van de Velde, (8/5/1956); "La VIIIe Foire Internationale de Liège," *Acier - Stahl - Steel* 21, no. 6 (1956); Lucien Kroll and Charles Vandenhove, "Mobilier V K," ibid., no. 10; "Les Meubles VK," *La*

Maison 11, no. 6 (1955); "Esthétique Industrielle. Architecte L. Kroll et C. Vandenhove,» *Habitat et Habitation. Documents d'architecture et d'urbanisme* 16, no. 8 (1956); Pierre-Louis Flouquet, «A la Foire Commerciale de Liège 1956. Le 3me Salon de l'Industrial Design,» *La Maison* 12, no. 6 (1956); «Le Salon d'esthétique industrielle Liège 1956,» *La Maison* 12, no. 7 (1956). "Studiedag op de internationale jaarbeurs van Luik 1956," *Berichten van het Instituut voor Industriële Vormgeving* 1, no. 1 (1957).

2 Both architects first studied architecture at the Institut Saint-Luc in Liège and then at La Cambre. For more information on Lucien Kroll and Charles Vandenhove see Geert Bekaert and Francis Strauven, *Bouwen in België 1945-1970* (Brussels: Nationale Confederatie van het Bouwbedrijf, 1971), 328; Geert Bekaert, *Hedendaagse architectuur* (Tielt: Lannoo, 1995), 95-108; Mil De Kooning, Fredie Floré, and Iwan Strauven, *Hedendaags design. Alfred Hendrickx en het fifties-meubel in België* (Mechelen: Stedelijke Musea, 2000), 136-44.

3 This international jury was advised by Paul Mahlberg of Industrieform Essen.

4 This portrait was a picture by Willy Kessels (1898-1974). It was equally published on the occasion of Van de Velde's 70th birthday in 1933 in the magazines *La Cité* and *Kunst*. *La Cité* mentions that the picture was taken around the time of its publication in April 1933. "Editorial," *La Cité* 11, no. 5/6 (1933); A. Vermeylen, "Bij den zeventigsten verjaardag van Henry van de Velde," *Kunst. Maandblad voor oude en jonge beeldende bouw en sierkunsten*, no. 10-11-12 (1933).

5 "La présidence d'honneur de cet Institut a été proposée à Henry van de Velde: j'ai donc apposé son portrait en grand dès l'entrée du stand, un peu comme une incantation... Son 'humanisme' pouvait nous protéger d'une professionnalisation trop commerciale et aussi humaniser les modèles 'Bauhaus' qui étaient les plus actuels mais encore confusément, les plus abstraits..." See Lucien Kroll, email message to author, October 20, 2014.

6 Javier Gimeno Martínez addresses this issue in an unpublished conference paper: Gimeno-Martínez, Javier, "Van de Velde after Van de Velde: Design Pioneers and the Promotion of Industrial Design in the 1950s," The Annual Design History Society conference "Writing Design," University of Hertfordshire, Hatfield (Verenigd Koninkrijk), 2009.

7 Ernesto Rogers wrote this in the introduction to the special issue (237, March 1960) dedicated to Henry van the Velde. See Ernesto N. Rogers, "Henry van de Velde o dell'evoluzione," *Casabella Continuità. Rivista Internationale di Architettura e di Urbanistica*, March 1960. See also Silvia Malcovati, "The Notion of the Total Work of Art and Italian Building Culture after World War II," in *The Death and Life of the Total Work of Art. Henry van de Velde and the Legacy of a Modern Concept*, ed. Carsten Ruhl, Chris Dähne, and Rixt Hoekstra (Berlin: Jovis, 2015).

8 Javier Gimeno-Martínez, "The Introduction and Dissemination of the English Word 'Design' in the Belgian Context," in *Words for Design. Comparative Etymology and Terminology of Design and Its Equivalents. The First Step Towards a Full Collection of Words for Design from All over the World*, ed. Haruhiko Fujita (Japan Society for the Promotion of Science, 2009).

9 Brussels, Archives & Musée de la Litérature (AML), Fonds Henry van de Velde, 211/14: Secretary Desamory and Vice-President Paquay, Letter to Henry van de Velde, (7/01/1956); Brussels, Archives & Musée de la Litérature (AML), Fonds Henry van de Velde, 211/14: Secretary Desamory and Vice-President Paquay, Letter to Henry van de "Velde, (28/01/1956). This process is also described by Javier Gimeno Martínez in an unpublished conference paper: Gimeno Martínez, Javier, "Van de Velde after Van de Velde: Design Pioneers and the Promotion of Industrial Design in the 1950s," The Annual Design History Society conference "Writing Design," University of Hertfordshire, Hatfield (Verenigd Koninkrijk), 2009.

10 Around January 1956 Louis Desamory visited Henry van de Velde in Oberägeri (Switserland) to discuss the erection of a "section d'esthétique industrielle" at the Royal Academy of Fine Arts in Liège. See Brussels, Archives & Musée de la Litérature (AML), Fonds Henry van de Velde, 211/14: Letter from Louis Desamory to Henry van de Velde, (07/01/1956); Brussels, Archives & Musée de la Litérature (AML), Fonds Henry van de Velde, 211/14: Louis Desamory, *Academie Royale des Beaux-arts de Liège. Création d'une section d'Esthétique Industrielle* (10/9/1955).

Next to this contact in real life, Maurice Paquay and Louis Desamory, as representatives of the Institute for Industrial Design, maintained a correspondence with Henry van de Velde. See Brussels, Archives & Musée de la Litérature (AML), Fonds Henry van de Velde, 211/14; Brussels, Archives & Musée de la Litérature (AML), Fonds Henry van de Velde, 488.

11 Lucien Kroll remembers his visit to Henry van de Velde after the establishment of the Institute together with Maurice Paquay and Charles Vandenhove. See Interview by the author with Lucien Kroll on October 21, 2014, Brussels. This is also mentioned in the book: Bekaert and Strauven, *Bouwen in België 1945-1970*, 328. The trip itself is discussed in some correspondence: Brussels, Archives & Musée de la Litérature (AML), Fonds Henry van de Velde, 629/2: Letter of Paquay to Henry van de Velde, (15/06/1956).

12 Nikolaus Pevsner, *Pioneers of Modern Design: From William Morris to Walter Gropius* (New Haven: Yale University Press, 2005).

13 Kathleen James-Chakraborty, "Henry van de Velde and Walter Gropius," in *Bauhaus Culture. From Weimar to the Cold War*, ed. Kathleen James-Chakraborty (Minneapolis/London: University of Minnesota Press, 2006).

14 It was a photographic portrait taken by Marcel Baugniet during one of his visits to Oberägeri, Van der Velde's Swiss postwar place of residence. See *Berichten van het Instituut voor Industriële Vormgeving*, no. 1 (1957).

15 James-Chakraborty, "Henry van de Velde and Walter Gropius," 28.

16 Virginie Devillez, "Een Belgische Werkbund? Het streven naar een nationale kunstnijverheid," in *Art Nouveau & Design. Sierkunst van 1830 tot Expo 58*, ed. Claire Leblanc (Tielt/Brussels: Lannoo/Racine, 2005).

17 Huib Hoste, "Henry van de Velde," *Bouwkundig Weekblad. Orgaan van de Maatschappij tot Bevordering der Bouwkunst Bond van Nerderlandsche Architecten* 45, no. 27 (1924); "Hulde aan Henry van de Velde bij zijn 90oste verjaardag," *Bouwen en wonen. Maandblad voor architectuur, stedenbouw, woningpolitiek, beeldende kunst, binnenhuiskunst en industriële vormgeving* 4, no. 11 (1957).

18 "Een deel der belgische pers was tegen mij opgehitst geworden. Zij gebruikte de lafste en hatelijkste verdachtmakingen om mijne inspanningen voor de oprichting van dit Hooger Instituut te doen mislukken." See Jean van de Voort, "Henry van de Velde, de leider, in het verleden, heden en toekomst," *Kunst. Maandblad voor oude en jonge beeldende bouw en sierkunsten*, no. 10–11-12 (1933): 351. We can see this sensitivity to this blackening of his reputation in a later letter of 1936 where Henry van de Velde thanks Marcel Schmitz for supporting him as one of the first journalists after his return to Belgium after the First World War: "Et c'est bien ce que Vous, qui avez été un des premiers à me soutenir depuis que je suis rentré en Belgique, ce que Vous et tous ceux qui se sont ranges à mes côtés attendez de moi." Brussels, Archives & Musée de la Litérature (AML), Fonds Henry van de Velde, 211/15: copy of a letter of Henry van de Velde to Marcel Schmitz, (October 1936).

19 Huib Hoste, "Henry van de Velde," *Opbouwen. Halfmaandelijks Tijdschrift voor Architectuur tevens Orgaan der Vlaamse Architectuurvereniging*, no. 3 (1933); Tanja Boon, "Ter Kameren (1927–1936)" (Katholieke Universiteit Leuven, 2007).

20 Roger Avermaete, "Henry van de Velde, Een groote figuur van onzen tijd," *Kunst. Maandblad voor oude en jonge beeldende bouw en sierkunsten*, no. 10–11-12 (1933).

21 "Maar, Van de Velde had af te rekenen met jaloerse rivalen. hij werd aangevallen – niet omdat de heren bekwaam waren het beter te doen dan hij, – maar hij werd het mikpunt van een polemiek. Hij werd voorgesteld als anti-Belg, – en als een van Duitsland verslaafd personage. Kortom, hij ging door voor een politiek verrader [...] Toen ik minister werd, in 1925, heb ik droogweg op die vuiligheid geantwoord. Ik heb hem de leiding van het splinternieuwe instituut van Terkameren toevertrouwd, en hem benoemd als professor op de Universiteit van Gent. Zo werd hij ook lid der Academie." See Camille Huysmans, "In Memoriam. Henry van de Velde,"

38 "Les paroles qu'il prononçait en 1892, conservent en 1923 un aspect d'immédiate vérité: en 1952, leur actualité pourrait-elle être diminuée?" See Henry van de Velde, *Les formules de la beauté architectonique moderne* (Brussels: Equerre, 1923).

39 Pierre Puttemans, *Modern Architecture in Belgium* (Brussels: Marc Vokaer Publishers, 1976), 82.

40 "Van de Velde ignore sans doute la dualité de l'instinct et de l'intelligence: chez lui, ces deux forces, naturellement sont alliées, et ce que propose son intuition était secrètement souhaité par sa raison." See van de Velde, *Les formules de la beauté architectonique moderne.*

41 Elie Haddad, "On Henry van de Velde's "Manuscript on Ornament"," *Journal of Design History* 16, no. 2 (2003).

42 "Het meest afdoende middel om de lelijkheid te bestrijden is: de terugkeer tot de rede, tot de strikte logica der vormen van alle dingen." Henry van de Velde, "De Heilige Weg," *Kunst. Maandblad voor oude en jonge beeldende bouw en sierkunsten* 10-11-12(1933).

43 "En l'exprimant avec passion, il a réhabilité la raison, accusée de froideur et de prosaïsme." Maurice Casteels, *Henry van de Velde* (Bruxelles: Editions des Cahiers de Belgique, 1932), 16.

44 The other architects are Otto Wagner (1841–1918), Adolf Loos (1870–1933), Louis Sullivan (1856–1924), Frank Lloyd Wright (1867–1959). See Pevsner, *Pioneers of Modern Design: From William Morris to Walter Gropius.*

45 The Belgian architectural magazine *La Cité* called him the "defender of reason" in 1933 on the occasion of his 70th birthday. Also Maurice Casteels portrayed Henry van de Velde's architecture rather as rationalist in his 1930 book *L'art moderne primitif* and his 1932 monography on Henry van de Velde. See "Editorial," 21; Verpoest and Van Impe, "Henry van de Velde. Brieven van architecten. Bezoekersgids; Casteels, *Henry van de Velde.*

46 Henry van de Velde and Hermann Muthesius, "Werkbund Theses and Antitheses," in *The Design History Reader*, ed. Grace Lees-Maffei and Rebecca Houze (Oxford/New York: Berg, 2010); Jonathan Woodham, *Twentieth-Century Design* (Oxford/New York: Oxford University Press, 1997), 21–23; Kenneth Frampton, *Moderne architectuur. Een kritische geschiedenis*, trans. Henk Hoeks, 4th ed. (Nijmegen: Sun, 2001), 120–24; John V. Maciuika, "Wilhelmine Precedents for the Bauhaus," in *Bauhaus Culture. From Weimar to the Cold War*, ed. Kathleen James-Chakraborty (Minneapolis/London: University of Minnesota Press, 2006).

47 See Ulrich Conrads, ed. *Programs and Manifestoes on 20th-Century Architecture* (Cambridge, Massachusetts: The MIT Press, 1971), 29, also cited in James-Chakraborty, "Henry van de Velde and Walter Gropius."

48 Geert Bekaert, *Verzamelde opstellen. Deel 1: Stapstenen 1950-1965* (Brussel: Stichting Monumenten- en Lanschapszorg, 1985), 246.

49 K.N. Elno, "Henry van de Velde. Genie zonder meesterwerken," *De Linie*, 29 maart 1963.

50 Stynen, «Nécessité de l'enseignement de l'esthétique industrielle en Belgique. Intervention de l'architecte Léon Stynen au récent colloquium d'industrial design.»

51 "[...] tandis que l'Amérique allait s'inspirer des expériences déployées en Allemagne par Van de Velde et les traduire sur le plan des réalités. De telle sorte que nous trouvons aujourd'hui dans cette situation paradoxale - je m'excuse de la rapidité de ce schéma - ou, le jeu de la concurrence jouant pleinement, nous importons une foule de produits américains qui sont consommés de préférence à des produits belge équivalent, grâce à des qualités dont nous aurions pu enrichir depuis longtemps non propre produits." See ibid.

52 Brussels, Archives et Musée de la Litérature (AML), Fonds Henry van de Velde, 211/15: Maurice Paquay, *Association Belge des Industrial Designers. Historique des débuts de l'association.*

53 "[...] l'arrivée au pouvoir des nazis entraîna la fin du Bauhaus. Gropius, Breuer, Mies van der Rohe et quelques autres quittèrent l'Allemagne pour aller travailler et enseigner en Amérique. Nantis de moyens financiers exceptionnels, il fallut alors peu d'années aux Américains pour accroître leurs productions de façon extraordinaire. Aujourd'hui, par le jeu des importations, nous achetons aux U.S.A. une quantité importante de produits dus aux techniques ressortant de la doctrine de Van de Velde et cela, de préférence aux produits belges.» See Brussels, Archives & Musée de la Litérature (AML), Fonds Henry van de Velde, 211/15: Maurice Paquay, *Association Belge des Industrial Designers. Historique des débuts de l'association.*

54 Greg Castillo, "The Bauhaus in Cold War Germany," in *Bauhaus Culture. From Weimar to the Cold War*, ed. Kathleen James-Chakraborty (Minneapolis/London: University of Minnesota Press, 2006); Kathleen James-Chakraborty, "From Isolationism to Internationalism," ibid.

55 Paul Betts, "Science, Semiotics and Society: The Ulm Hochschule für Gestaltung in Retrospect,» *Design Issues* 14, no. 2 (1998); Jane Pavitt, "Design and the Democratic Ideal," in *Cold War Modern: Design 1945-1970*, ed. David Crowley and Jane Pavitt (London: V&A Publishing, 2008), 86-89; Castillo, "The Bauhaus in Cold War Germany; Jens Müller and René Spitz, «HfG Ulm, kurze Geschichte der Hochschule für Gestaltung: Anmerkungen zum Verhältnis von Design und Politik = a Brief History of the Ulm School of Design: Notes on the Relationship between Design and Politics " (Zürich: Lars Müller, 2014); Philipp Oswalt, ed. *Bauhaus Conflicts, 1919-2009. Controversies and Counterparts* (Ostfildern: Hatje Cantz, 2009).

56 See the manuscripts in which he develops these thoughts on the four citadels of modernism: Brussels, Archives et Musée de la Litérature (AML),

Fonds Henry van de Velde, 46: Henry van de Velde, Les Citadelles (manuscript describing his school in Weimar (1907), the Bauhaus (1919), the ISAD (1926) and Ulm (1955) as the four citadels), (1955).

57 Schulte, "Henry van de Velde. De levensreis 1863–1957," 46.

58 See the correspondence between Henry van de Velde and Max Bill. The first letters date from 1948, and the Ulm project is mentioned for the first time in a letter of April 21, 1950: Brussels, Archives & Musée de la Littérature (AML), Fonds Henry van de Velde, 241. From 1952, Henry van de Velde corresponded also with the Scholl-Stiftung: Brussels, Archives & Musée de la Littérature (AML), Fonds Henry van de Velde, 708. Max Bill invites Henry van de Velde in 1955 to become a member of the HfG *Ehrencollegium*: Brussels, Archives & Musée de la Littérature (AML), Fonds Henry van de Velde, 203.8: Membre du "Ehrencollegium" de la "Hochschule für Gestaltung" à Ulm (1955), (3/10/1955).

Henry van de Velde also maintained a correspondence with Inge Aicher Sholl and the Geschwister-Sholl-Stiftung from 1952. See: Brussels, Archives & Musée de la Littérature (AML), Fonds Henry van de Velde, 708.

59 In this text, Bill divided the pioneers of modern architecture into two groups. The first group – including Henry van de Velde, Adof Loos and Walter Gropius – were pioneers in the morality and social responsibility in architecture, and the second – including Frank Lloyd Wright, Mies van der Rohe and Le Corbusier – were rather deemed important as creators of "forms." See the text of his 1953 Sao Paulo speech: Brussels, Archives & Musée de la Littérature (AML), Fonds Henry van de Velde, 241: Max, *Conférence tenue à la Faculté d'architecture et d'urbanisme de l'Université de Sao Paulo, le 9 juin 1953. L'architecte, l'architecture et la société.*

60 For the relation between Van de Velde and Gropius and their respective claims on the Bauhaus legacy see James-Chakraborty, "Henry van de Velde and Walter Gropius." See also Paul Betts, "The Bauhaus as Cold War Legend: West German Modernism Revisited," *German Politics and Society* 14, no. 2 (1996); Castillo, "The Bauhaus in Cold War Germany; Kathleen James-Chakraborty, "From Isolationism to Internationalism," ibid.

61 "Ich sah, dass ich bei den Anhängern einer neuen vernuftgemässen Architecktur und der in Europa enstandenen Tendenz zu reiner, primärer Formgestaltung, die übrigens auch in den Vereinigten Staaten in Erscheinung getreten war, als "Vater" die neuen Stils angesehen wurde," citation also mentioned in Schulte, „Henry van de Velde. De levensreis 1863–1957.»

62 J.-J. T, «Industrial Design,» *L'Ossature Métallique* 19, no. 6 (1954); Stynen, «Nécessité de l'enseignement de l'esthétique industrielle en Belgique. Intervention de l'architecte Léon Stynen au récent colloquium d'industrial design; Albert Bontridder, «L'art de rendre une forme humaine à la machine et ses produits,» *Architecture 55. Revue*

bimestrielle d'architecture et d'urbanisme, no. 14 (1955); Leon-Louis Sosset, "De lessen der Xe Driejaarlijkse van Milaan," *Kunstambachten en Kunstnijverheden. Maandelijks Inlichtingsblad der Provinciale Commissies voor Kunstambachten en Nijverheidsbedrijven*, no. 60 (1954).

63 Greg Castillo, *Cold War on the Home Front: The Soft Power of Midcentury Design* (Minneapolis: University of Minnesota Press, 2010); David Crowley and Jane Pavitt, eds., *Cold War Modern: Design 1945–1970* (London: V&A Publishing, 2008); Gay McDonald, "The 'Advance' of American Postwar Design in Europe: MoMA and the Design for Use, USA Exhibition 1951–1953," *Design Issues* 24, no. 2 (2008); "The Modern American Home as Soft Power: Finland, MoMA and the 'American Home 1953' Exhibition," *Journal of Design History* 23, no. 4 (2010); "Selling the American Dream: MoMA, Industrial Design and Post-War France," *Journal of Design History* 17, no. 4 (2004).

64 Marc Reynebeau, *Een geschiedenis van België* (Tielt: Lannoo, 2009), 218.

65 See chapter one.

66 «Cependant nous, Belges, avons l'honneur de compter parmi nos citoyens l'un des pionniers de l'esthétique industrielle: le Maître Henry van de Velde. C'est lui qui, dès 1902 fondait à Weimar une école ou pour la première fois l'étude des formes industrielles s'orienta vers la recherche d'une esthétique fonctionnelle. L'industrie belge de l'époque ne comprit guère la portée de l'efficacité de cette science nouvelle et ce n'est qu'à la longue, sur la foi des inquiétantes réussites acquises à l'étranger, qu'elle en vint a en éprouver toute la nécessité." See Brussels, Archives & Musée de la Litérature (AML), Fonds Henry van de Velde, 211/14: *A.B.I.D. L'esthétique industrielle en Belgique*

67 Sosset, "Gaat de "Industrial Design" in België wortel schieten? in K.N. Elno, "Belgische pointering 1956," *Scheppend Ambacht* 7, no. 6 (1956); "Ontstaan en groei der industriële vormgeving," in *Industriële Vormgeving (1957–1958)*, ed. Stichting Lodewijk De Raet (Brussels: Stichting Lodewijk De Raet, 1959); *Ruimte en beelding. Beschouwingen over architektuur, plastische kunsten, fotografie en typografie*, vol. 157, Vlaamse Pockets (Hasselt: Heideland, 1965), 65–81; Avermaete, "Henry van de Velde, Een groote figuur van onzen tijd; Stan Leurs, "Henri van de Velde, een vlaamsch architect?!," ibid.

68 "La présidence d'honneur de cet Institut a été proposée à Henry van de Velde: j'ai donc apposé son portrait en grand dès l'entrée du stand, un peu comme une incantation... Son 'humanisme' pouvait nous protéger d'une professionnalisation trop commerciale et aussi humaniser les modèles 'Bauhaus' qui étaient les plus actuels mais encore confusément, les plus abstraits..." See Lucien Kroll, email message to author, October 20, 2014.

69 "Henry van de Velde est mort. Sa mort sera-t-il aussi vaine, chez nous, que les conséquences de son enseignement. Qu'avons fait de son message? Nous l'avons ignoré, caché ou mieux défiguré... Il a sombré sous la

puissance des intérêts et le poids des habitudes." See Lucien Kroll, «La leçon de Henry van de Velde,» *La Maison*, no. 12 (1957).

70 David Crowley, «Europe Reconstructed, Europe Devided,» in *Cold War Modern: Design 1945-1970*, ed. David Crowley and Jane Pavitt (London: V&A Publishing, 2008), 58. See for example also the case of Jaap Bakema in the Netherlands and his concept of the *homo ludens*: Cornelis Wagenaar, "Jaap Bakema and the Fight for Freedom," in *Anxious Modernisms. Experimentation in Postwar Architectural Culture*, ed. Sarah Williams Goldhagen and Réjean Legault (Montréal, Cambridge, London: Canadian Centre for Architecture, MIT Press, 2000).

71 "Je lui avais suggéré en quelques mots qui l'avaient frappé, que la continuité de son effort vers la beauté allait devoir viser cette fois, non plus le machinisme devenu trop brutal, mais le respect de la personne et son droit à la différence, car le binôme 'Finances et Techniques' était devenu l'ennemi déclaré." See Interview by the author with Lucien Kroll, 21/10/2014, Brussels; Brussels, Archives et Musée de la Litérature (AML), Fonds Henry van de Velde, 629/2: Letter of Paquay to Henry van de Velde (15/06/1956).

72 It was a photographic portrait taken by Marcel Baugniet during one of his visits to the Swiss Oberägeri. See *Berichten van het Insituut voor Industriële Vormgeving*, no. 1 (1957).

73 For more information on his artistic work and activities in Latin America see María Amalia García and Alejandro Gabriel Crispiani, *Tomás Maldonado in Conversation with María Amalia García*, ed. Gabriel Perez-Barreiro, Conversations (New York/Caracas: Fundación Cisneros, 2010); Verónica Devalle, "The Role of the 'Vision' Course in the Creation of the Design Programmes in Argentina" (Paper presented at the 9th International Committee Design History and Design Studies. Tradition, transition, trajectories: major or minor influences?, Aveiro, 2014); Patricia Amorim and Virginia Cavalcanti, "Modern Design Meets Latin America: The Role of Pioneering Design Magazines Habitat and Nueva Visión in Brazil and Argentina," in *Design Frontiers - Territories, Concepts, Technologies, 8th Conference of the International Committee of Design History and Design Studies*, ed. Farias Priscila, et al. (Sao Paulo: blucher, 2012).

74 At the HfG Ulm, Maldonado was teacher (1954-1966) vice-rector (1954-1956, 1962-1964), member of the governing board (1956-1962) and rector (1964-1966). For more information see Gui Bonsiepe and John Cullars, "The Invisible Facets of the HfG Ulm," *Design Issues* 11, no. 2 (1995); Paul Betts, "Science, Semiotics and Society: The Ulm Hochschule für Gestaltung in Retrospect," ibid.14(1998); Silvia Fernández, "The Origins of Design Education in Latin America: From the HfG in Ulm to Globalization," ibid.22, no. 1 (2006); Cornelie Leopold, "Precise Experiments: Relations between Mathematics, Philosophy and Design at Ulm School of Design," *Nexus Network Journal* 15, no. 2 (2013); Müller

and Spitz, "HfG Ulm, kurze Geschichte der Hochschule für Gestaltung: Anmerkungen zum Verhältnis von Design und Politik = a Brief History of the Ulm School of Design: Notes on the Relationship between Design and Politics".

75 Gui Bonsiepe, email message to author, July 9, 2015; Tomas Maldonado, "Kunst en industrie. Vrij heterogene beschouwingen over een bijna legendarisch thema," *West-Vlaanderen* 13, no. 78 (1964); "Art et Industrie. Considérations presque hétérodoxes sur un theme presque légendaire,» *Informatieblad van het Instituut voor Industriële Vormgeving* 5, no. 6 (1962).

76 Provost was already involved as administrator in the activities of the Institute from 1958. See Ghent, Archief Adelbert Van de Walle, Museumarchief Design Museum Gent, Mapje instituut industriële vormgeving B.-L.: L. Desamory (Secrétaire) and Ph. Questienne (Président), *Assemblee Generale du Jeudi 13 Fevrier 1958. Procès-verbal*, (13/02/1958); "Naar de toekomst...(2)," *Berichten van het Instituut voor Industriële Vormgeving*, no. 2 (1959); Eva Van Regenmortel, "Pol Provost (1907–1990)," (2012).

77 Pol Provost, "Boodschap," *Berichten van het Instituut voor Industriële Vormgeving*, no. 4 (1959).

78 «Les tâches de l'Institut d'Esthétique Industrielle,» *Supplément à l'Écho de la Bourse*, September 9, 1959.

79 "Algemene Vergadering," *Berichten van het Instituut voor Industriële Vormgeving*, no. 3 (1960).

80 Brighton, University of Brighton, Design Archives, ICSID, 9-11-1: Josine des Cressonnières, Letter to Misha Black (Vice-President ICSID), (5/08/1959).

81 Michel Olyff's logo was later also adopted by the Design Centre.

82 "L'Institut d'Esthétique Industrielle, vient de subir une profonde réorganisation et deviant aujourd'hui le Centre de toutes les activités d'industriel design en Belgique. C'est à lui que le Ministre des Affaires Economiques confie l'ensemble des credits à repartir entre les diverses actions en faveur du design." See Brussels, Koninklijke Musea voor Schone Kunsten van België, Archief voor Hedendaagse Kunst in België, Fonds Léon-Louis Sosset, Caisse l'Esthétique industrielle: RB, *Compte-rendu de l'activité du Signe d'Or pendant l'année 1960,* (March 1961).

83 "Het leven van het Instituut," *Informatieblad van het Instituut voor Industriële Vormgeving* 5, no. 3 (1961).

84 Brussels, Koninklijke Musea voor Schone Kunsten van België, Archief voor Hedendaagse Kunst in België, Fonds Léon-Louis Sosset, Caisse l'Esthétique industrielle: *Prévisions d'activité du Signe d'Or pour 1961.*

85 The Institute, for example, refused to fund the educational seminars at the Lodewijk de Raet Foundation anymore because of the diverged vision. In the new Ulm fashion, the Institute seems to be allergic to the word crafts and wanted to keep the disciplines of design and crafts separated

from each other as best they could. The president of the Lodewijk de Raet Foundation, Maurice Van Haegendoren, consequently accused the institute of completely ignoring the social and cultural aspects of design. At meetings of the Institute he stated: "industrial design is not only a question of 'sales promotion' but also about the education of the consumer taste." See Ghent, Archief Adelbert Van de Walle, Museumarchief Design Museum Gent, unnumbered: *Proces-Verbaal van de 10de vergadering der werkgroep voor de studie van de Industriële Vormgeving, gehouden in het Instituut voor Industriële Vormgeving op 7 december 1960*; Brussel, Archief en Museum voor het Vlaams Leven te Brussel, BE AMVB 168, 1604/SLDR: M. van Haegendoren, Letter to Josine des Cressonnières [General Secreatry of the Institute for Industrial Design], (5/06/1961); Brussels, Archief en Museum voor het Vlaams Leven te Brussel, BE AMVB 168, 1604/SLDR: Josine des Cressonnières [General Secreatry of the Institute for Industrial Design], Letter to to M. van Haegendoren, (2/10/1961); Ghent, Archief Adelbert Van de Walle, Museumarchief Design Museum Gent, Mapje Lodewijk de Raet: Maurits van Haegendoren, Letter to Pol Provost, (15/10/1960).

86 The original name was Council of Societies of Industrial *Designers*, but it was soon changed in 1961 to Council of Societies of Industrial *Design*. See Brighton, University of Brighton, Design Archives, ICSID, 3-1-1 (Minutes of the IInd G.A.): *Second General Assembly. Venice, Italy: 14 to 16 September, 1961*, (December 1961), p. 3; «Le Congrès de l'I.C.S.I.D.,» *Esthetique Industrielle*, no. 52/53 (1961).

87 For a complete list of the international jury of the Signe d'Or, see Javier Gimeno-Martínez, "The Signe d'Or Award Scheme from 1956 to 1960: The Economic Reasons for "Good Design"," *Konsthistorisk Tidskrift/Journal of Art History* 79, no. 3 (2010).

88 Brighton, University of Brighton, Design Archives, ICSID, 9-11-1: Josine des Cressonnières, Letter to Peter Muller-Munk (president of the ICSID), (8/4/1959).

89 From its beginning the Belgian design scene was involved in this international endeavor. The Belgian art historian Robert Delevoy was at the Paris meeting of April 1956 that established the International Coordination Committee of Industrial Design and became a member of the provisional executive board with Peter Müller-Munk as president, Misha Black as vice-president, Pierre Vago as secretary and himself as treasurer. This Committee worked on the official establishment of the Council of Societies of Industrial Designers that eventually took place in June 1957. See «'Fondation du Comité de Liaison Internationale d'Esthétique Industrielle. Dîner d'acceuil des représentants étrangers',» *Esthetique Industrielle* VI, no. 22 (1956); Arthur J. Pulos, *The American Design Adventure, 1940-1975* (Cambridge: MIT Press, 1988), 216. The Institute gained the status of associate member already in 1958. In 1961, it became a full member. See Brighton, University of Brighton, Design Archives,

ICSID, 9-11-1: Peter Muller-Munk, Letter to Louis Desamory (secretary Institute for Industrial Design), (2/4/1958).

90 "Het leven van het Instituut," *Informatieblad van het Instituut voor Industriële Vormgeving*, no. 5 (1962); "Het ICSID congres te Venetië," *Informatieblad van het Instituut voor Industriële Vormgeving*, no. 5 (1962).

91 These were Robert Delevoy (founder of the Industrial Design program at La Cambre), architect and designer Walter Bresseleers, F. Van Espen (representative of the Belgian industry as adjunct of the commercial department of M.B.L.E.), H. Steenwinckel (director of the Ecole Supérieur des Beaux-Arts St. Luc in Liège) and F. Brossé of Flemish television. See "Het leven van het Instituut."

92 Josine des Cressonnières succeeded the German Mia Seeger in this position. The complete ICSID Executive Board elected in 1961 was: Sigvard Bernadotte as President, Misha Black as Past-President, Jay Doblin and Alberto Rosselli as Vice-Presidents and Josine des Cressonnières as Secretary General. See "Het ICSID congres te Venetië."; Brighton, University of Brighton, Design Archives, ICSID, 3-1-1 Minutes of the IInd G.A.: *Second General Assembly. Venice, Italy: 14 to 16 September, 1961*, (December 1961), p. 2, 5.

93 Betts, "Science, Semiotics and Society: The Ulm Hochschule für Gestaltung in Retrospect; Castillo, «The Bauhaus in Cold War Germany; Bonsiepe and Cullars, «The Invisible Facets of the HfG Ulm; Silvia Fernández, «The Origins of Design Education in Latin America: From the HfG in Ulm to Globalization,» ibid.22, no. 1 (2006); Leopold, «Precise Experiments: Relations between Mathematics, Philosophy and Design at Ulm School of Design; Müller and Spitz, «HfG Ulm, kurze Geschichte der Hochschule für Gestaltung: Anmerkungen zum Verhältnis von Design und Politik = a Brief History of the Ulm School of Design: Notes on the Relationship between Design and Politics "; Raimonda Riccini, "Tomás Maldonado and the Impact of the Hochschule für Gestaltung Ulm in Italy,» in *Made in Italy. Rethinking a Century of Italian Design*, ed. Grace Lees-Maffei and Kjetil Fallan (London/New York: Bloomsbury, 2014); Cornelie Leopold, "The Mathematical Approach at Ulm School of Design," in *Imagine Maths 4. Between Culture and Mathematics*, ed. Michele Emmer, Marco Abate and Marcela Villarreal (Unione Matematica Italiana, 2015).

94 Lecture at the Brussels World's Fair held on September 18, 1958. See Brussels, State Archives, BE-A0545/Fic 2000, Archive Design Centre, 731: Tomas Maldonado, *Design Education;* "Echo's en Berichten. Voordracht," *Berichten van het Instituut voor Industriële Vormgeving*, no. 4 (1958). *According to designer Jean-Jacques Stiefenhofer, Josine des Cressonnières got to know Tomas Maldonado and his ideas on this occasion. See Interview by the author with Jean-Jacques Stiefenhofer on 11/10/2011, 17/10/2011 and 3/5/2012, Antwerp.*

8 One exception is the article by Javier Gimeno-Martínez on the Signe d'Or (1956–1960): see Javier Gimeno-Martínez, "The Signe d'Or Award Scheme from 1956 to 1960: The Economic Reasons for 'Good Design'," *Konsthistorisk Tidskrift/Journal of Art History* 79, no. 3 (2010). See also the authors' publication: Katarina Serulus, "Au-delà des frontières: La promotion du design belge pendant la guerre froide/De grenzen voorbij: De promotie van Belgisch design tijdens De Koude Oorlog/Beyond the Borders: The Promotion of Belgian Design During the Cold War," in *Panorama: A History of Modern Design in Belgium*, ed. Katarina Serulus and Javier Gimeno-Martínez (Brussels: CFC Editions, 2017).

9 Jonathan Woodham, *Twentieth-Century Design* (Oxford/New York: Oxford University Press, 1997), 171–77; "Formulating National Design Policies in the United States: Recycling the "Emperor's New Clothes"?," *Design Issues* 26, no. 2 (2010); Artemis Yagou, "Unwanted Innovation. The Athens Design Centre (1961–1963)," *Journal of Design History* 18, no. 3 (2005); Kjetil Fallan, "How an Excavator Got Aesthetic Pretentions - Negotiating Design in 1960s' Norway," ibid. 20, no. 1 (2007); Christopher Thompson, "Modernizing for Trade: Institutionalizing Design Promotion in New Zealand," ibid. 24, no. 3 (2011).

10 Gimeno-Martínez, "The Signe d'Or Award Scheme from 1956 to 1960: The Economic Reasons for "Good Design"," 133.

11 «Quatre ans d'action économique,» ed. Ministère des Affaires Economiques (Brussels: Ministère des Affaires économiques, 1958), 41.

12 "[...] een permanent tentoonstellingscentrum, waarvan de doelstelling is, hetzij het prestige van de Belgische nijverheid, bijzonderlijk op de buitenlandse markten, te bevorderen [...]" See: "N. 477", Annexe au Moniteur Belge. Associations sans but lucratif et établissements d'utilité publique/Bijlage tot het Belgisch Staatsblad. Verenigingen zonder winstgevend doel en instellingen van openbaar nut [Attachment to the Belgian Official Journal. Not-for-profit Organizations and Public Utility Foundations], 1 February 1962, p. 212–214.

13 See the archival documents on Luxembourg's participation in the Design Centre: Brussels, State Archives, BDOP/OBAP, 65: Josine des Cressonnières, Letter to R. Talpaert with regard to Luxembourg's participation (December 30, 1963); Brussels, State Archives, BDOP/OBAP, 65: Josine des Cressonnières, *Compte rendu du Comité de Direction du Design Centre* (20 Nov. 1963).

14 Penny Sparke, *An Introduction to Design and Culture. 1900 to the Present*, 2 ed. (London/New York: Routledge, 2004), 198–206.

15 Widar Halén and Kerstin Wickman, eds., *Scandinavian Design Beyond the Myth. Fifty Years of Design from the Nordic Countries* (Stockholm: Arvinius Förlag/Form Förlag, 2003).

16 In his capacity as European commissioner, Jean Rey also supported the initiative of the Liaison Committee for Industrial Design in the Common Market.

17 Gimeno-Martínez, "The Signe d'Or Award Scheme from 1956 to 1960: The Economic Reasons for 'Good Design'."

18 Ibid., 141.

19 Brussels, Koninklijke Musea voor Schone Kunsten van België, Archief voor Hedendaagse Kunst in België, Fonds Léon-Louis Sosset, Caisse l'Esthétique industrielle: *Previsions d'Activité du Signe d'Or pour 1961;* Beveren, State Archives, the BKM collection (Archives of the Department of Fine Arts and Museums, Ministry of the Flemish Community/Transfer 2005 [1945–2004]): BKM 2013: *Rapport au Conseil d'Administration du Signe d'Or Industriel*, (17.9.1959), fol. 147r-150r; *Assemblée générale statutaire de l'A.S.B.L. "Le Signe d'Or Industriel" du jeudi 25 février 1960*, fol. 123–126; Rapport au Conseil d'Administration, (23/08/1960), fol. 111r-112r. *Commission «Design» 18.10.1960*, fol. 106r-109r; BKM 2013: *Previsions d'activité du Signe d'Or pour 1961*, fol. 87–88; BKM 2013: *Commission «Design». Compte-rendu de la réunion du 13.1.61*, fol. 95r-96r.

20 "Il semble que nous arrivions maintenant au point où le Signe d'Or a 'écrémé' toute la production et signalé ses meilleures réalisations. La rythme de la sélection annuelle commence à devenir essoufflant, et il serait souhaitable de repenser notre activité en tenant compte de ce fait nouveau." See Beveren, State Archives, BKM collection (Archives of the Department of Fine Arts and Museums, Ministry of the Flemish Community/Transfer 2005 [1945–2004]): BKM 2013: *Compte-rendu de l'activité du Signe d'Or pendant l'année 1960*, (mars 61) fol. 83r.

21 "La participation des firmes hollandaises a été plus forte cette année, et leur concurrence s'est confirmé redoutable" See Beveren, State Archives, BKM collection (Archives of the Department of Fine Arts and Museums, Ministry of the Flemish Community/Transfer 2005 [1945–2004]): BKM 2013: *Compte-rendu de l'activité du Signe d'Or pendant l'année 1960*, (mars 61) fol. 83r.

22 "Le Signe d'Or celui-ci devra être adapte aux nécessitées nouvelles, soit élargi à l'échelon européen (Marché Commun), soit ré-étudié, au point de vue national." See: Beveren, State Archives, BKM collection (Archives of the Department of Fine Arts and Museums, Ministry of the Flemish Community/Transfer 2005 [1945–2004]): BKM 2013: *Commission "Design". Réunion 1. Du 18-10-60*, fol. 106r.

23 "M. De Poerck demande des précisions quant aux 'projets de collaboration avec les grandes entreprises de distribution', signalés dans les prévisions d'activité pour 1960. Mme des Cressonnières explique qu'il y a d'abord une idée de base: inclure les grands magasins dans une politique d'expansions du design en Belgique, parce qu'ils peuvent y jouer un rôle très importante (témoin l'exposition actuellement en cours au

Bon Marché). Un exemple précis: l'expérience faite aux Etats-Unis par un petit jury de designers, qui ont sélectionné dans les grands magasins et au moment des fêtes, des produits bien dessinés et bon marché. Le public était surpris et frappé de découvrir, réunis dans une petite exposition, tous ces 'well designed products under 10 dollars'. Cette expérience, qui concourt à l'éducation du public, pourrait être réalisée dans les chaînes des grands magasins européens, avec l'aide des groupements d'esthétique industrielle." See Beveren, State Archives, BKM collection (Archives of the Department of Fine Arts and Museums, Ministry of the Flemish Community/Transfer 2005 [1945-2004]): BKM 2013: *Assemblée general statutaire de l'A.S.B.L. "Le Signe d'Or Industriel" du jeudi 25 février 1960, à 17 h.*, fol. 126r. Josine des Cressonnières also made an inquiry in 1959 from Peter Muller-Münk about the MoMA Good "Design program." Peter Muller-Munk informed her that this program died a natural death. See Brighton, University of Brighton, Design Archives, ICSID, 9-11-1: Peter Muller-Münk, Letter entitled 'Re: Bon Marché' to Josine des Cressonnières, (27/01/1959).

24 For information on the "Useful Objects" series and the "Good Design" program see Terence Riley and Edward Eigen, "Between the Museum and the Marketplace: Selling Good Design," in *The Museum of Modern Art at Mid-Century. At Home and Abroad*, ed. John Elderfield, *Studies in Modern Art* (New York: The Museum of Modern Art, 1994).

25 Ibid., 110–21; Arthur J. Pulos, *The American Design Adventure, 1940–1975* (Cambridge: MIT Press, 1988).

26 The exhibition was originally planned from 5 September until 28 September 1959, but the opening was eventually delayed to February 18, 1960 and eventually ran until March 5, 1960. The exhibition was under the patronage of the ICSID and the Ministry of Cultural Affairs, Public Education, Economic Affairs and Foreign Trade. See Beveren, State Archives, BKM collection (Archives of the Department of Fine Arts and Museums, Ministry of the Flemish Community/Transfer 2005 [1945-2004]): BKM 2013: *Compte Rendu de l'activité du Signe d'Or de février à août 1960*, fol. 113r; Beveren, State Archives, BKM collection (Archives of the Department of Fine Arts and Museums, Ministry of the Flemish Community/Transfer 2005 [1945-2004]): BKM 2013: *Bilan de l'activité du Signe d'Or Industriel pour l'exercice 1959*, fol. 127r-130r; Brighton, University of Brighton, Design Archives, ICSID, 9-11-1: *Exposition internationale d'Industrial Design 5 au 28 septembre 1959 au Bon Marché* [leaflet]; "Tentoonstellingen," *Berichten van het Instituut voor Industriële Vormgeving* 4, no. 2 (1960).

27 The exhibition was also under patronage of the Ministries of Cultural Affairs, Public Education, Economic Affairs and Foreign Trade.

28 In the preparation of the exhibition, Josine des Cressonnières especially maintained a correspondence with the first executive board: President Peter Muller-Munk, the Vice-President Misha Black, Vice-President

Enrico Peressutti and Secretary Treasurer Pierre Vago. See the correspondence at Brighton, University of Brighton, Design Archives, ICSID, 9-11-1.

29 For an exhaustive list of the annual national and international jury see Gimeno-Martínez, "The Signe d'Or Award Scheme from 1956 to 1960: The Economic Reasons for 'Good Design'."

30 Brighton, University of Brighton, Design Archives, ICSID, 9-11-1: Josine des Cressonnières, Letter to Peter Muller-Munk (president of the ICSID) (April 8, 1959); Brighton, University of Brighton, Design Archives, ICSID, 9-11-1: Misha Black (Vice-President of the ICSID), Letter to Raymond Vaxelaire (director of the Au Bon Marché department store) (February 20, 1959).

31 Brighton, University of Brighton, Design Archives, ICSID, 9-11-1: P. Yvanoff (Manager of the Au Bon Marché department store), Letter to Misha Black (Vice-President ICSID) (March 3, 1959).

32 Pulos, *The American Design Adventure, 1940–1975*, 216.

33 The participating foreign design bodies were: Rat für Formgebung (Germany), the Council of Industrial Design (Great Britain), Österreichisches Institut für Formgebung (Austria), National Industrial Design Council (Canada), Selskabet for Industriele Formgivning (Denmark), American Society of Crafts and Design (USA), American Society of Industrial Designers (USA), Finnish Society of Crafts and Design (Finland), Institut d'Esthétique Industriel "formes utiles" (France), Instituut voor Industriële Vormgeving (Netherlands), Premio "Compasso d'Oro" (Italy), Associazione per il Disegno Industriale (Italy), Norsk Gruppe for Industriell Formgivning (Norway), Form Design Center A B. Swedish Institute for Cultural Relations (Sweden), Schweizerischer Werkbund (Switzerland). See Ghent, Archief Adelbert Van de Walle, Museumarchief Design Museum Gent, unnumbered: folder *Au bon Marché*.

34 Concerning his appointment see the letter: Brighton, University of Brighton, Design Archives, ICSID, 9-11-1: P. Yvanoff (Manager of the Au Bon Marché department store), Letter to Misha Black (Vice-President ICSID) (March 3, 1959). His candidature was welcomed by the ICSID. The Vice-President Misha Black stated in this regard: "I have seen examples of his work and have no doubt that he is one of the outstanding Belgian designers in this field." See Brighton, University of Brighton, Design Archives, ICSID, 9-11-1: Misha Black (Vice-President ICSID), Letter to P. Yvanoff (Manager of the Au Bon Marché department store) (September 3,1959).

35 Ghent, Archief Adelbert Van de Walle, Museumarchief Design Museum Gent, unnumbered: folder *Industrial design 1959 in Gent*; Ghent, Archief Adelbert Van de Walle, Museumarchief Design Museum Gent, unnumbered: folder *Expo Signe d'Or in Gent*; Ghent, Archief Adelbert

Van de Walle, Museumarchief Design Museum Gent, unnumbered: folder *Internationale Jaarbeurs Vlaanderen 1959*; Leon-Louis Sosset, "Drie belangrijke tentoonstellingen te Gent," *Kunstambachten en Kunstnijverheden. Maandelijks Inlichtingsblad der Provinciale Commissies voor Kunstambachten en Nijverheidsbedrijven*, no. 118 (1959); Jacques Van der Schueren et al., "Nationaal Salon voor Industriële Vormgeving," ed. Ministerie van Economische Zaken. Administratie van de Handel. Dienst Tentoonstellingen en Jaarbeurzen (1959).

36 The display is analyzed and described on the basis of pictures preserved in personal archives and the GIB (Grand Bazar Innovation Bon Marché group) archive: Brussels, ULB, GIB archives, ARCHIULB/25Z, DOC001/238: Dossier Centenaire du Bon Marché 1960.

37 Josine des Cressonnières, «L'esthétique industrielle et le Signe d'Or 1959,» *Industrie* 1959, 664.

38 "D'autre part, nous sommes en pourparler avec les organismes d'esthétique industrielle des pays du Marché Commun, afin d'établir avec eux une collaboration. Celle-ci pourrait se traduire, sur le plan pratique, par l'organisation d'une sélection à l'échelon européen des meilleurs produits choisis dans chaque pays par les 'Design Centers', Signe d'Or, Compasso d'Oro, Beauté-France, etc... Ce projet se complèterait par un échange d'objets et d'idées, un effort commercial commun, et peut-être plus tard, par une unification des critères et des règlements de sélection dans les divers pays." See: Beveren, State Archives, BKM collection (Archives of the Department of Fine Arts and Museums, Ministry of the Flemish Community/Transfer 2005 [1945–2004]): BKM 2013: *Previsions d'activité du Signe d'Or pour 1961*, (mars) fol. 88r.

39 The name of the organization varied in different publications from "Comité de Liaison pour l'Industrial Design dans la C.E.E." to "Comité de Liaison pour l'Industrial Design dans les Pays du Marché Commun." "De industrial design en de gemeenschappelijke markt; "Informations Belgiques," *Esthetique Industrielle*, no. 51 (1961).

40 "Voor een goed overleg tussen de autoriteiten van de E.E.G. en de gezamelijke Instituten voor Industriële Vormgeving in de E.E.G.-landen, is een koördinerend orgaan nodig, een verbindingsdienst, die de gemeenschappelijke belangen van industriële vormgeving voor alle lid-staten behartigt en representeert bij de Europese Gemeenschap." See "De Industrial Design en de gemeenschappelijke markt," 6.

41 Ibid.

42 This second project (the study of European companies with regard to industrial design) was mentioned by Josine des Cressonnières in her lecture on December 15, 1962 at Fabrimetal. See Brighton, University of Brighton, Design Archives, ICSID, 9-23-2: *Exposé fait par Madame des Cressonnières, Secrétaire Générale de l'Institut d'Esthétique Industrielle, dans Le cadre du Séminaire Du 15 décembre 1962, à Fabrimetal.*

43 "Congressen en studiedagen. Bij de gemeenschappelijke markt," *Informatieblad van het Instituut voor Industriële Vormgeving* 6, no. 10–11 (1962).

44 The speakers at the symposium were: Pol Provost (Chairman of the *Institut d'Esthétique Industrielle*, Belgium), Tomás Maldonado (Rector of the *Hochschule f r Gestaltung* in Ulm, Germany), Augusto Morello (Director of the Design Department of the *Rinascente*, Secretary of the *Compasso d'Oro* award, board member of the *Associazione per il Disegno Industriale*, Italy), N.H. Benninga (Vice-Chairman of the *Raad voor Industriële Vormgeving* and chairman of the *Design Centrum* in Amsterdam, the Netherlands), Mia Seeger (board member and Secretary of the *Rat für Formgebung*, Germany), G. Combet (Chairman of the *Institut d'Esthétique Industrielle*, France). This program was announced on the invitation preserved in the archives of the Belgian Office for Increasing Productivity: see Brussels, State Archives, BDOP/OBAP, 65: *Comité de Liaison pour l'Industrial Design dans les Pays du Marché Commun* [Invitation to the symposium addressed to R. Talpaert].

45 "Het Leven van het Instituut," *Informatieblad van het Instituut voor Industriële Vormgeving* 5, no. 3 (1961): 5. In 1960, the "Design" Commission of the Institute for Industrial Design already expressed the intention to expand the activities of the Signe d'Or award to a European level. The organization of the *European Selection of Industrial Design* by the Institute for Industrial Design together with the Signe d'Or was also mentioned as one of the reasons why the Signe d'Or was not awarded in 1961. This information was found at the State Archives in Beveren in the BKM collection: see Archives of the Department of Fine Arts and Museums, Ministry of the Flemish Community/Transfer 2005 [1945–2004]: BKM 2013: *Commission "Design" 18.10.1960*, fol. 106r-109r; BKM 2013: *Previsions d'activité du Signe d'Or pour 1961*, fol. 87–88; BKM 2013: *Commission "Design". Compte-rendu de la réunion du 13.1.61*, fol. 95r-96r.

46 "Het leven van het Instituut," *Informatieblad van het Instituut voor Industriële Vormgeving* 6, no. 8–9 (1962).

47 "Het doel van het 'Comité de Liaison' is een panorama samenstellen van de Europese design en hierdoor een 'aspect van Europa' bekendmaken." See: ibid., 2.

48 Greg Castillo, "Marshall Plan Modernisme in Divided Germany," in *Cold War Modern: Design 1945–1970*, ed. David Crowley and Jane Pavitt (London: V&A Publishing, 2008); *Cold War on the Home Front: The Soft Power of Midcentury Design* (Minneapolis: University of Minnesota Press, 2010); Gay McDonald, "The 'Advance' of American Postwar Design in Europe: MoMA and the Design for Use, USA Exhibition 1951–1953," *Design Issues* 24, no. 2 (2008); "The Modern American Home as Soft Power: Finland, MoMA and the 'American Home 1953' Exhibition," *Journal of Design History* 23, no. 4 (2010); "Selling the American Dream: MoMA, Industrial Design and Post-War France," *Journal of Design History*

17, no. 4 (2004); Per Lundin and Thomas Kaiserfeld, eds., *The Making of the European Consumption. Facing the American Challenge* (Basingstoke: Palgrave Macmillan, 2015); David Crowley and Jane Pavitt, eds., *Cold War Modern: Design 1945-1970* (London: V&A Publishing, 2008); Robin Schuldenfrei, ed. *Atomic Dwelling. Anxiety, Domesticity, and Postwar Architecture* (London and New York: Routledge, 2012); Paul Betts, *The Authority of Everyday Objects: A Cultural History of West German Industrial Design* (Berkeley: University of California Press, 2004); Kjetil Fallan, "Cursing 'the Chromium-Plated Calf': Norwegian Fears and Desires of American Industrial Design in the 1950s," in *Appropriating America* (Amsterdam: ESF Inventing Europe, 2009); Mary Nolan, *The Transatlantic Century. Europe and America, 1890-2010*, ed. William Beik, T.C.W. Blanning and Brendan Simms, New Approaches to European History (New York: Cambridge University Press, 2012).

49 Per Lundin, "Introduction," in *The Making of the European Consumption. Facing the American Challenge*, ed. Per Lundin and Thomas Kaiserfeld (Palgrave Macmillan, 2015); Dmitry Azrikan, "VNIITE, Dinosaur of Totalitarianism or Plato's Academy of Design?," *Design Issues* 15, no. 3 (1999); Ruth Oldenziel and Adri Albert de la Bruhèze, "Europe's Mediation Junction: Technology and Consumer Society in the 20th Century," *History and Technology* 21, no. 1 (2005); Crowley and Pavitt, *Cold War Modern: Design 1945-1970*; Jack Masey and Conway Lloyed Morgan, *Cold War Confrontations. US Exhibitions and Their Role in the Cultural Cold War* (Baden: Lars Müller, 2008).

50 "[...] van zijn nijverheid op het hoogste niveau, waarin schranderheid, originaliteit, logica, durf, eerlijkheid en economie de synthese vormen van hetgeen Europa bijbrengt aan de wereldproductie." See "Het leven van het Instituut," 2.

51 This idea was especially developed in the talk by Pol Provost and Tomás Maldonado at the November 1962 symposium: see "De industrial design en de gemeenschappelijke markt." Similar ideas prevailed in the talk by Josine des Cressonnières in December 1962 at Fabrimetal. See Brighton, University of Brighton, Design Archives, ICSID, 9-23-2: *Exposé fait par Madame des Cressonnières, Secrétaire Générale de l'Institut d'Esthétique Industrielle, dans Le cadre du Séminaire du 15 décembre 1962, à Fabrimetal.*

52 Ibid., 6.

53 Nolan, *The Transatlantic Century. Europe and America, 1890-2010*, 203-04.

54 Ibid., 171.

55 A summary of this lecture is found in "Industriële Vormgeving," *Informatieblad van het Instituut voor Industriële Vormgeving* 5, no. 3 (1961).

56 This Janus-faced reception of US material culture in Europe is described by many design historians. In Germany, as design historian Paul Betts points out, American design, except for Eames or Knoll furniture, was

rejected because of its dishonest and irresponsible nature. Design historian Kjetil Fallan illustrates the European ambivalent attitude towards design with examples from Scandinavia, Great Britain and Italy. In Belgium, the reaction is similar: most of the leftish, humanistic and artistic critics that rejected the "American" design were palmed off with MoMA's "International style." As early as 1934 the MoMA rejected the styling of objects and principles such as streamlining with its canonical exhibition *Machine Art*. See McDonald, "The 'Advance' of American Postwar Design in Europe: MoMA and the Design for Use, USA Exhibition 1951–1953." Betts, *The Authority of Everyday Objects: A Cultural History of West German Industrial Design*, 10. Fallan, "Cursing "the Chromium-Plated Calf": Norwegian Fears and Desires of American Industrial Design in the 1950s;" "Love and Hate in Industrial Design: Europe's Design Professionals and America in the 1950s," in *The Making of the European Consumption. Facing the American Challenge*, ed. Per Lundin and Thomas Kaiserfeld (Palgrave Macmillan, 2015); Riley and Eigen, "Between the Museum and the Marketplace: Selling Good Design; Jennifer Jan Marshall, "In Form We Trust: Neoplatonism, the Gold Standard, and the "Machine Art" Show, 1934," *The Art Bulletin* 90, no. 4 (2008).

57 Luuk van Middelaar, *De passage naar Europa. Geschiedenis van een begin*, 4 ed. (Groningen: Historisch Uitgeverij, 2011), 435; Rik Coolsaet, *België en zijn buitenlandse politiek 1830–2000*, 3rd ed. (Leuven: Van Halewyck, 2001).

58 Nolan, *The Transatlantic Century. Europe and America, 1890–2010*, 204.

59 "De grote ondernemingen zouden bijvoorbeeld kunnen proberen de vindingrijke traditie van de vroegere nijverheid over te nemen en alzo de concurrentie te herwinnen die gebaseerd was op de integrale hernieuwing van het product [...]" See "De industrial design en de gemeenschappelijke markt," 8.

60 "In plaats van ons aan te zetten tot buitensporige verkwisting van onze energie- en materiaalvoorraad, in plaats van zekere behoeften prestige en opzichtigheid te voldoen, zal de industrieel design, indien ze goed opgevat wordt, bijdragen tot een betere verdelen van deze energie en materialen en aldus de mens helpen in zijn kulturele en sociale vooruitgang" See "De industrial design en de gemeenschappelijke markt," 8.

61 The editorial board of the French magazine *Esthétique Industrielle* similarly advocated in a reaction to the short account of the American Designer Muller-Münk a harmonized European production where the national would become neutralized "Ces créations s'inscrivent dans l'esprit des productions contemporaines sans que l'on puisse leur attribuer de nationalité précise si ce n'est celle de l'Europe." See Muller Münk, «Situation de l'esthétique industrielle dans l'Europe du Marché Commun.»

62 Sparke, *An Introduction to Design and Culture. 1900 to the Present*, 206.

63 "Même avec la standardization croissante il n'y a pas de raison de s'attendra à ce que des produits de nationalité différents deviennent absolument indiscernables. Par exemple, dans l'Europe du moyen âge, on a peint des madones pendant des siècles. Le convictions religieuses fixaient des limitations très rigides sur la compositions et même la couleur (souvent la robe de la madone devait être bleu). Malgré cela les madones peintes par les artistes de chaque pays restaient tout à fait différentes les unes des autres. Il existe actuellement au moins autant 'influences nationales et elles prouveront vraisemblablement leur importance en conservant aux produits manufactures de l'avenir l''aspect national' de leur esthétique. En fait, la sorte de concurrence et les autres conditions qui vont se développer à l'intérieur du Marché Commun et dans la zone adjacente dite des 7 Nations (Grande-Bretagne, Suisse, Autriche, Portugal et Pays Scandinaves) devront aider les caractéristiques des produits de chaque pays à être définies plus clairement" See Muller Münk, «Situation de l'esthétique industrielle dans l'Europe du Marché Commun.»

64 A similar case is the example of the automobile industry: see Sparke, *An Introduction to Design and Culture. 1900 to the Present*, 206–07.

65 Els De Vos, "The American Kitchen in Belgium? A Story of Countering, Reversing, Selective Appropriation and Slidelining," in *The Tensions of Europe/Inventing Europe Working Papers Series* (2009).

66 "La photographie de notre page de couverture est à cet égard caractéristique. Sur un plateau Formica, matériau international par excellence, sont disposées des assiettes suédoises, des couverts français et de la verrerie sarroise. Le lecteur non prévenu ne le devinera pas, mais fout-il le regretter? Pour notre part nous prendrions volontiers place à cette table européenne" See Muller Münk, «Situation de l'esthétique industrielle dans l'Europe du Marché Commun.»

67 It features in the book of photographs entitled *Europa Unita*, published in 1966 under the auspices of the EEC. Karel van Milleghem, ed. *Europa Unita* (Tielt/Den Haag: Lannoo, 1966).

68 This is illustrated in a 1968 broadcast for the French-speaking Community of Belgium on the real effects of the customs union. While the voice-over proclaims "L'Europe fait vendre" (Europe sells), the camera shows stills of neon signs and placards of shops and companies whose names were inspired by the alluring idea of Europe. See Fred Van Besien, «Marché Commun ou demain l'Europe (Partie 1),» in *9 millions* (Belgium: RTBF (Radio Télévision Belge Francophone), 1968). Rik Coolsaet also mentions that European integration became a synonym for economic expansion and affluence in Belgium, see: Coolsaet, *België en zijn buitenlandse politiek 1830-2000*, 437.

69 Katarina Serulus, "Meurop in Holland 1959-1960. The Mediation of Popular Design" (VU University Amsterdam, 2011); "Het plastic meubel in België (1954-1973)" (Katholieke Universiteit Leuven, 2010);

Javier Gimeno-Martínez, "Redefining Social Design in 1970s Belgium: Affordable Design vs. Elite Design," *Interiors* 2, no. 2 (2011).

70 Jean Rey and Emile Bernheim maintained a correspondence on European and economic issues. Bernheim even asked Rey for advice when he was invited to London to give a talk to British industry and commerce representatives. He wanted to take advantage of this situation to advocate the European idea. See Brussels, ULB, Jean Rey (mainly 1921–1982), ARCHIULB/126PP, box 88. Bernheim also published a study entitled "Pour créer une mentalité européenne" for the Société Belge d'Études & d'Expansion in 1951. See Brussels, ULB,GIB archives, ARCHIULB/25Z, MON000/760: E. Bernheim, *Pour créer une mentalité européenne*, 1951.

71 Victoria de Grazia, *Irresistible Empire. America's Advance through 20th Century Europe* (Cambridge/London: The Belknap Press of Harvard University Press, 2005), 400.

72 Brussels, ULB, Jean Rey (mainly1921–1982), ARCHIULB/126PP, box 88: *Un sondage européen en Belgique.*

73 Grands Magasins Au Bon Marché, *Les Grands Magasins À l'Innovation 1959–1960* (Brussels1960), 46. Les Grands Magasins "À l'Innovation" were very much concerned with the liberalization of the Market. Already in 1948 together with the Grands Magasins du Bijenkorf in the Netherlands it organized an exhibition called *Benelux. Documentaire et educative.* See Brussels, ULB, GIB archives, ARCHIULB/25Z, MON002/752: *Declaration faite par Monsieur Bernheim, Administrateur Delegue des Grands Magasins* "À l'*Innovation" à la conference de presse du 21 mai 1948, consacrée à l'Exposition Benelux.*

74 Brussels, ULB, GIB archives, ARCHIULB/25Z, PHO001/867: Album photo «BM Expo Europe 1956.»

75 "Belgische Industrial Design-selektie 1963," *Instituut voor Industriële Vormgeving Informatieblad* 6, no. 2 (1963).

76 "Een Design Centre te Brussel," *Informatieblad van het Instituut voor Industriële Vormgeving* 6, no. 2 (1963).

77 Its Dutch and French equivalents that were mostly used in the archival documents: Belgisch Instituut voor Industriële Vormgeving/Institut Belge pour le Développement de l'Industrial Design.

78 This name change coincided with a change in its statute. Previously a not-for-profit organization, it was turned into a public utility organization in 1969. See "N. 835", Annexe au Moniteur Belge. Associations sans but lucratif et établissements d'utilité publique/Bijlage tot het Belgisch Staatsblad. Verenigingen zonder winstgevend doel en instellingen van openbaar nut [Attachment to the Belgian Official Journal. Not-for-profit Organizations and Public Utility Foundations], February 13, 1969, p. 411–412.

79 «Belgische Industrial Design-selektie 1963.»

80 Haerens and des Cressonnières, «Design in België 1940–1984.»

81 "N. 477", Annexe au Moniteur Belge. Associations sans but lucratif et établissements d'utilité publique/Bijlage tot het Belgisch Staatsblad. Verenigingen zonder winstgevend doel en instellingen van openbaar nut [Attachment to the Belgian Official Journal. Not-for-profit Organizations and Public Utility Foundations], February 1, 1962, p. 212–214.

82 This part of the essay is based on various reports of the Design Centre, currently preserved in Brussels. See Brussels, State Archives: BDOP/OBAP, 1081: Procès-*verbal de la réunion du Conseil d'Administration du 30 novembre 1967*; BDOP/OBAP, 1081: *Séance plénière du Comité de Direction, 3 juillet 1968*; BDOP/OBAP, 1082: *Procès-verbal de la réunion de l'Assemblée Générale 26 mars 1970*; BDOP/OBAP, 1083: *Compte rendu de la réunion du Conseil d'Administration du 14 septembre 1972.*

83 "Le Design Centre belge étant le mieux organisé parmi ceux des six pays de la Communauté Economique Européenne, il pourrait se trouver en position de devenir le premier Design Centre «Marché Commun». Il importe dès lors d'étudier à fond le problème." See Brussels, State Archives, BDOP/OBAP, 1081: Procès-*verbal de la réunion du Conseil d'Administration du 30 novembre 1967.*

84 Brighton, University of Brighton, Design Archives, ICSID, 9-12-2: *Definition of a Design Centre.*

85 "La quatrième difficulté à signaler vient du statut national du Design Centre par rapport à un marché en voie rapide d'internationalisation. Il devient souvent difficile de dire si un produit est belge: tantôt sa conception l'est, mais sa réalisation est fragmentée entre diverses usines réparties dans le Marché Commun, tantôt c'est le contraire. Le fait national n'est plus adapté aux structures économiques de demain. Les associations d'entreprises, par-dessus les frontières, rationalisent la production en la spécialisant – mais le Design Centre se voit parfois contraint de refuser tel objet parce que, par souci de rationalisation, il est désormais fourni par la société-sœur (française ou allemande) d'une usine belge qui, en retour, se spécialise dans un secteur voisin. Vis-à-vis du public, il apparaît que les confrontations internationales sont d'un intérêt très vif. Le Design Centre le fait périodiquement, lors d'échanger d'expositions avec certains Centres étrangers. Mais on peut imaginer que l'avenir s'oriente vers une confrontation permanente, c'est-à-dire un type de Design Centre international.» See Josine des Cressonnières, «Le Design Centre a cinq ans,» *Industrie* 1969, 362.

86 Ibid.

87 «Le président résume le problème: il devient de plus en plus difficile de déterminer avec précision la nationalité de nombreux produits, vu l'internationalisation des complexes de production.» See Brussels, State Archives, BDOP/OBAP, 1083: *Compte rendu de la réunion du Conseil d'Administration du 14 septembre 1972.*

88 «Les membres se prononcent unanimement pour un principe de liberté. Il faut cependant tenir compte du fait que le DC est financièrement aidé par l'Office Belge du Commerce Extérieur, et doit de ce fait demeurer un outil de progrès et de promotion national.» See Brussels, State Archives, BDOP/OBAP, 1083: *Compte rendu de la réunion du Conseil d'Administration du 14 septembre 1972.*

89 Guy Rouckaerts, "15 jaar Design-promotie. Design Centre: Lukse of noodzaak?," *Komunikatief* 1979.

90 Josine des Cressonnières and Jean-Jacques Stiefenhofer, «Dossier: Het Design Centre: Tien jaar onderweg,» *A+*, no. 9/10 (1974).

91 "Meubelen uit België. Van Henry van de Velde tot heden/Mobilier de Belgique de Henry van de Velde à nos jours," ed. Design Centre (Brussels1978).

92 "Design uit Nederland," ed. Visual Arts Office for Abroad (Amsterdam: Visual Arts Office for Abroad, 1980); Joana Ozorio de Almeida Meroz, "The International as National: The Role of International Cultural Policy in the Construction of Dutch Design as Conceptual" (paper presented at the 9th International Committee Design History and Design Studies. Tradition, transition, trajectories: major or minor influences?, Aveiro, 2014).

Chapter IV: Design Exchanges between Brussels and Moscow in the Early 1970s

1 This chapter is based on the author's article published in a special issue of the design and culture journal "Design as Object of Diplomacy Post-1945" edited by dr. Harriet Atkinson and dr. Verity Clarkson: Katarina Serulus, ""Well-Designed Relations": Cold War Design Exchanges between Brussels and Moscow in the Early 1970s," *Design and Culture* 9, no. 2 (2017). First presented as a paper at the symposium *From 'Soft' to 'Hard' Power? Changing Visions of Diplomacy by Design, 1945 Onwards* at the University of Brighton, November 26, 2016. The author wishes to thank the editors dr. Harriet Atkinson and dr. Verity Clarkson for their invaluable comments on the first drafts of this article. Many thanks are also extended to Veronika Belik for helping to find relevant files within the VNIITE archive.

2 After its première in Brussels, the VNIITE exhibition was shown in Germany, Finland and India: Dmitry Azrikan, "VNIITE, Dinosaur of Totalitarianism or Plato's Academy of Design?," *Design Issues* 15, no. 3 (1999): 64.

3 Per Lundin and Thomas Kaiserfeld, eds., *The Making of the European Consumption. Facing the American Challenge* (Basingstoke: Palgrave Macmillan, 2015); Kjetil Fallan, "Love and Hate in Industrial Design:

Europe's Design Professionals and America in the 1950s," in *The Making of the European Consumption. Facing the American Challenge*, ed. Per Lundin and Thomas Kaiserfeld (Palgrave Macmillan, 2015); Gay McDonald, "Selling the American Dream: MoMA, Industrial Design and Post-War France," *Journal of Design History* 17, no. 4 (2004); "The 'Advance' of American Postwar Design in Europe: MoMA and the Design for Use, USA Exhibition 1951–1953," *Design Issues* 24, no. 2 (2008); "The Modern American Home as Soft Power: Finland, MoMA and the 'American Home 1953' Exhibition," *Journal of Design History* 23, no. 4 (2010); Greg Castillo, "Domesticating the Cold War: Household Consumption as Propaganda in Marshall Plan Germany," *Journal of Contemporary History* 40, no. 2 (2005); Victoria de Grazia, *Irresistible Empire. America's Advance through 20th Century Europe* (Cambridge/London: The Belknap Press of Harvard University Press, 2005).

4 Mary Nolan, *The Transatlantic Century. Europe and America, 1890–2010*, ed. William Beik, T.C.W. Blanning and Brendan Simms, *New Approaches to European History* (New York: Cambridge University Press, 2012).

5 For example: Fredie Floré, "Architect-Designed Interiors for a Culturally Progressive Upper-Middle Class: The Implicit Political Presence of Knoll International in Belgium," in *Atomic Dwelling. Anxiety, Domesticity, and Postwar Architecture*, ed. Robin Schuldenfrei (London and New York: Routledge, 2012).

6 Susan E. Reid and Rosalind P. Blakesley, eds., *Russian Art and the West. A Century of Dialogue in Painting, Architecture, and the Decorative Arts* (Dekalb: Northern Illinois University Press, 2007); Verity Clarkson, "'Sputniks and Sideboards': Exhibiting the Soviet 'Way of Life' in Cold War Britain, 1961–1979," in *A People Passing Rude: British Responses to Russian Culture*, ed. Anthony Cross (Cambridge: Open Book Publishers, 2012); Peter Romijn, Giles Scott-Smith and Joes Segal, eds., *Divided Dreamworld? The Cultural Cold War in East and West* (Amsterdam: Amsterdam University Press, 2012); Sandrine Kott, "Par-delà la guerre froide. Les organisations internationales et les circulations est-ouest (1947–1973),» *Vingtième Siècle Revue d'histoire*, no. 109 (2011).

7 Reid and Blakesley, *Russian Art and the West. A Century of Dialogue in Painting, Architecture, and the Decorative Arts*, 15.

8 Javier Gimeno-Martínez, "The Signe d'Or Award Scheme from 1956 to 1960: The Economic Reasons for "Good Design"," *Konsthistorisk Tidskrift/Journal of Art History* 79, no. 3 (2010): 127.

9 Arthur J. Pulos, *The American Design Adventure, 1940–1975* (Cambridge: MIT Press, 1988), 209–21.

10 Tania Messel, "International Norms and Local Design Research: ICSID and the Promotion of Industrial Design in Latin America, 1970–1979" (paper presented at the DRS2016: Design + Research + Society - Future-Focused Thinking, Brighton, 2016).

11 Pulos, *The American Design Adventure, 1940–1975*, 210.

12 The design journalist Christoph Junck refutes the idea that the London Design Centre was the first. In his series "Die Design Centres der Welt" in the magazine *Moebel Interior Design*, he claims that the Sammlung Industrieform in Essen was actually the first since it opened as early as in 1955, a year before the London example. See Christoph Junck, "Die Design Centres der Welt," *Moebel Interior Design*, no. 1 (1967).

13 Brighton, University of Brighton Design Archives, ICSID, 03-3-1 Minutes and texts: *ICSID Vth General Assembly. Ottawa - Canada - 11 & 12 september 1967* [Internal Report].

14 Yuri Soloviev, *My Life in Design* (Moscow1994); E. Susan Reid, "This Is Tomorrow! Becoming a Consumer in the Soviet Sixties," in *The Socialist Sixties. Crossing Borders in the Second World*, ed. Anne E. Gorsuch and Diane P. Koenker (Bloomington: Indiana University Press, 2013), 61; Raymond Hutchings, "The Weakening of Ideological Influences Upon Soviet Design," *Slavic Review* 27, no. 1 (1968).

15 For more information on the VNIITE see Tom Cubbin, "The Domestic Information Machine. Futurological Experiments in the Soviet Domestic Interior, 1968–76," *Home Cultures* 11, no. 1 (2014); Andres Kurg, "Prehistory of Contemporary Environment: Multimedia Program in the ICSID'75 Congress in Moscow," in *The Islands of Yuri Sobolev*, ed. Anna Romanova and Galina Metelichenko (Moscow: The Moscow Museum of Modern Art, 2014); Yulia Karpova, "Accommodating 'Design': Introducing the Western Concept into Soviet Art Theory in the 1950s-60s," *European Review of History: Revue européenne d'histoire* 20, no. 3 (2013); "'A Glass without a Bottom': Neodecorativism in Late 1960s Soviet Design," *Journal of Design History* 30, no. 1 (2016); Azrikan, "VNIITE, Dinosaur of Totalitarianism or Plato's Academy of Design?; Raymond Hutchings, "Soviet Design: The Neglected Partner of Soviet Science and Technology," *Slavic Review* 37, no. 4 (1978).

16 Brighton, University of Brighton Design Archives ICSID, 09-11-1 Correspondence (1957–59): Muller-Münk, Peter, [Letter addressed to Josine des Cressonnières], (January 22, 1959).

17 Brussels, State Archives, BE ARA. BDOP (510-2227), I504, no. 65: Josine des Cressonnières, Letter to the administrators of the Design Centre (September 19).

18 Katarina Serulus, "Take up Position: Defining Industrial Design in Postwar Belgium (1954–1970)," in *9th International Committee Design History and Design Studies. Tradition, Transition, Trajectories: Major or Minor Influences?*, ed. Helena Barbosa and Anna Calvera (Aveiro: UA Editora, 2014).

19 Fallan, "Love and Hate in Industrial Design: Europe's Design Professionals and America in the 1950s; McDonald, "The 'Advance' of American Postwar Design in Europe: MoMA and the Design for Use, USA

Exhibition 1951–1953; Paul Betts, *The Authority of Everyday Objects: A Cultural History of West German Industrial Design* (Berkeley: University of California Press, 2004), 10; Terence Riley and Edward Eigen, "Between the Museum and the Marketplace: Selling Good Design," in *The Museum of Modern Art at Mid-Century. At Home and Abroad*, ed. John Elderfield, *Studies in Modern Art* (New York: The Museum of Modern Art, 1994).

20 "Een Design Centre te Brussel," *Informatieblad van het Instituut voor Industriële Vormgeving* 6, no. 2 (1963): 9.

21 "Het leven van het Instituut. Dhr. Paul Reilly op bezoek te Brussel," *Informatieblad van het Instituut voor Industriële Vormgeving* 6, no. 2 (1963).

22 Josine des Cressonnières and Georges Burnay, "Le Design Centre," *Industrie* 19, no. 1 (1965): 30.

23 Paul Betts, "Science, Semiotics and Society: The Ulm Hochschule für Gestaltung in Retrospect,» *Design Issues* 14, no. 2 (1998).

24 Dmitry Azrikan, "VNIITE, Dinosaur of Totalitarianism or Plato's Academy of Design?," ibid.15, no. 3 (1999): 63.

25 Ibid., 48; Karpova, "Accommodating 'Design': Introducing the Western Concept into Soviet Art Theory in the 1950s-60s," 635; Soloviev, *My Life in Design*.

26 Azrikan, "VNIITE, Dinosaur of Totalitarianism or Plato's Academy of Design?," 63–4.

27 Josine des Cressonnières and Jean-Jacques Stiefenhofer, «Dossier: Het Design Centre: Tien jaar onderweg,» *A+*, no. 9/10 (1974): 18. Not all Design Centers were as active or well situated in the network: see for example the short-lived Greek Design Centre: Artemis Yagou, "Unwanted Innovation. The Athens Design Centre (1961–1963)," *Journal of Design History* 18, no. 3 (2005).

28 Brussels, State Archives, BE ARA. BDOP (510–2227), I504, no. 1081: *Rapport d'activités 1969*, p. 6; Antwerp, Collection Flanders Architecture Institute - Collection Flemish Community, Archive Jean-Jacques Stiefenhofer (1943–2013), Doos 24.2: Josines des Cressonnières, *Rapport concernant la mission effectuee au bresil par mme des cressonnières, 4–30 avril 1973*.

29 Josine des Cressonnières' maiden name was Naus. She was the widow of Denis des Cressonnières (1919–1952). Interview by the author with Geraldine des Cressonnières on November 8, 2013, Coutisse; «La pionnière de la pêche aux nouveautés,» *Entre Nous. Le Magazin de l'Inno* 32(1954).

30 "N. 477". 1962. Annexe au Moniteur Belge. Associations sans but lucratif et établissements d'utilité publique/Bijlage tot het Belgisch Staatsblad. Verenigingen zonder winstgevend doel en instellingen van openbaar nut [Attachment to the Belgian Official Journal. Not-for-profit Organizations and Public Utility Foundations] (1 February): 212–214.

31 Richard Carr, "Overseas. Design Centre in Brussels," *Design*, no. 185 (1964); des Cressonnières and Burnay, "Le Design Centre."

32 Azrikan, "VNIITE, Dinosaur of Totalitarianism or Plato's Academy of Design?"

33 For example, the Brussels Design Centre prepared a Belgian design exhibition for the Belgian Office for Foreign Trade in the frame of the Belgian Weeks in Budapest (1966), Prague (1966) and Warsaw (1968). This eye-catching exhibition was part of the commercial campaign to promote Belgian products on the Eastern markets. Brussels, State Archives, BE ARA. BDOP (510-2227), I504, no. 1081: *Rapport d'activités 1968.*

34 Brighton, University of Brighton Design Archives, Council of Industrial Design, Box 157: Reilly, Paul, [Letter to Josine des Cressonnières], (18/08/1972).

35 Ibid.

36 Brighton, University of Brighton Design Archives, ICSID, 04-2 Board meeting 1971–1974: *Minutes Restricted Board Meeting n° 9 - Brussels 8/9 September 1973*, p. 2–4.

37 Brighton, University of Brighton Design Archives, ICSID, 04-2 Board meeting 1971–1974: Josine des Cressonnières, *Note for the consideration by the executive board*, (March 23, 1973); Brighton, University of Brighton Design Archives, ICSID, 04-2 Board meeting 1971–1974: *Recapitulation of opinions exchanged during the Executive Board meeting in Milan (13-17 July 1973) concerning the proposition of moving ICSID Secretariat to London.*

38 The tide turned in 1974, when Carl Aubock was chosen as the new ICSID President at the General Assembly in Kyoto. Under his chairmanship, Josine des Cressonières could continue with her ambition and transfer the ICSID seat from Paris to Brussels, which only strengthened her connections to the heart of the transnational design community. Brighton, University of Brighton, Design Archives, ICSID, 04-2 Board meeting 1971–1974: *Minutes Newly Elected Executive Board Meeting - Kyoto 13 October 1973*, p. 1.

39 Michael Ryckewaert, *Building the Economic Backbone of the Belgian Welfare State. Infrastructure, Planning and Architecture 1945–1973* (Rotterdam: 010, 2011), 175–97; Carola Hein, *The Capital of Europe: Architecture and Urban Planning for the European Union* (Westport/London: Praeger, 2004), 136–59.

40 Vincent Dujardin, "Go-Between: Belgium and Détente, 1961–73," *Cold War History* 7, no. 1 (2007); Maarten Van Alstein, "From Enigma to Enemy. Paul-Henri Spaak, the Belgian Diplomatic Elite, and the Soviet Union, 1944–1945," *Journal of Cold War Studies* 13, no. 3 (2011); Frank Gerits, "Taking Off the Soft Power Lens. The United States Information Service in Cold War Belgium (1950–1958)," *Journal of Belgian History* 4, no. XLII (2012).

41 Brussels, State Archives, Archief Belgische Dienst voor Buitenlandse Handel (BDBH): *Ontwerp van Bedrijvigheidsverslag* [Internal Report], 1971, p. 39–40.

42 Brussels, State Archives, Archief Belgische Dienst voor Buitenlandse Handel (BDBH): Notulen van de 125ste vergadering van de Raad van Beheer vrijdag 23 juni, 1972, p.6.

43 Michael L. Dockrill and Michael F. Hopkins, *The Cold War: 1945–1991*, Second Edition ed., Studies in European History (Basingstoke, New York: Palgrave Macmillan, 2006), 120.

44 Soloviev, *My Life in Design.*

45 Due to political difficulties with visas for South Africa and Taiwan, the General Assembly could not proceed in Moscow. Only the conference was held in the Soviet capital. The General Assembly was evacuated to Brussels and held some months later in 1976. Brighton, University of Brighton, Design Archives, ICSID, 04-3 Board meeting 1975–1976: des Cressonnières, Josine, *ICSID and Politics*, (January 1976).

46 Moscow, Moscow Technological University, VNIITE Archives, *Report n° 1001* [internal report], 1973.

47 Moscow, Moscow Technological University, VNIITE Archives, *Report n° 1001* [internal report], 1973; "Sovjetrussische industrie in Design Centre. Vormgeving wordt dikwijls van staatswege opgelegd," *De Standaard*, 29.3.73 1973; H.D.W., "Russische vormgeving in wereldpremière te Brussel," *Het Laatste Nieuws*, 29.3.73 1973.

48 For example, see Clarkson, "'Sputniks and Sideboards': Exhibiting the Soviet 'Way of Life' in Cold War Britain, 1961–1979." and Lewis Siegelbaum, "Sputnik Goes to Brussels: The Exhibition of a Soviet Technological Wonder," *Journal of Contemporary History* 47, no. 1 (2012).

49 H.D.W., "Russische vormgeving in wereldpremière te Brussel."

50 Jacques Poncin, «Echange d'expositions de design entre Moscou et Bruxelles. Le project de présentation des produits belges en U.R.S.S. doit se concrétiser dans le cadre de la visite de notre mission économique,» *Le Soir*, 29.5.73 1973.

51 E. Susan Reid, "Communist Comfort: Socialist Modernism and the Making of Cosy Homes in the Khrushchev Era," *Gender & History* 21, no. 3 (2009).

52 H.D.W., "Russische vormgeving in wereldpremière te Brussel."

53 Reid, "Communist Comfort: Socialist Modernism and the Making of Cosy Homes in the Khrushchev Era."

54 "Maar van een verassende frisheid getuigen een mini-scooter en een motorfiets, die de toets met de modernste Japanse machines glansrijk doorstaan." H.D.W., «Russische vormgeving in wereldpremière te Brussel.»

55 Poncin, «Echange d'expositions de design entre Moscou et Bruxelles. Le project de présentation des produits belges en U.R.S.S. doit se concrétiser dans le cadre de la visite de notre mission économique.»

56 "Sovjetrussische industrie in Design Centre. Vormgeving wordt dikwijls van staatswege opgelegd."

57 «Echange d'expositions de design entre Moscou et Bruxelles. Le project de présentation des produits belges en U.R.S.S. doit se concrétiser dans le cadre de la visite de notre mission économique.»

58 "[...] een nostalgieke getuigenis van een verleden dat door Tolstoi beschreven werd." H.D.W., "Russische vormgeving in wereldpremière te Brussel."

59 Ibid.

60 E. Susan Reid and David Crowley, eds., *Style and Socialism. Modernity and Material Culture in Post-War Eastern Europe* (Oxford/New York: Berg, 2000); Reid and Blakesley, *Russian Art and the West. A Century of Dialogue in Painting, Architecture, and the Decorative Arts*; Reid, "Communist Comfort: Socialist Modernism and the Making of Cosy Homes in the Khrushchev Era; "This Is Tomorrow! Becoming a Consumer in the Soviet Sixties.»

61 Poncin, «Echange d'expositions de design entre Moscou et Bruxelles. Le project de présentation des produits belges en U.R.S.S. doit se concrétiser dans le cadre de la visite de notre mission économique.»

Chapter V: In Search of a "National" Style (1964–1970)

1 Brussels, State Archives, BE-A0545/Fic 2000, Archive Design Centre, unnumbered: [document sans titre].

2 Foundation Interieur, ed. *Interieur. Design Biennale since 1968* (Kortrijk: Interieur Foundation, 2010); Monique Bucqoye, ed. *24th International Biennale Interieur 2014. The Book* (Kortrijk: Interieur Foundation, 2014); Katarina Serulus, "1968–2016, 25 Editions of Biennale Interieur," in *25th Silver Edition. Biennale Interieur 2016*, ed. Biennale Interieur (Kortrijk: Biennale Interieur NPO, 2016).

3 "N. 477", Annexe au Moniteur Belge. Associations sans but lucratif et établissements d'utilité publique/Bijlage tot het Belgisch Staatsblad. Verenigingen zonder winstgevend doel en instellingen van openbaar nut [Attachment to the Belgian Official Journal. Not-for-profit Organizations and Public Utility Foundations], February 1, 1962, pp. 212–214. The Design Centre was established by the trio of Pol Provost (1907–1990), Frans Wildiers (1905–1986) and Josine des Cressonnières (1926–1985).

4 The already existing institute and the new Design Centre were supposed to function in "tandem." The institute had to promote industrial design in Belgian industry, and the Design Centre had to make sure it was fruitful

through the export of Belgian design. However, the foundation of the Design Centre eventually eclipsed the Institute for Industrial Design that was in constant financial difficulties. See "Een Design Centre te Brussel," *Informatieblad van het Instituut voor Industriële Vormgeving* 6, no. 2 (1963).

5 "N. 477", Annexe au Moniteur Belge. Associations sans but lucratif et établissements d'utilité publique/Bijlage tot het Belgisch Staatsblad. Verenigingen zonder winstgevend doel en instellingen van openbaar nut [Attachment to the Belgian Official Journal. Not-for-profit Organizations and Public Utility Foundations], February 1, 1962, pp. 212–214.

6 The Belgian Office of Foreign Trade was established in 1948 at the initiative of Minister of Foreign Trade Straten-Waillet. The Office was a semi-governmental body that included representatives from the government as well as from trade and industry. In 1960, Minister Van Offelen redefined the role of the Office. It would be exclusively qualified for foreign trade through the organization of trade missions and trade fairs. It was also responsible for contacts with the private sector. In 1962 Prince Albert became its honorary president and became its figurehead for prestigious trade missions. See Rik Coolsaet, Vincent Dujardin, and Claude Roosens, *Buitenlandse Zaken in België. Geschiedenis van een ministerie, zijn diplomaten en consuls van 1830 tot vandaag* (Tielt: Lannoo, 2014), 286, 341.

7 "Een Design Centre te Brussel; Josine des Cressonnières and Georges Burnay, "Le Design Centre," *Industrie* 19, no. 1 (1965).

8 In the same vein as the Institute of Industrial Design, the Design Centre at the time of its opening employed a Belgo-Luxembourg scope. However, this transnational intention did not prevail in its policies which were explicitly focused on introducing industrial design in the Belgian industry to upsize its export. See «Le Design Centre.»

9 Henry Dreyfuss, «Marketing et industrial design sur le marché des Etats-Unis,» *Bulletin Commercial Belge* 66, no. 12 (1952). He also wrote a second article with a similar purpose: "L'industrial design et l'homme d'affaires," *Bulletin Commercial Belge* 67, no. 3 (1953).

10 Overseas markets were previously not of vital importance for Belgium. However, participating in the European market and the West-Atlantic bloc, new trade possibilities opened up. See Rik Coolsaet, *België en zijn buitenlandse Politiek 1830-2000*, 3rd ed. (Leuven: Van Halewyck, 2001), 419.

11 "La mention 'Holland', par exemple, évoque dans l'esprit de l'Américain moyen l'image d'un pays ou l'on trouve des tulipes en fleurs, des moulin à vent, des gens chaussés de sabots ou de patins, des fromages, et des canaux. Des images familières analogues viennent à l'esprit américain aux noms de 'France', 'England', 'Switzerland' et 'Spain.' Mais je doute qu'il en soit de même pour la Belgique. A plus forte raison, beaucoup de mes compatriotes ne connaissent-ils pas autant qu'ils le devraient les fins textiles, ni les machines d'excellente qualité fabriqués en Belgique." See

Dreyfuss, «Marketing et industrial design sur le marché des Etats-Unis,» 58.

12 The enquiry "Products and People" was undertaken at the request the Reader's Digest by local research organizations. In Belgium is was undertaken by "Société Belge d'Economie et de Mathématique Appliquées S.A. (SOBEMAP). Belgisch Instituut voor Industriële Vormgeving, *Industriële vormgeving, innovatietechniek & een instrument voor dialoog met de verbruiker. Verslag van het onderzoek over de industriële vormgeving in de Belgische industrie* (Belgisch Instituut voor Industriële Vormgeving, 1968).

13 Brighton, University of Brighton Design Archives, ICSID Archive, 04-7 Board meeting 1968-1975 (Cahiers): *Comments by J. des Cressonnières on the Future of ICSID,* (March 1970).

14 "Quant aux raisons pour lesquelles la Belgique a pris ce retard, elles sont historiques, culturelles, économiques et même géographiques. Culturelles d'abord. Parce que nous avons toujours eu le goût de ce que est 'riche', nous aimons tout ce qui est haut en couleur, nous aimons les belles matières, nous aimons ce qui fais cossu, nous aimons la bonne nourriture. En fait, une joie de vivre. Contradictoire avec les objectifs primaires du design, caractérisés par une très grande pureté et un dépouillement qui trouvaient leur origine dans la volonté des créateurs de concilier les produits et leur fabrication en masse, dans une période de grande crise - celle de 1929 aux Etats-Unis notamment." See Brussels, State Archives, BE-A0545/Fic 2000, Archive Design Centre, unnumbered: article *Josine des Cressonnières: Le design comme moyen d'innover*, 1-14 décembre 1983, n°22, p. 1882 - 1885 [journalistic interview with Josine des Cressonnières, auteur et magazine unknown].

15 *The Arts of Belgium 1920-1960*, (Liège: G. Thone, 1960).

16 Albert Bontridder, *Dialoog tussen licht en stilte* (Antwerpen: Helios, 1963), 7.

17 Geert Bekaert et al., *Design. Made in Belgium. 1900-1994* (Kortrijk: Stichting Interieur, 1994), 154-56.

18 Ibid.

19 Ibid.

20 Marnix Beyen, "Tragically Modern. Centrifugal Sub-Nationalisms in Belgium, 1830-2009" (paper presented at the Contemporary Centrifugal Regionalism: Comparing Flanders and Northern Italy, Brussels, June 19-20, 2009); "De natie, de stad en de wereld. Belgische, Vlaamse en Waalse identiteitsvorming als antwoorden op de moderniteit," *Deus ex Machina. Tijdschrift voor Literatuur en Kunst* 26, no. 100 (2002).

21 For example, *The Arts of Belgium 1920-1960*.

22 "Quant aux raison géographiques, il s'agit principalement de notre situation à la croisée des chemins européens. Le fait que nous ayons été continuellement envahis a eu pour conséquence que nous avons toujours dû nous débrouiller. En fait, nous continuons à nous débrouiller." See

Brussels, State Archives, BE-A0545/Fic 2000, Archive Design Centre, unnumbered, article *Josine des Cressonnières: Le design comme moyen d'innover*, December 1–14, 1983, n°22, p. 1882 - 1885 [journalistic interview with Josine des Cressonnières, author and magazine unknown].

23 Bontridder, *Dialoog tussen licht en stilte*, 7.

24 "Nee, ik geloof dat wij in zo'n klein land leven en dat de invloedsferen van buitenaf zo groot zijn, dat wij geen Belgisch design met een typisch Belgisch gezicht hebben. Deze invloedsferen - o.a. door tijdschriften, bladen, gespecialiseerde revues - zijn zó groot en komen dus van buitenaf maar niet van het land zelf." See Brussels, State Archives, BE-A0545/Fic 2000, Archive Design Centre, 238: *BRT 2 Splinternieuws, dinsdag, 16 september 1980 - 16h.*

25 "Au point de vue économique, le risque reste grand de voir de trop nombreuses entreprises ne pas investir dans le risque de la création mais de ne regarder que ce qui se vend bien pour faire quelque chose "dans ce genre-là", c'est une attitude typiquement belge qui est due aussi au fait que nous sommes un petit pays et que la créativité représente en effet un risque. [...] Une autre explication est que nous sommes traditionnellement un pays d'ingénieurs, un pays d'industrie lourde: nous sommes un pays d'acier, un pays de grosses machines plus qu'un pays d'objets mobiliers fins. Nous sommes un pays ou ce sont des ingénieurs qui sont devenus des chefs d'entreprises en on mené ces entreprises. Les critères du design, c'est à dire le souci de la forme, le souci de la mode, il les ont considérés comme très secondaires." See Brussels, State Archives, BE-A0545/Fic 2000, Archive Design Centre, unnumbered: article *Josine des Cressonnières: Le design comme moyen d'innover*, 1–14 décembre 1983, n°22, p. 1882 - 1885 [journalistic interview with Josine des Cressonnières, author and magazine unknown].

26 Brussels, State Archives, BE-A0545/Fic 2000, Archive Design Centre, 238: *BRT 2 Splinternieuws, dinsdag, 16 september 1980 - 16h*; Belgisch Instituut voor Industriële Vormgeving, *Industriële vormgeving, innovatietechniek & een instrument voor dialoog met de verbruiker. Verslag van het onderzoek over de industriële vormgeving in de Belgische industrie.*

27 Bekaert et al., *Design. Made in Belgium. 1900–1994*, 160–61.

28 Gent, Archief Adelbert Van de Walle, Museumarchief Design Museum Gent, unnumbered, folder comm. national esth. industr: *compte-rendu de la 9ème renurion du groupe de travail pour l'étude de l'industrial design, tenu le 21 janvier 1960 au Ministère des Affaires économiques.*

29 "Les lauréats du 'Signe d'Or' nous prouvent que c'est possible. Sur les marches extérieurs ils nous défendent. A l'assemblée générale de Stockholm, ils ont fait mieux: ils ont aide à recréer un "visage" à la Belgique. Et c'est là un atout auquel les pays concurrents attachment une énorme importance. Quand les Etats-Unis exposent à la Triennale de Milan, ils n'en attendant guère de conséquences immédiates, mais

ils exposent le "visage" des Etats-Unis; ils vendent la manière de vivre américaine, le standing américain. Quel est à l'étranger le visage de la Belgique? Trop souvent, il est resté celui des dentelles de Bruge et du Manneken Pis! Mais si Olivetti, Farina, Gio Ponti peuvent faire à l'Italie un visage moderne, les "Signe d'Or" peuvent témoigner que la Belgique, elle aussi, travaille pour s'inscrire dans la grand linge de la meilleur production du vingtième siècle. See Josine des Cressonnières, «L'Esthétique Industrielle et le Signe d'Or 1959,» *Industrie* 1959, 664.

30 Only in the mid-1980s would the Design Centre include some references to Henry van de Velde as the father of Belgian industrial design. It was not so much the actuality of his legacy that was underlined - as was the case in the 1950s - but rather his historical importance. See Jo Haerens and Josine des Cressonnières, "Design in België 1940-1984," ed. Museum voor Moderne Kunst van Brussel (Brussel: Koninklijke Musea voor Schone Kunsten van België, 1984).

31 This analysis of the Design Centre's interior features is based on Katarina Serulus, "Het interieur van het Design Centre in Brussel (1964-1986). Belgisch design gekaderd," in *Vormgevingserfgoed in Vlaanderen in 1945*, ed. Ellen Van Impe and Eva Van Regenmortel (Antwerpen: Art Paper Editions & CVAa, 2016).

32 Marthe Blanpain, *Le patrimoine monumental de la Belgique. Bruxelles 1C.* (Brussels: Mardaga, 1989), 163-64.

33 Beveren, State Archives, BKM collection (Archives of the Department of Fine Arts and Museums, Ministry of the Flemish Community/ Transfer 2005 [1945-2004]): BKM 2013: *Compte-rendu de la réunion de la Commission Design du 15-2-63*, fol. 40r.

34 "Nous avons voulu faire du Design Centre une oeuvre d'Industrial Design en soi," des Cressonnières and Burnay, «Le Design Centre.»

35 Gertjan Hoste, *Constantin Brodzki. Architecte* (Sprimont: Mardaga, 2004).

36 Brussels, Archives et Bibliothèques d'Architecture de l'Université Libre de Bruxelles, Fonds Brodzki, Classeur "Affaires déjà terminées": *Curriculum Vitae*; Stephanie van de Voorde, Sofie De Caigny, and Ronny De Meyer, *Isia Isgour 1913-1967*, ed. Sofie De Caigny, et al., (Antwerpen: Centrum Vlaamse Architectuurarchieven, 2008). Brodzki testified that through the personal contacts of his mother with the cabinet of politician Paul-Henri Spaak, a key figure in the European unification and the first president of the United Nations, he could start as a trainee at the construction site of the headquarters of the United Nations. More study trips to the United States would follow in 1949 and 1968. See Constantin Brodzki, interview by the author on October 31 and November 19, 2013.

37 Ibid., 34-37, 58.

38 Some examples are those for the Belgian Union of the mechanical glass trade and the Belgian Institute for packaging. These fair stands were characterized by a highly modernistic esthetic and the use of American

furniture. Brodzki also collaborated in 1953 in the design of the model home of the Belgian Society of Modernist Urban designers and Architects (SUBAM) exhibited at the Brussels Fine-Arts center. On this occasion Knoll furniture was on display for the first time in Europe. See Geert Bekaert, *Hedendaagse architectuur* (Tielt: Lannoo, 1995), 124–25, 31; Constantin Brodzki, "Architectures. Interieurs d'aujourd'hui. Un cadre adapté à la vie actuelle,» *Les Beaux-Arts. Hebdomadaire d'Information Artistique*, no. 624 (1953); Sasson, «La maison modèle au Palais des Beaux-Arts,» *Architecture 55. Revue bimestrielle d'architecture et d'urbanisme*, no. 10 (1955); Hoste, *Constantin Brodzki. Architecte*, 48–49.

39 See chapter one for more information on the *First Industrial Design Show* of 1954.

40 The showroom was the first in the Benelux and was located in the Rue Royale in Brussels. The construction was a collaboration between Florence Knoll, Knoll Planning Unit, Marcel Breuer, Constantin Brodzki and De Coene. See Syracuse, Syracuse University Libraries, Marcel Breuer Digital Archive, Project Knoll Showroom (Brussels) Professional Papers, available online at http://breuer.syr.edu; Fredie Floré, "Architect-Designed Interiors for a Culturally Progressive Upper-Middle Class: The Implicit Political Presence of Knoll International in Belgium," in *Atomic Dwelling. Anxiety, Domesticity, and Postwar Architecture*, ed. Robin Schuldenfrei (London and New York: Routledge, 2012). The story goes that Hans Knoll asked for Brodzki's professional help after his visit to the Brussels bookshop the interior of which was designed by Brodzki and his usual partner in crime, Corneille Hannoset. It was the *La Librairie de l'Edition Universelle* bookshop in the Rue Royale, the same street where the Knoll showroom would be built. See Constantin Brodzki, interview by the author on October 31 and November 19, 2013; "Librairie à Bruxelles. Architecte Constantin Brodzki et décorateur Corneille Hannoset,» *Habitat et Habitation. Documents d'architecture et d'urbanisme* 15, no. 12 (1955); Hoste, *Constantin Brodzki. Architecte*, 29, 31; Sasson, «La Librairie de l'Édition Universelle,» *Architecture 55. Revue bimestrielle d'architecture et d'urbanisme*, no. 10 (1955).

41 Brussels, Archives et Bibliothèques d'Architecture de l'Université Libre de Bruxelles, Fonds Brodzki, Classeur "Affaires déjà terminées": *Curriculum Vitae.*

42 André De Poerck was the architect of many early exhibitions on industrial design: the Belgian section of Milan 1954, *National Show of Industrial Design* Ghent in 1959 and the *International Industrial Design Exhibition* at the Brussels *Bon Marché* in 1960. He designed a wooden beam for De Coene which got a diploma of honour at the Milan Triennial in 1954. From 1952 until 1954 he was a member of the editorial staff of the magazine *Architecture. Revue bimestrielle d'architecture et d'urbanisme* and from 1953 became the chief editor together with Robert Delevoy and Roger Thirion. In 1955, he became one of the first instructors in Industrial

Design at La Cambre. He was also closely involved in the first design institutionalization in Belgium. He became a member of the Institute of Industrial Design in 1957 and an associate member of the Signe d'Or in 1958.

43 Ghent, Archief Adelbert Van de Walle, Museumarchief Design Museum Gent, unnumbered, folder Instituut Industriële Vormgeving B.-L.: Correspondance André De Poerck ca. 1957.

44 André De Poerck, «L'objet, son existence et sa présentation,» *Architecture 54. Revue bimestrielle d'architecture et d'urbanisme*, no. 10 (1954).

45 Pierre Génicot and Daniël Duqué started to work together in 1955. Their design bureau existed for seven years. See Brussels, State Archives, BDOP/OBAP (5010–2227), I504, 1081: *Rapport d'activités et prévision de l'evolution du Design Centre,* p.2; «...Voorbeelden wekken. Daniël Duqué," *Infordesign. Informatieblad van het Instituut voor Industriële Vormgeving*, September 1965.

46 See Michel Olyff, «Questions graphiques. Enjeux et contraintes,» *Académie royale de Belgique. Bulletin de la Classe des Beaux-Arts* 6e série - tombe XII, no. 7–12 (2001).

47 Michel Olyff, interview by the author, October 10, 2013.

48 Constantin Brodzki, interview by the author on October 31 and November 19, 2013.

49 Geert Bekaert and Francis Strauven, *Bouwen in België 1945–1970* (Brussels: Nationale Confederatie van het Bouwbedrijf, 1971), 282.

50 He became a member of the Institute for Industrial Design in 1958. In the 1960s, he was also a member of the "Design Doctrine and Philosophy" working group of the ICSID. See Ghent, Design Museum Gent, Archief Van de Walle, unnumbered, folder Inst. Esth. Industri. Membres: *Volmakende lijst der leden van 14 februari tot 21 mars 1958*; ibid.

51 Constantin Brodzki, interview by the author on October 31 and November 19, 2013; des Cressonnières and Burnay, "Le Design Centre."

52 "[...] savant assemblage d'éléments préfabriqués." Hoste, *Constantin Brodzki. Architecte*, 78.

53 Constantin Brodzki, interview by the author on October 31 and November 19, 2013; ibid., 32.

54 Philippe Neerman (1930–2011), interview by the author on April 29, 2010, Kortrijk.

55 "[...] dans ma mémoire, s'est assimilé à l'ambiance d'un orchestra de jazz, avec un quartette compose de Constantin Brodzki au piano, Pierre Génicot à la contrebasse, Philippe Neerman à la batterie et Michel Olyff à la guitare. Bresseleers et De Poerck, surajouté à l'équipe, se sont contenté de batre la mesure." See Constantin Brodzki, lecture at the MAD Brussels fashion and design center in its series *Le Leçon Design*, June 11, 2014. The

small seminar also included lectures from the author and graphic designer Michel Olyff.

56 This part is based on the paper 'Design and the State. A Theoretical Account of the Role of the State in the Creation of Design Cultures' which I wrote in collaboration with dr. Javier Gimeno-Martínez for the *Raison d'être 2: Crafting a Global History of Interior Decoration and Design* international symposium on March 9 and 10 in New Orleans (Louisiana, US) and Gimeno Martínez's doctoral dissertation: Javier Gimeno-Martínez, "The Role of the Creative Industries in the Construction of Regional/European Identities (1975–2002)" (Katholieke Universiteit Leuven, 2006).

57 des Cressonnières and Burnay, «Le Design Centre.»

58 Ibid; Josine des Cressonnières, «Le Design Centre a cinq ans,» ibid. 1969; Josine des Cressonnières and Jean-Jacques Stiefenhofer, «Dossier: Het Design Centre: tien jaar onderweg,» *A+*, no. 9/10 (1974); «Le Design Centre a 20 ans,» *Design News* 10, no. mars (1984).

59 „Hochschule für Gestaltung Ulm. Hogeschool Voor Vormgeving Ulm," ed. Stedelijk Museum Amsterdam (Amsterdam1965); Jens Müller and René Spitz, „HfG Ulm, kurze Geschichte der Hochschule für Gestaltung: Anmerkungen zum Verhältnis von Design und Politik = a Brief History of the Ulm School of Design: Notes on the Relationship between Design and Politics „ (Zürich: Lars Müller, 2014).

60 Gui Bonsiepe and John Cullars, "The Invisible Facets of the HfG Ulm," *Design Issues* 11, no. 2 (1995): 15; "Hochschule für Gestaltung Ulm, Hogeschool voor Vormgeving Ulm."

61 «Le Design Centre a 20 ans.»

62 For more information on des Cressonnières' run for ICSID president see: Katarina Serulus, "Design & Politics. The Public Promotion of Industrial Design in Postwar Belgium (1950–1986)" (University of Antwerp, 2016), 141–42.

63 Gimeno-Martínez, "The Role of the Creative Industries in the Construction of Regional/European Identities (1975–2002)," 22–23, 28, 77.

64 Joseph Nye, *Soft Power. The Means to Success in World Politics* (New York: PublicAffairs, 2004), 1–32; Greg Castillo, Cold War on the Home Front: The Soft Power of Midce.

65 Castillo, Cold War on the Home Front: The Soft Power of Midcentury Design, 38–41.

66 Alison Clarke, Tupperware: The Promise of Plastic in 1950s America (Washington: Smithsonian Institution Press, 1999), 53.

67 Royal support towards design promotion was also noticeable in Sweden, the Netherlands and Great Britain. See: des Cressonnières, "Le Design Centre a cinq ans."

68 Brussels, Archief van het Koninklijk Paleis, Secretariaat Koningin Fabiola, hofdames; Baudouin D'hoore (archivist Archief van het Koninklijk Paleis), email message to the author and the archivist, September 3, 2015.

69 Constantin Brodzki, interview by the author on 31 October and 19 November 2013; Géraldine des Cressonnières, interview by the author on 8 November 2013 and 30 April 2015.

70 Sint-Genesius-Rode, Private archive Nanette Blondel: *Rapport concernant l'exposition "Le Design Centre Belge en Visite à Londre" - 25 mars - 25 avril 1969.*

71 Richard Carr, "Overseas. Design Centre in Brussels," *Design*, no. 185 (1964); "Belgisch Design Centre - pas geopend - heeft eigen karakter," *Maandbericht van het Instituut voor Industriële Vormgeving*, March/April 1964.

72 "Overseas. Design Centre in Brussels." And as shortly described by Jonathan Woodham: Jonathan Woodham, *Twentieth-Century Design* (Oxford/New York: Oxford University Press, 1997), 171.

73 Carr, "Overseas. Design Centre in Brussels."

74 "De propaganda is niet op een zo groot mogelijk publiek gericht, maar op de 'opinieleiders', op de handel (vooral export) en op het prestige dat de fabrikanten kunnen ontlenen aan het uitverkoren-zijn voor expositie in deze top-showroom." See "Belgisch Design Centre - pas geopend - heeft eigen karakter."

75 des Cressonnières and Stiefenhofer, "Dossier: Het Design Centre: tien jaar onderweg," 36–38.

76 K.N. Elno, "Bekvechten met een nieuw museum," *De Nieuwe*, 5 juni 1964.

77 "De technische kommentaar beluisteren. Esthetische deuntjes neuriën. Wikken en wegen. Met geciviliseerde distantie bewonderen. En er niet op letten dat dit apparaat tussen u en het scheren ligt te liggen en dat ge nu ook nog over al de klippen van die industriële vormgeving moet klauteren alvorens het u toegelaten is zoiets eenvoudigs en menselijk en zondeloos te verrichten als het scheren van een glorende baard," ibid.

78 "Van deze mogelijkheid wordt in de design centres zelden of nooit gebruik gemaakt en dat geeft al een aanduiding van hun vaak zeer verkeerde gerichtheid: ze dragen willens nillen bij tot de fictie dat een werkelijke industrieel product iets exclusiefs zou kunnen (mogen) zijn en alleen reeds daardoor blijft een design centre zoals dat te brussel meer een status-centre en een oord waar men de industriële techniek poogt in te lijven in een anti-moderne en anti-democratische levensvisie." See "Industriële bewustwording door plastische schepping. Vormgeving," *De Nieuwe*, June 24, 1966.

79 Unless otherwise referenced, the following part is based on the annual reports on the activities of the Design Centre preserved in Brussels at the State Archives: BDOP/OBAP (5010-2227), I504, 1081.

80 "Gezicht van de onderneming," *Infordesign*, no. 9 (1966).

81 "Er bestaan in België te weinig producten van esposerenswaardig niveau om een Design Centre op permanente wijze te bevoorraden." des Cressonnières, "Le Design Centre a cinq ans."

82 K.N. Elno, "Belgische pointering 1956," *Scheppend Ambacht* 7, no. 6 (1956); Leon-Louis Sosset, "De lessen der Xe Driejaarlijkse van Milaan," *Kunstambachten en Kunstnijverheden. Maandelijks Inlichtingsblad der Provinciale Commissies voor Kunstambachten en Nijverheidsbedrijven*, no. 60 (1954).

83 «De lessen der Xe Driejaarlijkse van Milaan; Elno, «Belgische pointering 1956;» Pierre-Louis Flouquet, «Le Salon d'esthétique industrielle Liège 1956,» *La Maison* 12, no. 7 (1956).

84 Javier Gimeno-Martínez, "The Signe d'Or Award Scheme from 1956 to 1960: The Economic Reasons for "Good Design"," *Konsthistorisk Tidskrift/Journal of Art History* 79, no. 3 (2010).

Chapter VI: C'est Belge Ça?

1 Alison Clarke, "'Actions Speak Louder.' Victor Papanek and the Legacy of Design Activism," *Design and Culture* 5, no. 2 (2013); Pauline Madge, "Design, Ecology, Technology: A Historiographical Review," *Journal of Design History* 6, no. 3 (1993); Javier Gimeno-Martínez, "Redefining Social Design in 1970s Belgium: Affordable Design vs. Elite Design," *Interiors* 2, no. 2 (2011); Alison Clarke, "Design for Development, ICSID and UNIDO: The Anthropological Turn in 1970s Design," *Journal of Design History* (2015).

2 Alice Twemlow, "I Can't Talk to You if You Say That: An Ideological Collision at the International Design Conference at Aspen, 1970," *Design and Culture* 1, no. 1 (2009).

3 Brussels, State Archives, BE-A0545/Fic 2000, Archive Design Centre, 20: Josine des Cressonnières, L'industrial design dans le monde et en Belgique [undated article published in the magazine *Europe 13*].

4 Brussels, State Archives, BDOP/OBAP, 1036: *The Instant City Manifesto.*

5 Ibid.

6 Victor Papanek, *Design for the Real World. Human Ecology and Social Change* (London: Thames & Hudson, 2006), IX.

7 Ibid., XIV.

8 Javier Gimeno-Martínez, Joana Ozorio de Almeida Meroz and Katarina Serulus, "Localizing Design/Designing Location: Creative Cities in the Low Countries," *Dutch Crossing* 37, no. 3 (2013).

9 Brussels, State Archives, *BDOP*, 1081: *Réunion du Comité de Direction du 21 april 1967.* Adelbert Van de Walle was no stranger in the world of design

promotion. From 1952 he was secretary of the Provincial Agency for Handicrafts and Industrial Design of East Flanders. He was also involved in the foundation of the Institute for Industrial Design for Belgium and the Grand Duchy of Luxembourg in 1956. See Fredie Floré, *Lessen in goed wonen. Woonvoorlichting in België 1945–1958* (Leuven: Universitaire Pers Leuven, 2010), 188–202.

10 Brussels, State Archives, *BDOP*, 1081: *Réunion du Comité de Direction du 21 april 1967;* Brussels, State Archives, *BDOP*, 1081: *Procès-verbal de la réunion du Comité de Direction du 26 april 1967*; Brussels, State Archives, *BDOP*, 1081: *Procès-verbal du Comité de Direction du 26 april 1967.*

11 For a similar reason the reformed Institute for Industrial Design refused any longer to finance the educational seminars on industrial design by the Lodewijk de Raet Foundation in the early 1960s. The president and secretary, respectively Pol Provost and Josine des Cressonnières, explained that this decision was guided by the different understandings of the nature of design. The president of the Lodewijk De Raet foundation, Maurice Van Haegendoren, understood design not in service of the industry, trade or professionals; no, all these had to be in the service of humankind. With these convictions, he accused the institute of completely ignoring the social and cultural aspects of design. On meetings of the Institute he stated: "industrial design is not only a question of 'sales promotion' but also about the education of the consumer taste." See Ghent, Archief Adelbert Van de Walle, Museumarchief Design Museum Gent, unnumbered: *Proces-Verbaal van de 10de vergadering der werkgroep voor de studie van de Industriële Vormgeving, gehouden in het Instituut voor Industriële Vormgeving op 7 december 1960;* Brussel, Archief en Museum voor het Vlaams Leven te Brussel, BE AMVB 168, 1604/SLDR: Letter of M. van Haegendoren to Josine des Cressonnières [General Secreatry of the Institute for Industrial Design], (5/06/1961); Brussel, Archief en Museum voor het Vlaams Leven te Brussel, BE AMVB 168, 1604/SLDR: Letter of Josine des Cressonnières [General Secreatry of the Institute for Industrial Design] to M. van Haegendoren, (2/10/1961); Ghent, Archief Adelbert Van de Walle, Museumarchief Design Museum Gent, unnumbered, folder Lodewijk de Raet: Letter of Maurits van Haegendoren to Pol Provost, (15/10/1960).

12 Tim Edensor, *National Identity, Popular Culture and Everyday Life* (Oxford/New York: Berg, 2002), 40, 108; Javier Gimeno-Martínez, "The Role of the Creative Industries in the Construction of Regional/European Identities (1975–2002)" (Katholieke Universiteit Leuven, 2006), 12, 22.

13 Belgian Ministry of Economic Affairs, *Belgium To-Day* (Brussels: Belgian Ministry of Economic Affairs, 1949).

14 Floré, *Lessen in goed wonen. Woonvoorlichting in België 1945–1958*, 188–202.

15 Adelbert van de Walle drafted this vision with the help of the later president baron Fernand van Ackere (1878–1958) and linguist Blancqueart. See Ghent, Archief Adelbert Van de Walle, Museumarchief Design Museum

Gent, unnumbered, folder Nationaal raad voor wetenschapsbeleid: Van de Walle, Adelbert, "Bepaling van de kunstambachten," (10/12/1959).

16 "Door hun ambachtelijk karakter staan zij dichter dan zgn. vrije kunsten bij het volk waaraan zij niet alleen een pittoreske broodwinning verzekeren, doch waarvan zij tevens de ziel weerspiegelen, alle grote internationale stromingen ten spijt [...] De kunstambachten kunnen tenslotte ook beschouwd worden als de wil van een volk om naast de zuivere 'nuttigheidsberoepen' ook andere uit te oefenen, waaraan een esthetisch en dikwijls kreatief aspect nauw verbonden is, doch waarbij de ambachtelijke productiemogelijkheden die dichter bij het volksgemoed staan dan de ver doorgedreven mechanisatie in ere gehouden worden." See Ghent, Archief Adelbert Van de Walle, Museumarchief Design Museum Gent, unnumbered, folder Nationaal raad voor wetenschapsbeleid: Van de Walle, Adelbert, "Bepaling van de kunstambachten," (10/12/1959).

17 A.M. Botte, "Kunstambachten en industriële vormgeving," *Informatieblad van het Instituut voor Industriële Vormgeving* 5, no. 5 (1962). Alphonse Botte was the representative for handicrafts on the executive committee of the Design Centre.

18 "A ceux qui craignent qu'une uniformité ne naisse de la multiplication de ces échanges, on peut répondre que les anciennes valeurs régionales tendront inéluctablement à se fondre dans une civilization qui devient planétaire, mais que le folklore peut être remplacé par l'homme, non pas l'homme seul, mais l'homme dans l'équipe, dans l'entreprise, dans le bureau de design.» See Josine des Cressonnières, «La signification d'un Design Centre,» *Industrie* 1965.

19 It was the Ministry of the Middle Class who paid the costs of the exhibition. See Christoph Junck, "Die Design Centres der Welt," *Moebel Interior Design*, no. 1 (1967); Josine des Cressonnières, "Le Design Centre a cinq ans," *Industrie* 1969.

20 Some Design Centers, such as the Greek, British and Finnish, included a great many crafted products. Others, such as the Dutch, excluded these.

21 Brussels, State Archives, *BDOP*, 1081: *Prévisions d'activités pour 1968.*

22 The exhibition was organized in collaboration with the city of Charleroi and the Belgian Chambers of Commerce and Industry. See Brussels, State Archives, *BDOP*, 1081: *Rapport d'activités 1969*; Brussels, State Archives, *BDOP*, 1081: *Entrevue avec Monsieur Hubeaux, burgemestre de Charleroi, le 20 juin 1968.*

23 The exhibition is also mentioned in: Els De Vos, "The Interieur Biennale and Domesticity," in *24th International Biennale Interieur 2014. The Book*, ed. Monique Bucqoye (Kortrijk: Interieur Foundation, 2014).

24 The exhibition was sponsored by the city of Kortrijk.

25 Brussels, State Archives, *BDOP*, 1081: *Rapport d'activités pour 1968 et prévisions d'activités pour 1969;* Brussels, State Archives, *BDOP*, 1081: *Séance plénaire du mercredi 14 février 1968;* Brussels, State Archives, *BDOP*, 1081:

Rapport d'activités pour 1968 et prévisions d'activités pour 1969; Brussels, State Archives, *BDOP*, 1081: *Rapport d'activités 1970*; Brussels, State Archives, *BDOP*, 1081: *Rapport d'activités 1972*.

26 Brussels, State Archives, *BDOP*, 1081: *Rapport d'activités 1972*. Brussels, State Archives, *BDOP*, 1081: *Prévisions d'activités pour l'exercice 1973*.

27 "Wat vindt u mooi, funktioneel en vernuftig? Belgische design in Antwerps grootwarenhuis," *Gazet van Antwerpen*, 12/01/1973 1973; "Inauguration de l'exposition d'objets 'design' ouverte au Grand Bazar jusqu'au 14 février," *La Wallonie*, 03/02/73 1973; A.W., "Au Grand Bazar de la Place Saint-Lambert. L'exposition d'objets 'design' permettra aux visiteurs de participer à un concours original (Ouverte jusqu'au 14 février),» *Le Monde du Travail*, 05/02/73 1973; André Pacquot, «Une importance plus grande à la qualité. Une exposition 'Design Center' à Liège. La culture contemporaine s'exprime à travers l'objet,» *La Dernière Heure*, 03/02/73 1973.

28 "Wat vindt u mooi, funktioneel en vernuftig? Belgische design in Antwerps grootwarenhuis."

29 Brussels, State Archives, BE-A0545/Fic 2000, Archive Design Centre: 204, 302; "Urbanizatie. Akelige lelijkheid,» *De Nieuwe*, 24/11 1972; «Pour une nouvelle culture urbaine,» *La Dernière Heure*, 16/10/72 1972; «Une intéressante exposition du Design Centre de Bruxelles: «Pour une nouvelle culture urbaine»,» *Le Jour*, 8/12/72 1972; «Une exposition sur une nouvelle culture urbaine circulera à travers du pays,» *Nord Eclair*, 11/10/72 1972; «Nieuwe stedelijke cultuur in Design Centre te Brussel,» *Gazet van Antwerpen*, 17/10/72 1972; «Inspraak van publiek in leefmilieu. Design Centre brengt goed idee niet aan de man," *De Standaard*, 31/10/72 1972.

30 "Eén van de belangrijkste kenmerken van onze naoorlogse generatie is het dwingende besef dat deze aarde – 'spaceship earth' – te allen prijze moet behouden blijven en dat dit ons allen aangaat. Dat environment – de 20m2 van onze kamer, het milieu van onze straat, onze wijk, onze stad, onze wereld ... dat leefmilieu gaat iedereen aan." See "Environnement-tentoonstelling in het Design Centre te Brussel," *Volksgazet*, 10/10/72.

31 "Inspraak van publiek in leefmilieu. Design Centre brengt goed idee niet aan de man."

32 Vincent Dujardin et al., eds., *Nieuwe geschiedenis van België. 1950-Heden*, vol. III (Tielt: Lannoo, 2009), 1443; Els Witte, Jan Craeybeckx, and Alain Meynen, *Politieke geschiedenis van België. Van 1830 tot heden* (Antwerpen: Standaard, 2005), 437–40.

33 "Wie tot op heden, gedeeltelijk terecht, de franstalige gevel van het niettemin angstvallig tweetalige Design Centre schuwde, moet toch even de moeite nemen en eens over de Belgische drempel heenstappen." See G.S., "Vormgeving: een souvenirshop moet niet noodzakelijk kitscherig zijn," *De Nieuwe*, 17 juli 1970.

34 For example this excerpt from a 1972 review in De *Nieuwe*: "Het is triest, weer te moeten vaststellen dat de Nederlandstalige burger hierbij niet aangesproken werd. Waar het Design Centre strikte tweetaligheid nastreeft, kwam het Nederlands van deze tentoonstelling wat over als een welwillende, maar niet volgehouden poging om een woordje uitleg aan de minus habentes te verstrekken, die zich het Frans nog niet konden eigen maken. Er kon wel een Nederlandstalige rondleiding aangevraagd worden, maar zoiets doet men als enkeling niet gemakkelijk. [...]" in "Urbanizatie. Akelige lelijkheid," ibid., November 24, 1972.

35 The regular column was the first in Flanders and ranged from half a page to two complete pages. In the first number he describes his mission statement as follows: "Maar er is in dit verband een specificatie nodig: die uitgave (dag- of week- of maandblad) moet in haar geheel ook nog geëngageerd zijn jegens een waaier aan waarden waarvan de zin in menselijk, boven-werkelijk, tijdsgebonden en situationistisch perspectief onbetwistbaar is. Waarmede ik bedoel dat, voor mij, voor ons, total design in totale publicistiek hetzelfde is als wijding-aan-vormgeving in een twintigste-eeuws en Vlaamsbewuste (1) geest, in een christelijk-socialistische hoop, in een denken dat zowel het 'alle Menschen werden Brüder' als het onaanraakbaar Vlamingschap (vervelend woord, natuurlijk, maar schept moed, vrienden, we zullen het te boven komen) omsluit.... die dus het gewone van ons aller dagelijkse leven voldoende raadselachtig en indrukwekkend vindt om het diep te beminnen." See interview by the author with Marc Grammens, Liedekerke, October 18, 2012; Fredie Floré, "Elno, K.-N., pseudoniem van Karel Horemans, architectuur-, design- en kunstcriticus," in *Nationaal biografisch woordenboek* (Brussels: Koninklijke Academiën van België, 2011); "Design Criticism and Social Responsibility: The Flemish Design Critic K.-N. Elno (1920–1993)," in *Writing Design: Words and Objects*, ed. Grace Lees-Maffei (Oxford: Berg, 2011).

36 In April 1968, Pol Provost would be succeeded in his position as president by Louis Ameye, then the president of the Federations of the Glass Industry.

37 Ludo Meyvis, *Markt en macht* (Tielt: Lannoo, 2004); Kristin Van der Wee and Edward de Maesschalck, *Vlaanderen ontmoet Nederland. Geschiedenis van de Orde van den Prince* (Tielt: Lannoo, 2005).

38 He was for example the first secretary of the Orde van den Prince. In 1966 he wrote in a confidential letter to Pol Provost: "Mijn intrede in het IIV werd door de Design Centre en OBCE-kringen beschouwd als een zog. Vervlaamsing van het IIV. De standpunten ter zake in die kringen zijn overbekend; denk aan de ééntalige toespraak van Mijnheer Burnay in aanwezigheid van Prins Albert en Uzelf. In dergelijke omstandigheden kan alleen een gelijkaardige steun van de vlaamsvoelende middens aan het IIV de nodige vleugelslag verlenen." See Brussels, State Archives, I 504 40–48: Hubert De Sy, Letter to Pol Provost, (9/05/1966).

39 Rik Coolsaet, Vincent Dujardin and Claude Roosens, *Buitenlandse Zaken in België. Geschiedenis van een ministerie, zijn diplomaten en consuls van 1830 tot vandaag* (Tielt: Lannoo, 2014), 304–09.

40 "Industriële vormgeving heeft veel te maken met taal, kulturele eigenwaarde en specifieke levensomstandigheden van een (echt-) nationale groep (want het gaat om KARAKTER in de vervaardiging en om begeleiding-door-producten van onvervalste degelijkheid) en dus ook met ... kulturele autonomie. Waarop wachten zijn om het I.I.V. bedaard en vastberaden TE FEDERALISEREN?" See K. N. Elno, "Tweemaal I.V.V.," *De Nieuwe*, January 1, 1965.

41 Dujardin et al., *Nieuwe geschiedenis van België. 1950-Heden*, 437–40; Witte, Craeybeckx and Meynen, *Politieke geschiedenis van België. Van 1830 tot heden*.

42 «B.70: 300 produits à l'image de la Belgique. 27 juin-2 septembre 1970," ed. Design Centre (Brussels: Design centre, 1970).

43 For the design and craft promotional actions of the Economic and Social Institute for the Middle Classes (ESIM) in the 1980s see Gimeno-Martínez, "The Role of the Creative Industries in the Construction of Regional/European Identities (1975–2002)," 81–87.

44 W. Degryse, "Het scheppend ambacht op 'B 70'," *Infordesign*, no. 34 (1970).

45 The Minister of the Middle Classes Charles Hanin opened the exhibition on June 26. It was on view until September 2. The French and Dutch exhibition titles were respectively: *B.70: 300 produits à l'image de la Belgique* and *B.70: 300 produits à l'image de la Belgique*. See «B.70: 300 produits à l'image de la Belgique. 27 Juin-2 September 1970.»

46 The exhibition was opened by the Brussels Mayor Louis Cooremans. The opening was graced by folkloristic music, boudin noir, fromage blanc and beer. The French and Dutch exhibition titles were respectively: *C'est belge!* and *Belgisch ook nog!* See "Exposition de produits et objets belges typiques," *Agefi*, 14/12/1971 1971. One of the numbers of *Infordesign* served as the catalog of the exhibition: "'En Belgisch ook nog': typische voorwerpen en producten," *Infordesign*, no. 42 (1971).

47 Marianne Burton, « »C'est ainsi que les touristes étrangers emporteront dans leurs bagages une certaine image de la Belgique,» *Libelle*, 18/08 1970; «Expositions. La Belgique des Frères Grimm," *Special*, 26/08 1970.

48 "Wie tot op heden, gedeeltelijk terecht, de franstalige gevel van het niettemin angstvallig tweetalige Design Centre schuwde, moet toch even de moeite nemen en eens over de Belgische drempel heenstappen. Het zal met genoegen vaststellen dat 'design' niet uitsluitend een mooie vorm rondom een handige mixer betekend, maar daarenboven bekommernis om het algemene beeld van de samenleving. Om de gezondheid ook van het industriële milieu. De enorme belangstelling voor het zgn. 'environment' mag zich niet beperken tot een ecologisch beschermingsgebaar.

Door mensenhanden vervaardigde milieu wordt minstens zo belangrijk als onze natuurlijk leefwereld." See G.S., "Vormgeving: Een souvenirshop moet niet noodzakelijk kitscherig zijn."

49 "Agréablement nationale. C'est probablement l'expression qui qualifie de la façon la plus exacte cette exposition [...]" See «Le temps des souvenirs: une exposition agréablement nationale,» *Le Soir*, 08/08 1970; «C'est belge dit le Design Centre en nous présentant des produits et objets typiquement 'à l›image de notre pays',» *Le Soir*, 7/12 1971.

50 «Expositions. La Belgique des Frères Grimm.»

51 For the complete list see the catalogue: «B.70: 300 produits à l'image de la Belgique. June 27 - September 2, 1970."

52 For more on foodstuff and national identity see Nelleke Teughels and Peter Scholliers, eds., *A Taste of Progress. Food at International and World Exhibitions in the Nineteenth and Twentieth Centuries* (Aldershot: Ashgate, 2015).

53 Walter Baeke, "Brussels Design Centre voert grootscheepse 'souveniraktie.' Vervangen hespen en worsten de gebruikelijke souvenirs," *Het Laatste Nieuws*, 23/07 1970.

54 «Design 'C'est Belge',» *Special*, 22/12 1971.

55 «Entendons-nous: IL Y A des choses ravissantes (anciennes très souvent, ou renouvelées de l'ancien). Oui, des choses ravissantes, inattendues, charmantes, qui font de très jolis " souvenirs de voyage". Mail il y a aussi un ramassis d'idées farfelues (quand elles ne sont pas tout simplement très choquantes pour le bon goût) et qui semblent d'autant plus déplacées que les voici exposées sous le couvert et avec l'apparente bénédiction du Design Centre, que nous avons toujours considéré comme une sorte de conservatoire du bon goût passé présent et à venir." See «C'est belge dit le Design Centre en nous présentant des produits et objets typiquement 'à l›image de notre pays'.»

56 "Selon les objets que vous choisirez à l'exposition voir pourrez vous dire que nous sommes très artistes, très originaux, très spirituels...ou très plats, très tourmenté, très complexé." See ibid.

57 "Les savons de Manneken-Pis, que je ne trouve pas drôles du tout". See ibid; "Handjes wassen met het Manneke?," *De Standaard*, 6/01 1972; "Design 'C'est Belge'."

58 "Manneke binnen kontekst," *De Standaard*, 27/1 1972. In that sense, the second exhibition seems to be rather an exponent of ESIM's selection policies - showing a broad and heterogeneous collection - than those of the Design Centre. See Gimeno-Martínez, "The Role of the Creative Industries in the Construction of Regional/European Identities (1975-2002)," 82–87.

59 "The Role of the Creative Industries in the Construction of Regional/European Identities (1975-2002)," 70–71.

60 "'En Belgisch ook nog': typische voorwerpen en producten," 5.

61 "The Role of the Creative Industries in the Construction of Regional/ European Identities (1975-2002)," 33-34, 36-38; Edensor, *National Identity, Popular Culture and Everyday Life*; Umut Özkirimli, *Theories of Nationalism. A Critical Introduction*, 2d ed. (Basingstoke: Palgrave Macmillan, 2010), 170-71.

62 Brighton, Design Archives, ICSID, 04-1 Board meeting 1962-1970: *Minutes of the Executive Board Meeting held in Brussels on 20th February 1964*; Brighton, Design Archives, ICSID, 04-7 Board meeting 1968-1975 (Cahiers): *Comments by J. des Cressonnières on the Future of ICSID,* (March 1970); Brighton, Design Archives, ICSID, 04-7 Board meeting 1968-1975 (Cahiers): *Notes by Josine des Cressonnières*, (May 1970); Brighton, Design Archives, ICSID, 04-2 Board meeting 1971-1974: *ICSID Executive Board Meeting n°3 - Kyoto 22-25/10/72;* Brighton, Design Archives, ICSID, 04-2 Board meeting 1971-1974: des Cressonnières, Josine, *ICSID's Programme for 1973-75 - J. des Cressonnières' Views*; Brighton, Design Archives, ICSID, 03-9-8 Nominations: Bojar, Ryszard and Josine des Cressonnières, *Statement of Intent*; Brighton, University of Brighton, Design Archives, ICSID, 04-4 Board meeting 1977-1979: des Cressonnières, Josine, *statement of intent* [first draft], (June 1977).

63 Brighton, University of Brighton, Design Archives, ICSID, 03-9-8 Nominations: des Cressonnières, Josine, *Statement of Intent;* Brighton, University of Brighton, Design Archives, ICSID, 03-9-8 Nominations: List of nominees for the election of the ICSID Executive Board 1977-1979; Brighton, University of Brighton, Design Archives, ICSID, 04-4 Board meeting 1977-1979: des Cressonnières, Josine, *statement of intent* [first draft], (June 1977).

64 Brighton, University of Brighton, Design Archives, ICSID, 03-9-8 Nominations: des Cressonnières, Josine, *Statement of Intent.*

65 Coutisse, Archives des Cressonnières: des Cressonnières, Josine, *Proposal for a fellowship to the Woodrow Wilson International Center for Scholars. September 78 (Washington). "Design and Public Authorities" A proposal for a global policy of industrial design.*

66 This plan was reminiscent of the actions of one of the ICSID working groups on design promotion - of which des Cressonnières was a member - that compiled a bunch of national case histories in the early 1970s. The subject of a state policy of design was well explored within the ICSID. One of the sessions at the Kyoto conference in 1973 was devoted to this theme. The 1975 conference in Moscou even took "Design as a state policy" as its main theme. See Brighton, University of Brighton, Design Archives, ICSID, 06-6bis Working Group VI: Promotion, des Cressonnières, Josine, *Minutes of Meeting Berlin, December 16 to 17, 1974, Session on "Design as a State Policy"*, (6/01/1975).

67 Brussels, State Archives, BE-A0545/Fic 2000, Archive Design Centre, 898: des Cressonnières, Josine, *"5 minutes pour convaincre", prononcé au CCI paris lors de l'assemblé général du ICSID du février 1980.*

68 "Monsieur le Ministre. Vous avez dit - écrit - qu'il est essentiel pour notre pays d'investir dans l'innovation et la créativité (qu'il l'ait dit ou non, c'est sans importance, il sera d'accord pour qu'on le lui attribue) - la creativité et l'innovation étant des plus précieux atouts qui puissent compenser, sur les marchés internes, le coût de notre main-d'oevre, de nos grèves, de notre train de vie, de nos charges sociales croissantes etc..." See Brussels, State Archives, BE-A0545/Fic 2000, Archive Design Centre, 898: des Cressonnières, Josine, *"5 minutes pour convaincre", prononcé au CCI paris lors de l'assemblé général du ICSID du février 1980.*

69 "On peut attendre un résultat supplémentaire de cette politique: une nouvelle image du pays, qualité éventuelle pour notre pouvoir à l'exportation. Que l'on pense par exemple combine l'image du Japon a changé en 20 ans sur nos marches! Un dernier point, Monsieur le Ministre, il n'y a aucun pays au monde où il existe à ce jour une politique complète et cohérente d'Industrial Design. Pourquoi ne serions-nous pas les premiers?" See Brussels, State Archives, BE-A0545/Fic 2000, Archive Design Centre, 898: des Cressonnières, Josine, *"5 minutes pour convaincre", prononcé au CCI paris lors de l'assemblé général du ICSID du février 1980.*

70 Paul Betts, *The Authority of Everyday Objects: A Cultural History of West German Industrial Design* (Berkeley: University of California Press, 2004), 249–64.

71 Tomas Maldonado, *Design, Nature and Revolution: Toward a Critical Ecology*, trans. Mario Domandi (New York: Harper & Row, 1972). The first edition was published in 1970 in Italian (*La Speranza Progettuale*) and the second revised version would be translated in 1972 into English, French, German and Spanish.

72 Katarina Serulus, "1968–2016, 25 Editions of Biennale Interieur," in *25th Silver Edition. Biennale Interieur 2016*, ed. Biennale Interieur (Kortrijk: Biennale Interieur npo, 2016); "Biennale Interieur, een halve eeuw designgeschiedenis," in *Designx50* (Veurne: Kannibaal, 2017).

73 Ludo Bekker and Kaat Dutreeuw, "10 jaar Design Center," in *Kortweg* (VRT (Vlaamse Radio- en Televisieomroeporganisatie), 1974).

74 Gui Bonsiepe was at that moment setting up a "Design Centre" in Chile and rather preferred the term "Zentrum für Produktentwicklung" (Centre for Product Development). This name better covered the needs in the periphery, which were not so much focused on design promotion, but design production "beyond the standard lamp-chair-ashtray candidates." In certain circles, there lived a certain aversion towards the word "design." Antwerp, Collection Flanders Architecture Institute - Collection Flemish Community, Archive Jean-Jacques Stiefenhofer (1943–2013), Doos 18.1: Gui Bonsiepe (Argentine), Letter to F. Burkhardt (Germany),

(20/02/1974); Bonsiepe, Gui, [email to the author], (09/07/2015); Katarina Serulus, "Van industrieel design tot productontwikkeling," in *Van academie tot universiteit. 350 jaar architectuur in Antwerpen*, ed. Els De Vos and Piet Lombaerde (Brussels: UPA, 2013).

75 "Des Ausdruck 'Designzentrum' weckt Konnotationen aus der Bewegung der Guten Form un des Guten Designs mit seinen bildungsbürgerlichen Ballast und der Hypertrophie der formalin oder ästhetischen Eigenschaften von Produkten; diese Form von Designauffassung und diese Form von Designförderung schert mit olympischer Unbekümmertheit an der Bedürfnissen der Peripherie vorei [...]" in Antwerpen, Collection Flanders Architecture Institute – Collection Flemish Community, Archive Jean-Jacques Stiefenhofer (1943–2013), Doos 18.1: Gui Bonsiepe (Argentine), Letter to F. Burkhardt (Germany), (20/02/1974).

76 "Design Centers were initially linked to the *Good Form Movement*, with socio-pedagogical intentions to educate consumers. They singled out formal-esthetic aspects of products considered to comply with a formal canon. Though they mentioned function and functionality as an important aspect to be taken into account when evaluating a product, "form" ranked as the outstanding criterion. When I headed a design centre in Argentina that depended of the National Institute for Industrial Technology, the government authorities were less concerned with design promotion than with developing examples of design beyond the standard lamp-chair-ashtray candidates." See Bonsiepe, Gui, [email to the author], (29/06/2015).

77 The International Design-Zentrum (IDZ) in 1974 organized (April 8–10) in cooperation with the ICSID an international meeting for Design Centres in Berlin. Antwerp, Collection Flanders Architecture Institute – Collection Flemish Community, Archive Jean-Jacques Stiefenhofer (1943–2013), Doos 18.1: Stiefenhofer, Jean Jacques; Loek van der Sande and François Burckhardt, *Internationales Treffen der Design-Zentren im IDZ Berlin vom 8. bis 10. April 1974*, (8/04/1974). Jean-Jacques Stiefenhofer, „Design promotors aller landen, verenigt U!," *A+*, no. 9/10 (1974).

78 François Burkhard – good friends with the philosopher Jean-François Lyotard and head of the IDZ – was determined to introduce the postmodernist viewpoint in the design discourse and this marked certainly the discussions and outcome. The convictions of François Burkhardt resound in his defense for a Design Centre in Zurich he wrote that same year. First of all, he considers the Design Centres' link with the industry logic and necessary (since there is no "free space" that is disconnected with the reality of the market economy). This did not make the goals he granted to Design Centres less utopic. He envisioned the ultimate and very broad function of a Design Centre: to find a way that results in a better industrial society. See Burkhardt, François, Beitrag für "werk" Nr. 3/März 1974. Warum nich auch ein Design-Zentrum in Zürich? Einige Gedanken zur Rolle der Design-Zentren in hochindustrialisierten Ländern, (15/1/1974).

79 Antwerpen, Collection Flanders Architecture Institute - Collection Flemish Community, Archive Jean-Jacques Stiefenhofer (1943–2013), Doos 18.1: Stiefenhofer, Jean Jacques; Loek van der Sande and François Burckhardt, *Internationales Treffen der Design-Zentren im IDZ Berlin vom 8. bis 10. April 1974. Zusammenfassende Auswertung der Fragebogen*, (8/04/1974).

80 Katarina Serulus, "Take up Position: Defining Industrial Design in Postwar Belgium (1954–1970)," in *9th International Committee Design History and Design Studies. Tradition, Transition, Trajectories: Major or Minor Influences?*, ed. Helena Barbosa and Anna Calvera (Aveiro: UA Editora, 2014).

81 Brussels, State Archives, BE-A0545/Fic 2000, Archive Design Centre, 898: 235.

82 This Commission was established on April 5, 1977 under the presidency of Robert De Paepe (1927–2011), general secretary of the Ministry of Public Works. The Commission was under the protection of Louis Olivier, the Minister of Public Works, Brussels, State Archives, BE-A0545/Fic 2000, Archive Design Centre, 67: *Interministeriële Designcommissie*, (October 1981). Louis Olivier, "Design: pour l'État, voor de Staat, für den Staat, for the State," ed. Design Centre and Interministerial Design Commission (Brussels1983).

83 Brussels, State Archives, BE-A0545/Fic 2000, Archive Design Centre, 898.

84 The exhibition was on show in the Design Centre from November 10, 1983 until February 1, 1984 under the patronage of the King Baudouin Foundation, ICSID, and the Belgian Ministry of Public Works. Olivier, "Design: pour l'État, voor de Staat, für den Staat, for the State."

85 "Belgium Is Design," http://www.belgiumisdesign.be/?page=about.

86 In the recent book *Belgium's Best Design*, for example, the author Hadewijch Ceulemans (2013) constructs a canon of Belgian design icons through top 10 lists of 44 design peers, while according to her "[...] there is no specific Belgian style, nothing particular that makes Belgians as a group stand out [....]." Hadewijch Ceulemans, *Belgium's Best Design* (Antwerpen: Luster, 2013), 4.

87 Justin McGuirk, "The Belgian Affair," http://www.domusweb.it/en/design/2013/03/20/the-belgian-affair.html.

Conclusion

1 "A ceux qui craignent qu'une uniformité ne naisse de la multiplication de ces échanges, on peut répondre que les anciennes valeurs régionales tendront inéluctablement à se fondre dans une civilization qui devient planétaire, mais que le folklore peut être remplacé par l'homme, non pas

l'homme seul, mais l'homme dans l'équipe, dans l'entreprise, dans le bureau de design.» See Josine des Cressonnières, «La signification d'un Design Centre,» *Industrie* 1965.

Bibliography

Archives

Algemeen Rijksarchief/Archives générales du Royaume, Beveren and Brussels

Archief Adelbert Van de Walle, Museumarchief Design museum Gent, Ghent

Archief en Museum voor het Vlaams Leven te Brussel (AMVB), Brussels

Archief van het Koninklijk Paleis, Brussels (email contact)

Archief voor Hedendaagse Kunst in België, Brussels

Archives & Musée de la Littérature, Brussels

Archives de l'Université Libre de Bruxelles, Brussels

Archives et Bibliothèques d'Architecture de l'Université Libre de Bruxelles

Achivio Fotografico della Triennale di Milano, Milan (online database)

Collection Flanders Architecture Institute — Collection Flemish Community, Antwerp

Documentatiecentrum Design Vlaanderen, Brussels

HfG Archiv, Ulm (email contact)

Les Arts Décoratifs, Paris

Letterenhuis, Antwerp

Norsk Designråds historiske arkiv, Oslo (email contact)

Special Collections Research Center Syracuse University Libraries, Syracuse (email contact and online database)

Stadsarchief Brussel, Brussels

The University of Brighton Design Archives, Brighton

VNIITE archive, Moscow (email contact)

VRT beeldarchief, Brussels

Private archives

Marie-Christine Baeyens and Frank Smout

Nanette Blondel

Constantin Brodzki

Géraldine des Cressonnières

Jean Keup

Michel Olyff

Guy Roeckaerts

Jean-Jaccques Stiefenhofer

Interviews

Constantin Brodzki (1924–), interview by the author on October 31, 2013 and November 19, 2013, Linkebeek.

Geraldine des Cressonnières (1952–), interview by the author on November 8, 2013, Coutisse.

Mark Grammens (1933–), interview by the author on October 18, 2012, Liedekerke.

Françoise Jollant, interview by the author on August 29, 2015, Paris.

Lucien Kroll (1927–), interview by the author on October 21, 2014, Brussels.

Philippe Neerman (1930–2011), interview by the author on April 29, 2010, Kortrijk.

Michel Olyff (1927–), interview by the author on October 10, 2013, Ittre.

Jean-Jacques Stiefenhofer (1943–2013), interview by the author on October 11, 2011, October 17, 2011 and May 3, 2012, Antwerp.

Email

Lucien Kroll, email message to author, October 20, 2014.

Gui Bonsiepe, email message to author, July 9, 2015.

Magazines

7 Arts

A+

Acier

Architecture

Berichten van het Instituut voor Industriële Vormgeving voor België en het Groothertogdom Luxemburg/Bulletin d'Information Institut d'Esthétique Industrielle pour la Belgique et le Grand-Duché de Luxembourg

Bouwen en Wonen

Bulletin Commercial Belge

Bulletin de la Foire Internationale de Liège

Bulletin hebdomadaire d'information et de liaison edité par la federation des entreprises de l'industrie des fabrications métalliques

Bulletin du Grand Liège

Casabella Continuità

Design

Design News

Distributie

Entre Nous

Esthétique Industrielle

Habitat et Habitation

Industrie

Infordesign

Informatieblad van het Instituut voor Industriële Vormgeving/Bulletin de l'Institut d'Esthétique Industrielle

Komunikatief

Kunst. Maandblad voor oude en jonge beeldende, bouw en sierkunsten

Kunstambachten en Kunstnijverheden

L'Ossature Métallique

La Cité

La Maison

Le Bulletin du Grand Liège

Les Beaux Arts

Lettres Mensuelle d'Informations de Formes Nouvelles

Libelle

Maandbericht van het Instituut voor Industriële Vormgeving

MD Moebel Interior Design

Opbouwen

Ruimte

Scheppend Ambacht

The Designer

Ulm. Zeitschrift der Hochschule für Gestaltung

West-Vlaanderen

Newspapers

Spectator

Volksgazet

De Standaard

Het Laatste Nieuws

De Linie

De Nieuwe

Vooruit

Gazet van Antwerpen

De Spectator
Agefi
La Wallonie
Le Soir
La Dernière Heure
Le Monde du Travail
Le Peuple
Le Jour
Echo de la Bourse
Nord Eclair

Published sources

http://jamesbond.wikia.com/wiki/List_of_firearms.

Adamson, Glenn, Giorgio Riello, and Sarah Teasley, eds., *Global Design History*. London/New York: Routledge, 2011.

Adriaenssens, Werner, "Art Deco. Evolutie en beïnvloeding van de interieurkunsten in België na het uitdoven van de Art Nouveau tot de Tentoonstelling van 1925," in *Art Nouveau & Design. Sierkunst van 1830 tot Expo 58*, edited by Claire Leblanc, 104–19. Tielt/Brussels: Lannoo/Racine, 2005.

——"Les arts décoratifs et l'etat belge de 1890 à 1940/De decoratieve kunsten en de Belgische Staat van 1890 tot 1940/The Decorative Arts and the Belgian State from 1890 to 1940," in *Panorama: A History of Modern Design in Belgium*, edited by Katarina Serulus and Javier Gimeno-Martínez, 89–118. Brussels: CFC Editions, 2017.

Amalia García, María, and Alejandro Gabriel Crispiani, *Tomás Maldonado in Conversation with María Amalia García*. Conversations. edited by Gabriel Perez-Barreiro. New York/Caracas: Fundación Cisneros, 2010.

Amelina, Anna, Thomas Faist, Nina Schiller and Devrimsel Nergiz, "Methodological Predicaments of Cross-Border Studies," in *Beyond Methodological Nationalism: Research Methodologies for Cross-Border Studies*, edited by Anna Amelina, Devrimsel Nergiz, Thomas Faist and Nina Schiller, 91–110. New York/London: Routledge, 2012.

Amelina, Anna, Devrimsel Nergiz, Thomas Faist and Nina Schiller, eds., *Beyond Methodological Nationalism: Research Methodologies for Cross-Border Studies*. New York/London: Routledge, 2012.

Amorim, Patricia, and Virginia Cavalcanti, "Modern Design Meets Latin America: The Role of Pioneering Design Magazines Habitat and Nueva Visión in Brazil and Argentina," in *Design Frontiers - Territories, Concepts,*

Technologies, 8th Conference of the International Committee of Design History and Design Studies, edited by Farias Priscila, Marcos Braga, Anna Calvera and Zuleica Schincariol, 499–502. Sao Paulo: Blucher, 2012.

Andrew, Nell, "Living Art: Arakova and the Belgian Avant-Garde," *Art Journal* 68, no. 2 (2009), 26–49.

Antonelli, Paola, and Jamer Hunt, eds., *Design and Violence*. New York: The Museum of Modern Art, New York, 2015.

Art et Industrie. Kunst en Nijverheid. Brussels: Art et Industrie. Kunst en Nijverheid, 1938.

The Arts of Belgium 1920-1960. Liège: G. Thone, 1960.

Atkinson, Harriet, *The Festival of Britain. A Land and Its People*. London/New York: I.B. Tauris, 2012.

Atkinson, Harriet, and Verity Clarkson, "Editors' Introduction," *Design and Culture* 9, no. 2 (2017), 117–22.

Avermaete, Roger, "Henry van de Velde, Pionier van een Nieuwe Stijl," *Mededelingen van de Koninklijke Vlaamse Academie voor Wetenschappen, Letteren en Schone Kunsten van België* 25, no. 2 (1963).

——"Kanttekeningen bij het werk van A.M. Hammacher 'De wereld van Henry van de Velde'," *Mededelingen van de Koninklijke Vlaamse Academie voor Wetenschappen, Letteren en Schone Kunsten van België* 29, no. 2 (1967).

Aynsley, Jeremy, *Nationalism and Internationalism. Design in the 20th Century*. London: Victoria & Albert Museum, 1993.

Azrikan, Dmitry, "VNIITE, Dinosaur of Totalitarianism or Plato's Academy of Design?," *Design Issues* 15, no. 3 (1999) 45–77.

«B.70: 300 Produits à l'image de la Belgique. 27 Juin-2 September 1970," edited by Design Centre. Brussels: Design Centre, 1970.

Bailes, Kendall E., "Soviet Design: A Comment and an Alternative View," *Slavic Review* 37, no. 4 (1978), 595-601.

Bailey, Christopher, "The Global Future of Design History [Introduction]," *Journal of Design History* 18, no. 3 (2005), 231–33.

Bártolo, Carlos, "The Importance of Being 'Fascist', a Design History of 'Bad' People," in *PhD Workshop/DHS Day Symposium "Design History in Practice: Theory, Method, and Materials", organized by the Department of Philosophy, Classics, History of Art and Ideas of the University of Oslo and the Design History Society*. Oslo, 2015.

Baugniet, Marcel, *Essai sur la psychologie des formes*. Dilbeek (Brussels): Editions de la Maison du Poète, 1963.

Bekaert, Geert, *Hedendaagse architectuur*. Tielt: Lannoo, 1995.

—— *Verzamelde opstellen. Deel 1: Stapstenen 1950-1965*. Brussels: Stichting Monumenten- en Lanschapszorg, 1985.

Bekaert, Geert, Andrea Branzi, François Burkhardt, Lieven Daenens, Christian Kieckens and Christine Smout-Baeyens, *Design. Made in Belgium. 1900-1994*. Kortrijk: Stichting Interieur, 1994.

Bekaert, Geert, and Francis Strauven, *Bouwen in België 1945-1970*. Brussels: Nationale Confederatie van het Bouwbedrijf, 1971.

Bekker, Ludo, and Kaat Dutreeuw, "10 Jaar Design Center," in *Kortweg*: VRT (Vlaamse Radio- en Televisieomroeporganisatie), 1974.

Belgian Ministry of Economic Affairs, *Belgium To-Day*. Brussels: Belgian Ministry of Economic Affairs, 1949.

"Belgium Is Design," http://www.belgiumisdesign.be/?page=about.

Bender, Thomas, "Writing National History in a Global Age," *Correspondence: an international review of culture and society*, no. 7 (2001), 13-14.

Bennett, Tony, and Patrick Joyce, eds., *Material Powers. Cultural Studies, History and the Material Turn*. New York: Routledge, 2010.

Betts, Paul, *The Authority of Everyday Objects: A Cultural History of West German Industrial Design*. Berkeley: University of California Press, 2004.

——"The Bauhaus as Cold War Legend: West German Modernism Revisited," *German Politics and Society* 14, no. 2 (1996), 82-83.

——"Science, Semiotics and Society: The Ulm Hochschule für Gestaltung in Retrospect," *Design Issues* 14, no. 2 (1998), 67-82.

Beyen, Marnix, "De natie, de stad en de wereld. Belgische, Vlaamse en Waalse identiteitsvorming als antwoorden op de moderniteit," *Deus ex Machina. Tijdschrift voor Literatuur en Kunst* 26, no. 100 (2002), 7-11.

——"Tragically Modern. Centrifugal Sub-Nationalisms in Belgium, 1830-2009," Paper presented at the Contemporary Centrifugal Regionalism: Comparing Flanders and Northern Italy, Brussels, June 19-20, 2009.

Blanpain, Marthe, *Le patrimoine monumental de la Belgique. Bruxelles 1c. Pentagone*. Vol. 1C. Brussels: Mardaga, 1989.

Bonsiepe, Gui, and John Cullars, "The Invisible Facets of the HfG Ulm," *Design Issues* 11, no. 2 (1995), 11-20.

Bontridder, Albert, *Dialoog tussen licht en stilte*. Antwerpen: Helios, 1963.

——*Gevecht met de rede. Léon Stynen: Leven en werk*. Antwerpen: Comité Léon Stynen, 1979.

Boon, Tanja, "Ter Kameren (1927-1936)," Katholieke Universiteit Leuven, 2007.

Böröcz, "A Remarkable Continuity between 1930s Ideas and Reconstruction after the Second World War. New Stained-Glass Windows in War-Damaged Churches in the Diocese of Namur (Belgium)," in *Living with History*, edited by Nicholas Bullock and Luc Verpoest, 174-87. Leuven: Leuven University Press, 2011.

Bouchez, Hilde, "'I Don't Care What It Is for, I Want It!': The Flow of Meaning of 'High Design' in Flanders from 1980–2010, through the Dynamics between Producer, Media and Consumer," KU Leuven, 2013.

Braun, Benoît, «Esthétique et industrie: l'usine est un merveilleux instrument aux mains de l'artiste,» *Les Beaux-Arts. Hebdomadaire d'Information Artistique*, no. 658 (1954), 12.

Bucqoye, Monique, ed., *24th International Biennale Interieur 2014. The Book.* Kortrijk: Interieur Foundation, 2014.

Bucqoye, Monique, Lieven Daenens and Mil De Kooning, *Forms from Flanders. From Henry van de Velde to Maarten Van Severen 1900–2000*. Ghent/ Amsterdam: Ludion, 2001.

Buelens, Frans, *Congo 1885–1960. Een financieel-economische geschiedenis.* Berchem: epo, 2007.

Calvera, Anna, "From the East Side of the Iberian Peninsula: Meanings of Design When It Arrived During the 50s and Early 60s to Barcelona," *The Radical Designist*, no. December (2015).

——"Local, Regional, National, Global and Feedback: Several Issues to Be Faced with Constructing Regional Narratives," *Journal of Design History* 18, no. 3 (2005), 371–83.

Casteels, Maurice, *Henry van de Velde*. Brussels: Editions des Cahiers de Belgique, 1932.

Castillo, Greg, "The Bauhaus in Cold War Germany," in *Bauhaus Culture. From Weimar to the Cold War*, edited by Kathleen James-Chakraborty, 171–93. Minneapolis/London: University of Minnesota Press, 2006.

——*Cold War on the Home Front: The Soft Power of Midcentury Design.* Minneapolis: University of Minnesota Press, 2010.

——"Marshall Plan Modernisme in Divided Germany," in *Cold War Modern: Design 1945–1970*, edited by David Crowley and Jane Pavitt, 66–71. London: V&A Publishing, 2008.

Ceulemans, Hadewijch, *Belgium's Best Design*. Antwerpen: Luster, 2013.

Charles Graffart, Liège: Département des arts et lettres du Service des affaires culturelles de la Province de Liège, 1981.

Clarke, Alison, "'Actions Speak Louder.' Victor Papanek and the Legacy of Design Activism," *Design and Culture* 5, no. 2 (2013), 151–68.

——"Design for Development, ICSID and UNIDO: The Anthropological Turn in 1970s Design," *Journal of Design History* (2015), 1–15.

——*Tupperware: The Promise of Plastic in 1950s America*. Washington: Smithsonian Institution Press, 1999.

Cockcroft, Eva, "Abstract Expressionism," *Artforum* 15, no. 10 (1974), 39–41.

Coirier, Lise, *Design en Belgique/in Belgium/in België 1945–2000*. Brussels: Racine, 2004.

«Compagnie d'Afrique pour l'Industrie et la Finance. Bilan et compte de profits et pertes exercise 1952/1953,» Brussels: Weissenburch: Compagnie d'Afrique pour l'Industrie et la Finance, 1953.

Conrads, Ulrich, ed., *Programs and Manifestoes on 20th-Century Architecture*. Cambridge, Massachusetts: The MIT Press, 1971.

Coolsaet, Rik, *België en zijn buitenlandse politiek 1830–2000*. 3d ed. Leuven: Van Halewyck, 2001.

Coolsaet, Rik, Vincent Dujardin and Claude Roosens, *Buitenlandse Zaken in België. Geschiedenis van een ministerie, zijn diplomaten en consuls van 1830 tot vandaag*. Tielt: Lannoo, 2014.

Crowley, David, "Europe Reconstructed, Europe Devided," in *Cold War. Modern Design 1945-1970*, edited by David Crowley and Jane Pavitt, 43–65. London: V&A Publishing, 2008.

Crowley, David, and Jane Pavitt, eds., *Cold War. Modern Design 1945-1970*. London: V&A Publishing, 2008.

Cubbin, Tom, "The Domestic Information Machine. Futurological Experiments in the Soviet Domestic Interior, 1968–76," *Home Cultures* 11, no. 1 (2014), 5–32.

de Grazia, Victoria, *Irresistible Empire. America's Advance through 20th Century Europe*. Cambridge/London: The Belknap Press of Harvard University Press, 2005.

Nouvelle Biographie Nationale. Académie Royale des sciences, de lettres et des beaux-arts de Belgique, 2007.

De Kooning, Mil, *Willy Van der Meeren*. Vol. 21–24, Damme: Laat-XXe-eeuws genootschap, 1993.

De Kooning, Mil, Fredie Floré and Iwan Strauven, *Hedendaags design. Alfred Hendrickx en het fifties-meubel in België*, Mechelen: Stedelijke Musea, 2000.

De Vos, Els, "The American Kitchen in Belgium? A Story of Countering, Reversing, Selective Appropriation and Slidelining," in *The Tensions of Europe/ Inventing Europe Working Papers Series*, 2009.

——*Hoe zouden we graag wonen? Woonvertogen in Vlaanderen tijdens de jaren zestig en zeventig*. Leuven: Universitaire Pers Leuven, 2012.

——"The Interieur Biennale and Domesticity," in *24th International Biennale Interieur 2014. The Book*, edited by Monique Bucqoye, 60–69. Kortrijk: Interieur Foundation, 2014.

Defour, Frans, *Belgische meubelkunst in de XXe Eeuw. Van Horta tot heden*. Tielt: Lannoo, 1979.

den Boer, Pim, Peter Brugge and Ole Wæver, *The History of the Idea of Europe*. What Is Europe? London/New York: Routledge, 1993.

"Design uit Nederland," edited by Visual Arts Office for Abroad. Amsterdam: Visual Arts Office for Abroad, 1980.

Devalle, Verónica, "The Role of the 'Vision' Course in the Creation of the Design Programmes in Argentina," Paper presented at the 9th International Committee Design History and Design Studies. Tradition, transition, trajectories: major or minor influences?, Aveiro, 2014.

Devillez, Virginie, "Een Belgische Werkbund? Het streven naar een nationale kunstnijverheid," in *Art Nouveau & Design. Sierkunst van 1830 tot Expo 58*, edited by Claire Leblanc, 158–73. Tielt/Brussels: Lannoo/Racine, 2005.

——*Kunst aan de orde. Kunst en politiek in België 1918–1945*. Brussels/Ghent: Dexia Bank/Snoeck, 2003.

Devos, Rika, "A Cold War Sketch: The Visual Antagonism of the US vs. the USSR at Expo 58," *Revue Belge de Philologie et d'Histoire* 87, no. 3–4 (2009), 723–42.

——"Modern at Expo 58: Discussions on Post-War Architectural Representation." University of Ghent, 2008.

Devos, Rika, Mil De Kooning and Stefaan Decraene, *Moderne architectuur op Expo 58: "Voor een humanere wereld"*. Brussels: Dexia Bank and Mercatorfonds, 2006.

A Dictionary of Modern Design. 2010 ed. Oxford: Oxford University Press, 2004.

Dockrill, Michael L., and Michael F. Hopkins, *The Cold War: 1945–1991*. Studies in European History. 2d ed. Basingstoke, New York: Palgrave Macmillan, 2006.

Dorfles, Gillo, Christian Kieckens and Eva Storgaard, "Pieter De Bruyne 1931–1987: Pionier van het Postmoderne," edited by Design Museum Gent. Brussels: ASP, 2012.

Dreyfuss, Henry, «L'industrial design et l'homme d'affaires,» *Bulletin Commercial Belge* 67, no. 3 (1953), 35–40.

——»Marketing et industrial design sur le marché des Etats-Unis,» *Bulletin Commercial Belge* 66, no. 12 (1952), 55–62.

Dujardin, Vincent, "Go-Between: Belgium and Détente, 1961–73," *Cold War History* 7, no. 1 (2007), 95–116.

Dujardin, Vincent, Michel Dumoulin, Marnix Beyen and Philippe Destatte, eds., *Nieuwe geschiedenis van België. 1950–heden*. Vol. III. Tielt: Lannoo, 2009.

Edensor, Tim, *National Identity, Popular Culture and Everyday Life*. Oxford/New York: Berg, 2002.

Elno, K.N., "Ontstaan en groei der industriële vormgeving," in *Industriële Vormgeving (1957–1958)*, edited by Stichting Lodewijk De Raet, 7–9. Brussels: Stichting Lodewijk De Raet, 1959.

——*Ruimte en beelding. Beschouwingen over architektuur, plastische kunsten, fotografie en typografie*. Vlaamse Pockets. Vol. 157, Hasselt: Heideland, 1965.

Epple, Angelika, "The Global, the Transnational and the Subaltern: The Limits of History Beyond the National Paradigm," in *Beyond Methodological Nationalism: Research Methodologies for Cross-Border Studies*, edited by Anna

Amelina, Devrimsel Nergiz, Thomas Faist and Nina Schiller, 155–75. New York/London: Routledge, 2012.

Fallan, Kjetil, "Annus Mirabilis: 1954, Alberto Rosselli and the Institutionalisation of Design Mediation," in *Made in Italy. Rethinking a Century of Italian Design*, edited by Grace Lees-Maffei and Kjetil Fallan, 255–70. London/New York: Bloomsbury, 2014.

——'Cursing 'the Chromium-Plated Calf': Norwegian Fears and Desires of American Industrial Design in the 1950s," in *Appropriating America*. Amsterdam: ESF Inventing Europe, 2009.

——*Design History. Understanding Theory and Method*. Oxford/New York: Berg, 2010.

——"How an Excavator Got Aesthetic Pretentions - Negotiating Design in 1960s' Norway," *Journal of Design History* 20, no. 1 (2007), 43–59.

——"Love and Hate in Industrial Design: Europe's Design Professionals and America in the 1950s," in *The Making of the European Consumption. Facing the American Challenge*, edited by Per Lundin and Thomas Kaiserfeld. Palgrave Macmillan Transnational History Series, 134–56, Basingstoke: Palgrave Macmillan, 2015.

——"Nordic Noir: Deadly Design from the Peacemongering Periphery," *Design and Culture* 7, no. 3 (2015), 377–402.

——"Our Common Future. Joining Forces for Histories of Sustainavle Design," *Technoscienza. Italian Journal of Science & Technology Studies* 2, no. 5 (2014), 15–32.

Fallan, Kjetil, and Grace Lees-Maffei, "Introduction: The History of Italian Design," in *Made in Italy. Rethinking a Century of Italian Design*, edited by Grace Lees-Maffei and Kjetil Fallan, 2–33. London/New York: Bloomsbury, 2014.

Fernández, Silvia, "The Origins of Design Education in Latin America: From the HfG in Ulm to Globalization," *Design Issues* 22, no. 1 (2006), 3–19.

Ferran-Wabbes, Marie, *Jules Wabbes. Architecte d'intérieur 1919–1974*. Tournai: La renaissance du livre/Dexia Banque, 2002.

Fierens, Paul, *Marcel Baugniet*. Edited by Paul Fierens and Luc Haesaerts. Brussels/Paris: Editions Marion, 1942.

Fiss, Karen, "Design in a Global Context: Envisioning Postcolonial and Transnational Possibilities," *Design Issues* 25, no. 3 (2009), 3–10.

Fiss, Karen, and Hazel Clark, "Special Issue - Design in a Global Context," *Design Issues* 35, no. 3 (2009).

Flood, Catherine, and Gavin Grindon, *Disobedient Objects*. London: V&A Publishing, 2014.

Floré, Fredie, "Architect-Designed Interiors for a Culturally Progressive Upper-Middle Class: The Implicit Political Presence of Knoll International in Belgium," in *Atomic Dwelling. Anxiety, Domesticity, and Postwar Architecture*, edited by Robin Schuldenfrei, 169–85. London/New York: Routledge, 2012.

———"The Belgian Participation in the Milan Triennials," in *Design Frontiers - Territories, Concepts, Technologies, 8th Conference of the International Committee of Design History and Design Studies*, edited by Farias Priscila, Marcos Braga, Anna Calvera and Zuleica Schincariol, 278–80. Sao Paulo: Blucher, 2012.

———"Design Criticism and Social Responsibility: The Flemish Design Critic K.-N. Elno (1920–1993)," in *Writing Design: Words and Objects*, edited by Grace Lees-Maffei, 33–44. Oxford: Berg, 2011.

———"Diplomatic Encounters. Jules Wabbes an the Production of American Dunbar Furniture in Brussels," Paper presented at the 9th International Committee Design History and Design Studies. Tradition, transition, trajectories: major or minor influences? Aveiro, 2014.

———"Elno, K.-N., pseudoniem van Karel Horemans, architectuur-, design- en kunstcriticus," in *Nationaal biografisch woordenboek*, 316–23. Brussels: Koninklijke Academiën van België, 2011.

———*Lessen in goed wonen. Woonvoorlichting in België 1945–1958*. Leuven: Universitaire Pers Leuven, 2010.

———»Sociaal modernisme. De designkritiek van K.-N. Elno (1920–1993)," *De Witte Raaf*, no. 89 (2001), 6–8.

———"Wabbes and the Milan Triennales: Representing the Belgian Nation," in *Jules Wabbes. Furniture Designer*, edited by Iwan Strauven and Marie Ferran-Wabbes, 24–33. Brussels: A+ editions/Bozar Books, 2012.

Floré, Fredie, and Rika Devos, "Modern Wood: De Coene at Expo 58," *Construction History* 24 (2009), 103–20.

Foire 6e Internationale De Liège: Mines, métallurgie, mécanique, *électricité, 24 Avril-9 Mai 1954. Catalogue Officiel.* Liège1954.

Foucault, Michel, and Paul Rabinow, "Space, Knowledge, and Power," in *Power*, edited by James Faubion. The Essential Works of Foucault 1954–1984, 349–64. New York: The New Press, 2001.

Frampton, Kenneth, *Moderne architectuur. Een kritische geschiedenis,* translated by Henk Hoeks. 4th ed. Nijmegen: SUN, 2001.

Fry, Tony, *Design as Politics*. Oxford/New York: Berg, 2011.

———*Design History. Australia. A Source Text in Methods and Resources*. Sydney: Hale & Iremonger, 1988.

Fry, Tony, Clive Dilnot and Susan Stewart, *Design and the Question of History*. London/New York: Bloomsbury, 2015.

Gerits, Frank, "Taking off the Soft Power Lens. The United States Information Service in Cold War Belgium (1950–1958)," *Journal of Belgian History* 4, no. XLII (2012), 10–49.

Gimeno-Martínez, Javier, "Industrial Design in the Museum. The Case of the FN Milking Machine (Ca. 1947)," *The Burlington Magazine* 152, no. 1290 (2010), 603–08.

——"The Introduction and Dissemination of the English Word 'Design' in the Belgian Context," in *Words for Design. Comparative Etymology and Terminology of Design and Its Equivalents. The First Step Towards a Full Collection of Words for Design from All over the World*, edited by Haruhiko Fujita, 53–59. Japan Society for the Promotion of Science, 2009.

——"Redefining Social Design in 1970s Belgium: Affordable Design vs. Elite Design," *Interiors* 2, no. 2 (2011), 149–68.

——"The Role of the Creative Industries in the Construction of Regional/European Identities (1975-2002)," Leuven: Katholieke Universiteit Leuven, 2006.

——"The Signe d'Or Award Scheme from 1956 to 1960: The Economic Reasons for 'Good Design'," *Konsthistorisk Tidskrift/Journal of Art History* 79, no. 3 (2010), 127–45.

Gimeno-Martínez, Javier, Joana Ozorio de Almeida Meroz and Katarina Serulus, "Localizing Design/Designing Location: Creative Cities in the Low Countries," *Dutch Crossing* 37, no. 3 (2013), 274–92.

Goodin, Robert, ed., *The Oxford Handbook of Political Science*. New York: Oxford University Press, 2009.

Goris, Jan-Albert, ed., *Belgian American Survey 1958*. New York: Belgian Chamber of Commerce, 1958.

Grands Magasins au Bon Marché, *Les Grands Magasins* À l›*Innovation 1959-1960*. Brussels,1960.

Grindon, Gavin, "Cruel Designs at Banksy's Dismaland," http://gavingrindon.net/?p=246.

Guilbaut, Serge, *How New York Stole the Idea of Modern Art. Abstract Expressionism, Freedom, and the Cold War*. Chicago and London: The University of Chicago Press, 1983.

Haddad, Elie, "On Henry van de Velde's 'Manuscript on Ornament'," *Journal of Design History* 16, no. 2 (2003), 119–38.

Haerens, Jo, and Josine des Cressonnières, "Design in België 1940–1984," edited by Museum voor Moderne Kunst van Brussel. Brussels: Koninklijke Musea voor Schone Kunsten van België, 1984.

Hague, Rod, and Martin Harrop, *Comparative Government and Politics*. 7th ed. Basingstoke/New York: Palgrave Macmillan, 2007.

Halén, Widar, and Kerstin Wickman, eds., *Scandinavian Design Beyond the Myth. Fifty Years of Design from the Nordic Countries*. Stockholm: Arvinius Förlag/Form Förlag, 2003.

Harmel, Pierre, *Culture et profession*. «Batir». Collection d'études pour la restauration de la société chrétienne. Tournai/Paris: Casterman, 1944.

Harmel, Pierre, and Lucien Christophe, «'Art et travail': L'art monumental dans les établissements publics et industriels,» edited by Palais des Beaux-Arts Bruxelles. Brussels: Editions de la Connaissance, 1952.

Hein, Carola, *The Capital of Europe: Architecture and Urban Planning for the European Union*. Westport/London: Praeger, 2004.

Helmreich, Jonathan E. *Belgium and Europe. A Study in Small Power Diplomacy*. Issues in Contemporary Politics. Historical and Theoretical Perspectives. The Hague, Paris: Mouton, 1976.

Herman, Frank, Terenja Van Dijk, and Marc Dubois, *Kortrijkse Kunstwerkstede Gebroeders De Coene. 80 jaar ambacht en industrie: meubelen, interieurs, architectuur.* Kortrijk: Groeninghe, 2006.

Heskett, John. *Industrial Design*. London: Thames and Hudson, 1980.

Heylighen, F., "Belgium: Society, Character and Culture. An Essay on the Belgian Identity," http://pespmc1.vub.ac.be/BelgCul2.html.

"Hochschule für Gestaltung Ulm. Hogeschool voor Vormgeving Ulm," edited by Stedelijk Museum Amsterdam. Amsterdam, 1965.

Hoste, Gertjan, *Constantin Brodzki. Architecte*. Sprimont: Mardaga, 2004.

Huppatz, D.J., "Globalizing Design History and Global Design History," *Journal of Design History* 28, no. 2 (2015), 182–202.

Hutchings, Raymond, "Soviet Design: The Neglected Partner of Soviet Science and Technology," *Slavic Review* 37, no. 4 (1978), 567–83.

——"The Weakening of Ideological Influences Upon Soviet Design," *Slavic Review* 27, no. 1 (1968), 71–84.

Huygen, Frederike, *Brits Design, Imago en Identiteit*. London/Rotterdam: Thames and Hudson/Museum, Boymans-Van Beuningen, 1989.

——. "Vechten tegen de bierkaai? De promotie van industriële vormgeving via instituten en overheid," in *Holland in vorm: Vormgeving in Nederland 1945–1987*, edited by Geert Staal and Hester Wolters, 76–86. s'-Gravenhage: Stichting Holland in Vorm, 1987.

Interieur, Foundation, ed., *Interieur. Design Biennale since 1968*. Kortrijk: Interieur Foundation, 2010.

Jaarboek Nieuwe Vormen 1955. Annuaire Formes Nouvelles 1955. Brussels: Formes Nouvelles/Nieuwe Vormen, 1955.

Jacobs, Steven. *Henry van de Velde. Wonen als een kunstwerk. Een woonplaats voor kunst*. 2d ed. Leuven: Van Halewyck, 2005.

James-Chakraborty, Kathleen, ed., *Bauhaus Culture. From Weimar to the Cold War*. Minneapolis/London: University of Minnesota Press, 2006.

——"From Isolationism to Internationalism," Chap. 8 in *Bauhaus Culture. From Weimar to the Cold War*, edited by Kathleen James-Chakraborty, 153–70. Minneapolis/London: University of Minnesota Press, 2006.

——"Henry van de Velde and Walter Gropius," Chap. 2 in *Bauhaus Culture. From Weimar to the Cold War*, edited by Kathleen James-Chakraborty, 26–42. Minneapolis/London: University of Minnesota Press, 2006.

Jerlei, Triin, "Socialist Elements in Soviet Design Ideology," Paper presented at the 9th International Committee Design History and Design Studies. Tradition, transition, trajectories: major or minor influences?, Aveiro, 2014.

Jones, Michelle, "Design and the Domestic Persuader. Television and the British Broadcasting," *Journal of Design History* 16, no. 4 (2003), 307–18.

Julier, Guy, "Re-Drawing the Geography of European Design: the Case of Transitional Countries", paper given at the 2d European Academy of Design Conference, "Contextual Design — Design in Context", Stockholm, April 1997, http://www.designculture.info/reviews/ArticleStash/GJReDrawing1997.pdf

——*The Culture of Design*. 2d ed. London: Sage, 2010.

——"Design," in *The Palgrave Dictionary of Transnational History*, edited by Akira Iriye and Pierre-Yves Saunier, 258–60. Basingstoke: Palgrave Macmillan, 2009.

——"Más Allá De Los Margines: La Historia Del Diseño, Transnacionalización Y Globalización," in *Diseño e Historia: Tiempo, Lugar y Discurso*, edited by I. Campi, 117–26. Barcelona/Mexico City: Designio, 2010.

Kantor, Sybil, *Alfred H. Barr, Jr., and the Intellectual Origins of the Museum of Modern Art*. Cambridge: MIT Press, 2003.

Karpova, Yulia, "Accommodating 'Design': Introducing the Western Concept into Soviet Art Theory in the 1950s-60s," *European Review of History: Revue européenne d'histoire* 20, no. 3 (2013), 627–47.

——"'A Glass without a Bottom': Neodecorativism in Late 1960s Soviet Design," *Journal of Design History* 30, no. 1 (2016), 1–15.

Kikuchi, Yuko, and Yunah Lee, "Special Issue: Transnational Modern Design Histories in East Asia," *Journal of Design History* 27, no. 2 (2014).

——"Transnational Modern Design Histories in East Asia: An Introduction," *Journal of Design History* 27, no. 4 (2014), 323–34.

Kirkham, Pat, and Susan Weber, eds., *History of Design. Decorative Arts and Material Culture, 1400-2000*. New York/New Haven and London: Bard Graduate Center: Decorative Arts, Design History, Material Culture/Yale University Press, 2013.

Kott, Sandrine, «Par-delà la guerre froide. Les organisations internationales et les circulations est-ouest (1947–1973),» *Vingtième Siècle Revue d'histoire*, no. 109 (2011): 143–54.

Kroll, Lucien, J. Hamels and Ministry of Economic Affairs, *XI.A Triennale di Milano. Sezione Belga*. Brussels: Ministère des Affaires Economiques/Service des Expositions et Foires, 1957.

Kurg, Andres, "Prehistory of Contemporary Environment: Multimedia Program in the Icsid'75 Congress in Moscow," in *The Islands of Yuri Sobolev*, edited by Anna Romanova and Galina Metelichenko. Moscow: The Moscow Museum of Modern Art, 2014.

«L'Esthétique Industrielle. La 7me Foire Internationale de Liège,» edited by Ministère des Affaires Economiques: Ministère des Affaires Economiques, 1955.

Lavrentiev, Alexander N., and Yuri V. Nasarov. *Russian Design. Tradition and Experiment 1920–1990*. Academy Editions,1995.

Le Boeuf, Jocelyne, "Jacques Viénot and the 'Esthétique Industrielle' in France (1920–1960)," *Design Issues* 22, no. 1 (2006), 46–63.

Leblanc, Claire, ed. *Art Nouveau & Design. Sierkunst van 1830 tot Expo 58*. Tielt/Brussels: Lannoo/Racine, 2005.

Lees-Maffei, Grace, "The Production-Consumption-Mediation Paradigm," *Journal of Design History* 22, no. 4 (2009), 351–76.

Lees-Maffei, Grace, and Kjetil Fallan, eds., *Made in Italy. Rethinking a Century of Italian Design*. London/New York: Bloomsbury, 2014.

Leopold, Cornelie, "The Mathematical Approach at Ulm School of Design," in *Imagine Maths 4. Between Culture and Mathematics*, edited by Michele Emmer, Marco Abate and Marcela Villarreal, 15–28. Unione Matematica Italiana, 2015.

——"Precise Experiments: Relations between Mathematics, Philosophy and Design at Ulm School of Design," *Nexus Network Journal* 15, no. 2 (2013), 363–80.

Loeffler, Jan, *The Architecture of Diplomacy: Building America's Embassies*. New York: Princeton Architectural Press, 1998.

Loze, Pierre, Thierry Aughuet, Michelle Feiner-Weinstein and Philippe Niels, *Christophe Gevers Designer*. Brussels: Editions d'Art Laconti, 2008.

Lundin, Per, "Introduction," in *The Making of the European Consumption. Facing the American Challenge*, edited by Per Lundin and Thomas Kaiserfeld. Palgrave Macmillan Transnational History Series. Basingstoke: Palgrave Macmillan, 2015.

Lundin, Per, and Thomas Kaiserfeld, eds., *The Making of the European Consumption. Facing the American Challenge,* edited by Akira Iriye and Rana Mitter, Palgrave Macmillan Transnational History Series. Basingstoke: Palgrave Macmillan, 2015.

Maciuika, John V., "Wilhelmine Precedents for the Bauhaus," Chap. 1 in *Bauhaus Culture. From Weimar to the Cold War*, edited by Kathleen James-Chakraborty, 1–25. Minneapolis/London: University of Minnesota Press, 2006.

Madge, Pauline, "Design, Ecology, Technology: A Historiographical Review," *Journal of Design History* 6, no. 3 (1993), 149–66.

Maguire, Patrick, and Jonathan Woodham, eds., *Design and Cultural Politics. The Britain Can Make It Exhibition of 1946*. London: Leicester University Press, 1997.

Malcovati, Silvia, "The Notion of the Total Work of Art and Italian Building Culture after World War II," in *The Death and Life of the Total Work of Art. Henry van de Velde and the Legacy of a Modern Concept*, edited by Carsten Ruhl, Chris Dähne and Rixt Hoekstra, 164–78. Berlin: Jovis, 2015.

Maldonado, Tomas, *Design, Nature and Revolution: Toward a Critical Ecology,* translated by Mario Domandi. New York: Harper & Row, 1972.

Margolin, Victor, *World History of Design*. Vol. 3. London: Bloomsbury Academic, 2015.

——"A World History of Design and the History of the World," *Journal of Design History* 18, no. 3 (2005), 235–43.

Marsh, David, and Gerry Stoker, eds., *Theory and Methods in Political Science,* edited by Guy Peters, Jon Pierre and Gerry Stoker. 3d ed., Political Analysis. Basingstoke and New York: Palgrave Macmillan, 2010.

Marshall, Jennifer Jan, "In Form We Trust: Neoplatonism, the Gold Standard, and the 'Machine Art' Show, 1934." *The Art Bulletin* 90, no. 4 (2008), 597–615.

Masey, Jack, and Conway Lloyed Morgan, *Cold War Confrontations. US Exhibitions and Their Role in the Cultural Cold War*. Baden: Lars Müller Publishers, 2008.

McDonald, Gay, "The 'Advance' of American Postwar Design in Europe: MoMA and the Design for Use, USA Exhibition 1951–1953," *Design Issues* 24, no. 2 (2008), 15–27.

——"The Modern American Home as Soft Power: Finland, MoMA and the 'American Home 1953' Exhibition," *Journal of Design History* 23, no. 4 (2010), 287–408.

——Selling the American Dream: MoMA, Industrial Design and Post-War France," *Journal of Design History* 17, no. 4 (2004), 397–412.

McGuirk, Justin, "The Belgian Affair," http://www.domusweb.it/en/design/2013/03/20/the-belgian-affair.html.

Mertens, Phil, and P. Roberts-Jones, "Naar een zuiver beelden. De eerste Belgische abstracten/1918–1930," edited by Koninklijke Musea voor Schone Kunsten van België. Brussels: Koninklijke Musea voor Schone Kunsten van België, 1972.

"Meubelen uit België. Van Henry van de Velde tot heden/Mobilier de Belgique de Henry van de Velde à nos jours," edited by Design Centre. Brussels, 1978.

Meyvis, Ludo. *Markt en macht*. Tielt: Lannoo, 2004.

Miller, Daniel, "Coca-Cola: A Black Sweet Drink from Trinidad," in *Material Cultures*, edited by Daniel Miller, 169–87. London: UCL Press/University of Chicago Press, 1997.

Moriarty, Catherine, "A Backroom Service? The Photographic Library of the Council of Industrial Design, 1945–1965," *Journal of Design History* 13, no. 1 (2000), 39–57.

Müller, Jens, and René Spitz, "HfG Ulm, kurze Geschichte der Hochschule für Gestaltung: Anmerkungen zum Verhältnis von Design und Politik = a Brief History of the Ulm School of Design: Notes on the Relationship between Design and Politics," Zürich: Lars Müller, 2014.

Narotzky, Viviana, "Our Cars in Havana," in *Autopia Cars and Culture*, edited by Peter Wollen and Joe Kerr, 169-76. London: Reaktion Books, 2002.

Nolan, Mary, *The Transatlantic Century. Europe and America, 1890-2010*. New Approaches to European History, edited by William Beik, T.C.W. Blanning and Brendan Simms. New York: Cambridge University Press, 2012.

Nye, Joseph, *Soft Power. The Means to Success in World Politics*. New York: PublicAffairs, 2004.

Oldenziel, Ruth, and Adri Albert de la Bruhèze, "Europe's Mediation Junction: Technology and Consumer Society in the 20th Century," *History and Technology* 21, no. 1 (2005), 107-39.

Oldenziel, Ruth, and Karin Zachmann, *Cold War Kitchen: Americanization, Technology, and European Users*. Cambridge: MIT Press, 2009.

Olivier, Louis, "Design: pour l'Etat, voor de Staat, für den Staat, for the State," edited by Design Centre and Interministerial Design Commission. Brussels, 1983.

Olyff, Michel, «Questions Graphiques. Enjeux et Contraintes,» *Académie royale de Belgique. Bulletin de la Classe des Beaux-Arts* 6e série - tombe XII, no. 7-12 (2001).

Oswalt, Philipp, ed., *Bauhaus Conflicts, 1919-2009. Controversies and Counterparts*. Ostfildern: Hatje Cantz Verlag, 2009.

Özkirimli, Umut, *Theories of Nationalism. A Critical Introduction*. 2d ed. Basingstoke: Palgrave Macmillan, 2010.

Ozorio de Almeida Meroz, Joana, "The International as National: The Role of International Cultural Policy in the Construction of Dutch Design as Conceptual," Paper presented at the 9th International Committee Design History and Design Studies. Tradition, transition, trajectories: major or minor influences? Aveiro, 2014.

Ozorio de Almeida Meroz, Joana, "Transnational Material Politics. Constructions of Dutch Design, 1970-2012" (Unpublished PhD Dissertation, Vrije Universiteit Amsterdam, 2018).

Paenhuysen, An,. *De nieuwe wereld. De wonderjaren van de Belgische avant-garde (1918-1939)*. Antwerpen: Meulenhoff Manteau, 2010.

Papanek, Victor, *Design for the Real World. Human Ecology and Social Change*. London: Thames & Hudson, 2006.

Pavitt, Jane, "Design and the Democratic Ideal," in *Cold War Modern: Design 1945-1970*, edited by David Crowley and Jane Pavitt, 73-91. London: V&A Publishing, 2008.

Pevsner, Nikolaus, *Pioneers of Modern Design: From William Morris to Walter Gropius*. New Haven: Yale University Press, 2005.

Piérard, Louis, and Adrien Van der Bruch, *Manifestation en l'honneur de M. Louis Piérard. Fondateur et Président de l'Association Art et Industrie. Organisée le 14 Mai 1938 au Cercle Artistique et Littéraire Bruxelles*. Liège: Georges Thone, 1938.

Prina, Daniela N., "Design in Belgium before Art Nouveau: Art, Industry and the Reform of Artistic Education in the Second Half of the Nineteenth Century," *Journal of Design History* 23, no. 4 (2010), 329–52.

——"Belgian Decorative Arts in the Later Nineteenth Century. Needs for a National Museum and Debates Surrounding Didactic Collections in Brussels," *Journal of the History of Collections* 24, no. 2 (2012), 257–74.

Pulos, Arthur J., *The American Design Adventure, 1940-1975*. MIT, 1988.

Putnam, Tim, "Foreword: Design and Polity in Comparative Perspective," *Journal of Design History* 20, no. 2 (2007), 89–92.

Puttemans, Pierre, *Modern Architecture in Belgium*. Brussels: Marc Vokaer Publishers, 1976.

«Quatre ans d'action économique,» edited by the Ministère des Affaires Economiques. Brussels: Ministère des Affaires Economiques, 1958.

Reid, E. Susan, "Communist Comfort: Socialist Modernism and the Making of Cosy Homes in the Khrushchev Era," *Gender & History* 21, no. 3 (2009), 465–98.

——"'Our Kitchen Is Just as Good': Soviet Responses to the American National Exhibition in Moscou, 1959," in *Cold War Modern: Design 1945-1970*, edited by David Crowley and Jane Pavitt, 154–61. London: V&A Publishing, 2008.

——"This Is Tomorrow! Becoming a Consumer in the Soviet Sixties," in *The Socialist Sixties. Crossing Borders in the Second World*, edited by Anne E. Gorsuch and Diane P. Koenker, 25–65. Bloomington: Indiana University Press, 2013.

——"Toward a New (Socialist) Realism. The Re-Engagement with Wester Modernism in the Khrushchev Thaw," in *Russian Art and the West. A Century of Dialogue in Painting, Architecture, and the Decorative Arts*, edited by Susan E. Reid and Rosalind P. Blakesley, 215–39. Dekalb: Northern Illinois University Press, 2007.

——"Cold War Cultural Transactions: Designing the USSR for the West at Brussels Expo '58," *Design and Culture* 9, no. 2 (2017), 123–145.

Reid, Susan E., and David Crowley, eds., *Style and Socialism. Modernity and Material Culture in Post-War Eastern Europe*. Oxford/New York: Berg, 2000.

Reid, Susan E., and Rosalind P. Blakesley, eds., *Russian Art and the West. A Century of Dialogue in Painting, Architecture, and the Decorative Arts*. Dekalb: Northern Illinois University Press, 2007.

Reynebeau, Marc, *Een geschiedenis van België*. Tielt: Lannoo, 2009.

Riccini, Raimonda, "Tomás Maldonado and the Impact of the Hochschule für Gestaltung Ulm in Italy," in *Made in Italy. Rethinking a Century of Italian Design*, edited by Grace Lees-Maffei and Kjetil Fallan, 89–105. London/New York: Bloomsbury, 2014.

Riley, Terence, and Edward Eigen, "Between the Museum and the Marketplace: Selling Good Design," in *The Museum of Modern Art at*

Mid-Century. At Home and Abroad, edited by John Elderfield. Studies in Modern Art, 151–75. New York: The Museum of Modern Art, 1994.

Rodgers, Daniel T., *Atlantic Crossings. Social Politics in a Progressive Age*. Cambridge/London: The Belknap Press of Harvard University Press, 1998.

Romijn, Peter, Giles Scott-Smith and Joes Segal, eds., *Divided Dreamworld? The Cultural Cold War in East and West*. Amsterdam: Amsterdam University Press, 2012.

Ryckewaert, Michael, *Building the Economic Backbone of the Belgian Welfare State. Infrastructure, Planning and Architecture 1945–1973*. Rotterdam: 010, 2011.

Saunier, Pierre-Yves, "Transnational," in *The Palgrave Dictionary of Transnational History*, edited by Akira Iriye and Pierre-Yves Saunier, 1047–54. Basingstoke: Palgrave Macmillan, 2009.

Schuldenfrei, Robin, ed., *Atomic Dwelling. Anxiety, Domesticity, and Postwar Architecture*. London and New York: Routledge, 2012.

Schulte, Birgit, "Henry van de Velde. De levensreis 1863–1957," in *Henry van de Velde. Een Europees kunstenaar in zijn tijd*, edited by Klaus-Jürgen Sembach and Birgit Schulte, 19–53. Antwerpen: Pandora, 1993.

Scott, James C., *Seeing Like a State. How Certain Schemes to Improve the Human Condition Have Failed*. New Haven and London: Yale University Press, 1998.

Senkevitch, Anatole, "Art, Architecture, and Design: A Commentary," *Slavic Review* 37, no. 4 (1978), 587–94.

Serulus, Katarina, "Het plastic meubel in België (1954–1973)," Leuven: Katholieke Universiteit Leuven, 2010.

——"Meurop in Holland 1959–1960. The Mediation of Popular Design," VU University Amsterdam, 2011.

——"Take up Position: Defining Industrial Design in Postwar Belgium (1954–1970)," in *9th International Committee Design History and Design Studies. Tradition, Transition, Trajectories: Major or Minor Influences?*, edited by Helena Barbosa and Anna Calvera, 575–79. Aveiro: UA Editora, 2014.

——"Van industrieel design tot productontwikkeling," in *Van academie tot universiteit. 350 Jaar architectuur in Antwerpen*, edited by Els De Vos and Piet Lombaerde, 168–73. Brussels: UPA, 2013.

——"Design & Politics. The Public Promotion of Industrial Design in Postwar Belgium (1950–1986)" (Unpublished PhD Dissertation, University of Antwerp, 2016).

——"1968–2016, 25 Editions of Biennale Interieur," in *25th Silver Edition. Biennale Interieur 2016*, edited by Biennale Interieur, 8–13. Kortrijk: Biennale Interieur npo, 2016.

——"Het interieur van het Design Centre in Brussel (1964–1986). Belgisch design gekaderd," in *Vormgevingserfgoed in Vlaanderen in 1945*, edited by Ellen Van Impe and Eva Van Regenmortel, 55–73. Antwerpen: Art Paper Editions & CVAa, 2016.

———"Au-delà des frontières: la promotion du design belge pendant la guerre froide/De grenzen voorbij: de promotie van Belgisch design tijdens de Koude Oorlog/Beyond the borders: the promotion of Belgian design during the Cold War," in *Panorama: A History of Modern Design in Belgium,* edited by Katarina Serulus and Javier Gimeno-Martínez, 153–180. Brussels: CFC Editions, 2017.

———"Biennale Interieur, Een halve eeuw designgeschiedenis," in *Designx50,* 12–15. Veurne: Kannibaal, 2017.

———"'Well-Designed Relations': Cold War Design Exchanges between Brussels and Moscow in the Early 1970s.," *Design and Culture* 9, no. 2 (2017), 147–65.

Serulus, Katarina, and Javier Gimeno-Martínez, eds., *Panorama: A History of Modern Design in Belgium.* Brussels: CFC Editions, 2017.

Soloviev, Yuri, *My Life in Design.* Moscow1994.

Souto, Helena Maria, "Editorial. Design Policies: Between Dictatorship and Resistance. Special Issue of the Radical Designist," *The Radical Designist*, no. December (2015).

Souto, Helena Maria, Anjoom Satar and Márcia Gomes, "Portuguese Design History (1960–1974). Shaping the Future: The Role of the Art, Industrial Architecture and Industrial Design Nuclei of the National Institute of Industrial Research (I.N.I.I.)," Paper presented at the 9th International Committee Design History and Design Studies. Tradition, transition, trajectories: major or minor influences?, Aveiro, 2014.

Sparke, Penny, *An Introduction to Design and Culture. 1900 to the Present.* 2d ed. London/New York: Routledge, 2004.

———*Japanese Design.* New York: Museum of Modern Art, 2009.

Strauven, Iwan, «Victor Bourgeois, 1897–1962: Radicaliteit en pragmatisme, moderniteit en traditie,» Université libre de Bruxelles, 2015.

Strauven, Iwan, and Maurice Culot, *De gebroeders Bourgeois. Architecture et plastique pure.* Brussels: AAM, 2005.

Strauven, Iwan, Fredie Floré and Ferran-Wabbes, Marie, *Jules Wabbes. Furniture Designer.* Brussels: A+ editions/Bozar Books, 2012.

Tavernier, Roger, *Russia and the Low Countries: An International Bibliography, 1500-2000.* Groningen: Barkhuis, 2006.

Teughels, Nelleke, and Peter Scholliers, eds., *A Taste of Progress. Food at Internatioanl and World Exhibitions in the Nineteenth and Twentieth Centuries.* Aldershot: Ashgate, 2015.

Theunis, Katrien, "De zoektocht naar een Belgisch woonproject, 1965–1975: Toenaderingen tussen ontwerpers en overheid in de praktijk van het private wonen." Leuven: K.U.Leuven, 2008.

Thiesse, Anne-Marie, *The Transnational Creation of National Arts and Crafts in 19th-Century Europe.* Nise Monographs. NISE, 2014.

Thompson, Christopher, "Modernizing for Trade: Institutionalizing Design Promotion in New Zealand," *Journal of Design History* 24, no. 3 (2011), 223–39.

Twemlow, Alice, "I Can't Talk to You If You Say That: An Ideological Collision at the International Design Conference at Aspen, 1970," *Design and Culture* 1, no. 1 (2009), 23–49.

Vago, Pierre, *Pierre Vago: Une vie intense*. Brussels: Archives d'Architecture Moderne, 2000.

Van Alstein, Maarten, "From Enigma to Enemy. Paul-Henri Spaak, the Belgian Diplomatic Elite, and the Soviet Union, 1944–1945," *Journal of Cold War Studies* 13, no. 3 (2011), 126–48.

Van Besien, Fred, «Marché Commun ou demain l'Europe (Partie 1),» in *9 millions*. Belgium: RTBF (Radio Télévision Belge Francophone), 1968.

Van de Velde, Henry, *Les formules de la beauté architectonique moderne*. Brussels: Equerre, 1923.

Van de Velde, Henry, and Hans Curjel, *Geschichte meines Lebens*. München: Piper, 1962.

Van de Velde, Henry, and Hermann Muthesius, «Werkbund Theses and Antitheses,» in *The Design History Reader*, edited by Grace Lees-Maffei and Rebecca Houze. Oxford/New York: Berg, 2010.

Van de Voorde, Stephanie, Sofie De Caigny, and Ronny De Meyer, *Isia Isgour 1913-1967*. Focus Architectuurarchieven, edited by Sofie De Caigny, Annelies Nevejans, Line Mertens and Bregje Provo. Antwerpen: Centrum Vlaamse Architectuurarchieven, 2008.

Van der Schueren, Jacques, Adelbert Van de Walle, André De Poerck, and J. Iwens, "Nationaal Salon voor Industriële Vormgeving," edited by Ministerie van Economische Zaken. Administratie van de Handel. Dienst Tentoonstellingen en Jaarbeurzen, 1959.

Van der Wee, Kristin, and Edward de Maesschalck, *Vlaanderen ontmoet Nederland. Geschiedenis van de Orde van den Prince*. Tielt: Lannoo, 2005.

Van Loo, Anne, Denis Laurent and Andy Jacobs, *Design(s) à La Cambre 54 04*. Brussels: Ecole nationale supérieure des arts visuels de La Cambre, 2006.

Van Middelaar, Luuk, *De passage naar Europa. Geschiedenis van een begin*. 4th ed. Groningen: Historisch Uitgeverij, 2011.

Van Milleghem, Karel, ed., *Europa Unita*. Tielt/Den Haag: Lannoo, 1966.

Van Regenmortel, Eva, "Paul Mariamé," in *ODIS*. Record last modified date: September 1, 2014. Available from World Wide Web: http://www.odis.be/lnk/PS_113086.

——"Pol Provost (1907–1990)," in *ODIS*. Record last modified date: December 17, 2012. Available from World Wide Web: http://www.odis.be/lnk/PS_35108.

Van Regenmortel, Eva, and Ellen Van Impe, eds., *Vormgevingserfgoed in Vlaanderen na 1945*. Art Paper Editions & CVAa, 2016.

Verbeek, Peter-Paul, "Politics at Issue. On Art and the Democratization of Things," *Open: Politics of Things. What Art & Design do in Democracy*, no. 24 (2012), 18–29.

Verdickt, Caterina, "The Case of Elisabeth De Saedeleer (1902–1972): The Influence of Welsh Hospitality in the Great War on Belgian Modernist Interior Design," *The British Art Journal* 15, no. 3 (2015), 93–98.

Verpoest, Luc, "Immigrant in het vaderland. De tweede Belgische periode van Van de Velde," in *Henry van de Velde. Een Europees kunstenaar in zijn tijd*, edited by Klaus-Jürgen Sembach and Birgit Schulte, 397–411. Antwerpen: Pandora, 1993.

Verpoest, Luc, and Ellen Van Impe, "Henry van de Velde. Brieven van architecten. Bezoekersgids," edited by Koninklijke Bibliotheek van België, Centrum Vlaamse Architectuurarchieven and Archives et Musée de la Littérature, 2013.

Wagenaar, Cornelis, "Jaap Bakema and the Fight for Freedom," in *Anxious Modernisms. Experimentation in Postwar Architectural Culture*, edited by Sarah Williams Goldhagen and Réjean Legault, 261–77. Montréal/Cambridge/London: Canadian Centre for Architecture, MIT Press, 2000.

Waldeck, Mila, "Typography and National Identity," Paper presented at the 9th International Committee Design History and Design Studies. Tradition, transition, trajectories: major or minor influences?, Aveiro, 2014.

Walker, John, *Design History and the History of Design*. London: Pluto Press, 1989.

Wharton, Annabel Jane, *Building the Cold War: Hilton International Hotels and Modern Architecture*. Chicago: University of Chicago Press, 2001.

Whitehouse, Denise, "The State of Design History as a Discipline," in *Design Studies. A Reader*, edited by Hazel Clark and David Brody, 54–63. Oxford/New York: Berg, 2009.

Wimmer, Andreas, and Nina Schiller, "Methodological Nationalism and Beyond: Nation-State Building, Migration and the Social Sciences," *Global Networks* 4, no. 2 (2002), 301–34.

Winner, Langdon, *The Whale and the Reactor. A Search for Limits in an Age of High Technology*. Chicago and London: The University of Chicago Press, 1986.

Witte, Els, Jan Craeybeckx and Alain Meynen, *Politieke geschiedenis van België. Van 1830 tot heden*. Antwerpen: Standaard, 2005.

Woodham, Jonathan, "Formulating National Design Policies in the United States: Recycling the 'Emperor's New Clothes'?". *Design Issues* 26, no. 2 (2010), 27–46.

——"Local, National and Global: Redrawing the Design Historical Map," *Journal of Design History* 18, no. 3 (2005). 257–367.

——*Twentieth-Century Design*. Oxford/New York: Oxford University Press, 1997.

Yagou, Artemis, "Unwanted Innovation. The Athens Design Centre (1961–1963)," *Journal of Design History* 18, no. 3 (2005), 269–83.